Astronomers' Universe

Series Editor

Martin Beech, Campion College, The University of Regina, Regina, SK, Canada

The Astronomers' Universe series attracts scientifically curious readers with a passion for astronomy and its related fields. In this series, you will venture beyond the basics to gain a deeper understanding of the cosmos—all from the comfort of your chair.

Our books cover any and all topics related to the scientific study of the Universe and our place in it, exploring discoveries and theories in areas ranging from cosmology and astrophysics to planetary science and astrobiology.

This series bridges the gap between very basic popular science books and higher-level textbooks, providing rigorous, yet digestible forays for the intrepid lay reader. It goes beyond a beginner's level, introducing you to more complex concepts that will expand your knowledge of the cosmos. The books are written in a didactic and descriptive style, including basic mathematics where necessary.

Jie Wang

Eye Beyond the Sky

27 Telescopes and Space Probes, from Hooker to JWST

Jie Wang
Shanghai, China

Translated by
Xiaoyan Huang
Shanghai, China

ISSN 1614-659X ISSN 2197-6651 (electronic)
Astronomers' Universe
ISBN 978-981-99-9817-3 ISBN 978-981-99-9818-0 (eBook)
https://doi.org/10.1007/978-981-99-9818-0

Jointly published with Shanghai Educational Publishing House Co., Ltd.

This Springer imprint is published by the registered company Springer Nature Singapore Pte Ltd.
The registered company address is: 152 Beach Road, #21-01/04 Gateway East, Singapore 189721, Singapore

Contents

1

Mount Wilson Observatory: A Surprising Discovery that Kept Einstein Awake at Night

Mount Wilson Observatory

It was early 1930. The great physicist Albert Einstein was eager to depart for California, USA. One thing had been far more important than a single trip. His second wife Elsa traveled with him (Isaacson 2007) (see Fig. 1.1).

It was not simple to travel from Germany to the United States on the other side of the world before the advent of commercial flying. From Germany, Einstein traveled a train to Antwerp, Belgium, where he boarded a ship for the transatlantic crossing to New York City. After a little stopover, Einstein set sail from New York and sailed south along the East Coast through the Panama Canal, via the Canal to the Pacific Ocean, and then north along the East Coast of the Pacific Ocean before docking in Los Angeles. He spent a whole month on the road.

Although this was Einstein's second trip to America (Wikipedians, Albert Einstein 2023a), he had never visited the West Coast before. On the invitation of the California Institute of Technology (Pais 1994), he traveled all the way to the Golden State for a short academic stay. Yet another significant motive for Einstein's travel was to meet a new person in a fresh location. Edwin Hubble (1889–1953) (Smith 1998), a great American astronomer, was the target of his visit, and the Mount Wilson Observatory in California was the location he wished to explore. Hubble's observations at Mount Wilson

Supplementary Information The online version contains supplementary material available at https://doi.org/10.1007/978-981-99-9818-0_1.

J. Wang, *Eye Beyond the Sky*, Astronomers' Universe,
https://doi.org/10.1007/978-981-99-9818-0_1

Fig. 1.1 Albert Einstein with his second wife Elsa (*Source* United States Library of Congress)

Observatory revitalized not only Einstein's but all of humanity's perspective on the cosmos (Livio 2014). As a result of this discovery, the Mount Wilson Observatory has also made a significant contribution to the annals of astronomical literature.

Known as the "Founder of modern solar observation," American astronomer George Ellery Hale (1868–1938) (Wikipedians, George Ellery Hale 2023b) traveled to Mount Wilson, northeast of Los Angeles (Wikipedians, Mount Wilson Observatory 2023c) in 1904. At first sight, he was smitten with this mountain of treasure. This is due to the presence of an inversion layer, a rare form of meteorological phenomenon, above the Mount Wilson region. In most cases, the temperature will drop as you gain altitude. Nevertheless, due to a variety of complex factors, there are some locations that experience the exact reverse, and this phenomenon is known as the inversion layer (Runge and Patterson 2007) (see Fig. 1.2).

The atmospheric stability afforded by the inversion layer is Mount Wilson's main advantage. One of the most stable atmospheres in all of North America may be found atop Mount Wilson, making it an ideal location for astronomical studies (Rottman 2007) (see Fig. 1.3).

Fig. 1.2 George Ellery Hale (*Left, Source* The World's Work: A History of Our Time) surveyed Mount Wilson. (*Right, Source* Carnegie/Huntington Library)

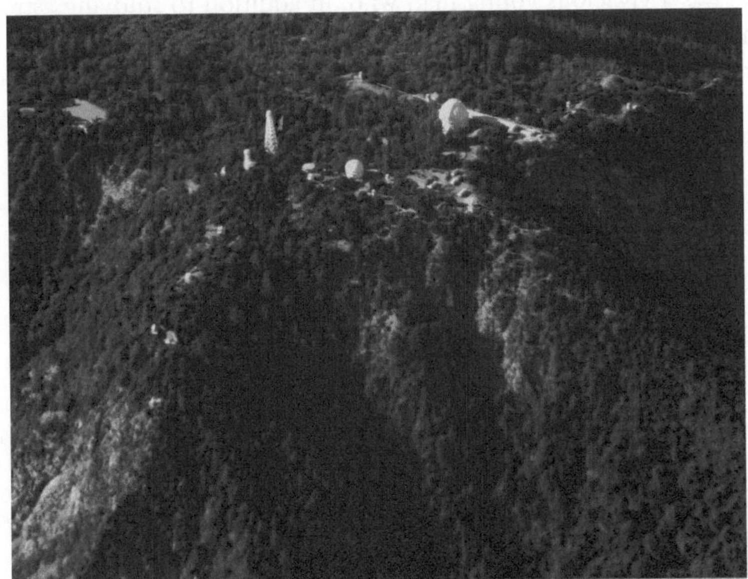

Fig. 1.3 Mount Wilson (*Credit* Jeffrey Beall)

In the immediate aftermath of Hale's discovery of this rich mountain, the most cutting-edge astronomical telescopes in the world were quickly relocated from the Yerkes Observatory to this new location (Osterbrock 1993). Therefore, the Mount Wilson Observatory was formally inaugurated.

The strong magnetic field of the Sun is concentrated in sunspots, which Hale and his team of scientists saw and studied after the Observatory was established. Furthermore, this was the first time that humans had ever been able to detect the magnetic field of a celestial body other than Earth. Yet,

Hale soon got concerned that the telescope's aperture was inadequate. When it came to telescope size, astronomers would never be satisfied (see Fig. 1.4).

Hale was talented in both scientific inquiry and the raising of funds. Donations to the Observatory came quickly after he approached affluent local merchants. It was at Mount Wilson Observatory in 1908 (Florence 2011) that the then-world's largest astronomical telescope was eventually built.

Mount Wilson Observatory was suddenly in the spotlight since it housed the world's most sophisticated telescope. Researchers from all over the flocked to this new site of astronomy. Among them was the legendary astronomer, Edwin Hubble (see Fig. 1.5).

Early in his career, Hubble conducted research at Yerkes Observatory (Christianson 1996). Once Hale relocated the telescope and constructed a new, larger one at Mount Wilson, Hubble began making regular trips there. Hubble was a vivacious young man who, in addition to studying astronomy and mathematics at the University of Chicago, was also a passionate boxer (information@eso.org n.d.). After finishing graduate school, he had several

Fig. 1.4 The world's largest astronomical telescope with an aperture of about 1.5 m (60 in.) in 1908 (*Source* The Huntington Library, Art Museum, and Botanical Gardens)

Fig. 1.5 Edwin Hubble (*Credit* NASA and ESA)

career options open to him: he could go into astrophysics, mathematics, or the professional boxing world.

Fortunately for astronomy, Hubble ultimately committed his life to the field. When Hubble decided to devote himself to astronomy, the prevailing academic view held that the Milky Way was the entirety of the cosmos (This month in physics history: Edwin Hubble Expands our View of the Universe n.d.), that the universe was no larger than the diameter of the galaxy, and that all heavenly bodies were either near the galactic core or on its periphery, like our solar system. In any case, there are dissenting opinions. Astronomers first spotted a spiral nebula in the night sky as early as the seventeenth century. Despite the fact that modern amateur astronomers are all well aware that the nebulae that look like snowflakes in telescopes are actually galaxies, this was not always the case.

It wasn't until the Milky Way's spiral structure was confirmed that some began to speculate that the spiral nebulae might actually be other spiral galaxies.

If this view was correct, then our universe would not only be the Milky Way. There were galaxies outside of the Milky Way, which was equal to subverting the human vision of the cosmos. It would be a tremendous event. However, extraordinary claims require extraordinary evidence. Just making outlandish predictions is not enough. In order to draw a scientific conclusion, evidence must be presented (see Fig. 1.6).

Hubble had been given the responsibility of figuring this out by the course of history. Where could he locate the evidence? Hubble couldn't shake off this lingering question.

Hubble used spectrum analysis as his approach (Hubble finds proof that the universe is expanding n.d.). After being dissected by the grating apparatus, every bright spot in the sky will reveal a colored spectrum. Hubble obtained and carefully studied as many spectra as possible from each nebula. The Andromeda Nebula, he discovered (NASA n.d.), has stellar properties and a spectrum very similar to the Sun's. Although William Huggins (1824–1910) (Wikipedians, William Huggins 2023d), a British astronomer, made the identical finding in 1864, the technology and data at the time were not nearly as precise. This time, Hubble and Hale knew for sure that the Andromeda Nebula was composed of many bright stars because Hale had

Fig. 1.6 Image of the Andromeda Nebula taken at the Yerkes Observatory in the early twentieth century (*Credit* Kippax, John R. (John Robert), 1849–1922)

independently confirmed the data collected by Hubble. Hubble also took photographs of the Andromeda Nebula using the Super Telescope at Mount Wilson Observatory (Hubble's Famous M31 VAR! plate n.d.).

Yet, establishing that the Andromeda Nebula is composed of stars does not indicate that it is a separate galaxy from our own. Due to the abundance of star clusters in the Milky Way, their finding was not particularly noteworthy. To Hubble, the key lies in pinpointing the Andromeda Nebula's distance. Given the circumstances, however, determining the distance to a dim star in the night sky was an extremely difficult job (see Fig. 1.7).

Larger telescopes, the bigger the better, were the only practical solution to this issue.

Hence, Hale reemerged, showcasing his fundraising prowess once again. He successfully invited Andrew Carnegie (1835–1919), a prominent philanthropist (Nasaw 2007), steel tycoon, and the world's second richest man at the time, to Mount Wilson Observatory, and then presented Carnegie with a popular scientific "lecture." Carnegie was finally impressed with him (Our history n.d.), and he donated $10 million, a staggering figure even by today's standards. Science popularization can play an important role in bringing in much-needed funding for scientific research, which is why almost all US research institutions place a premium on it.

Fig. 1.7 Image of the Triangulum Galaxy taken at Mount Wilson Observatory in August 1910 (*Source* GALERIE GADCOLLECTION)

When viewed from this perspective, having money makes a lot of things a lot simpler to accomplish. The world's largest telescope, in terms of aperture, was erected on November 1, 1917 (Building the 100-inch Telescope n.d.). British poet Alfred Noyes (1880–1958) commemorated this event in his epic *Watches of the Sky* by writing (The Observatory, by Alfred Noyes n.d.):

> The explorers of the sky, the pioneers
>
> Of science, now made ready to attack
>
> That darkness once again, and win new worlds.

The name "Hooker Telescope" was given to this particular instrument over time (Florence 2011). It was with it that Hubble eventually discovered a means to calculate the length of the journey to the Andromeda Nebula (Hubble 1929) (see Fig. 1.8).

Cepheids are a type of variable star that undergoes periodic variations in luminosity and can be seen in the night sky. Cepheid variable stars have the potential to serve as the cosmological "standard candle" (Leavitt 1979) due to the consistent link between their absolute luminosity and their variable light period. The brightness of the light is inversely proportional to the square of the distance (Halliday et al. 2013). Measuring the apparent brightness is the same as measuring the distance if the absolute luminosity of the light source is known (see Fig. 1.9).

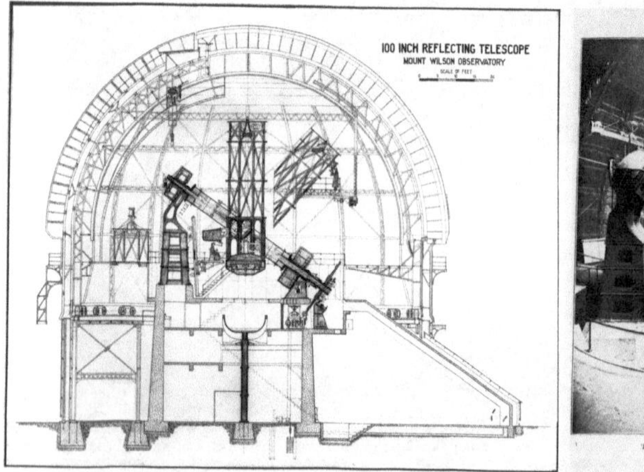

Fig. 1.8 Hooker telescope (*Left, Source* The Huntington Library, Art Museum, and Botanical Gardens) (*Right, Source* The History Trust of South Australian, South Australian Government Photo)

Fig. 1.9 Snapshots of the star that changed the universe (*Credit* NASA, ESA and Z. Levay (STScI). *Credit* NASA, ESA and the Hubble Heritage Team (STScI/AURA))

With the Super Hooker Telescope, Hubble obtained multiple images of the Andromeda Nebula, from which he identified 34 Cepheid variable stars. Then he meticulously spent over 2 years determining the absolute luminosity of these Cepheids by tracing their light cycle curves. He was able to determine the distance between these Cepheid variables and the Earth by basing his calculations on the apparent brightness of the variables in the telescope (see Fig. 1.10).

Hubble estimated that the Andromeda Nebula was roughly 900,000 light-years away (the most recent data puts it at 2.54 million light-years) (Hubble 1979), well beyond the Milky Way's diameter. The Andromeda Nebula is unequivocally not a part of our galaxy in any way.

At a conference of the American astronomical community in 1925, Hubble reported the results, which immediately caused a stir. Hubble's findings were rapidly accepted because they were so precise and carefully documented. It's fair to say that he single-handedly widened humanity's view of the cosmos (Bartusiak 2010).

Fig. 1.10 Hubble operated the Hooker telescope. (*Source* The Huntington Library, Art Museum, and Botanical Gardens)

Edwin Hubble and the Mount Wilson Observatory were at the forefront of astronomical discovery at the time. In addition, Hubble did not cease his investigation into the universe. He maintained the Hooker telescope's gaze on the faraway galaxies, and he was about to solve another incredible cosmic enigma.

Hubble discovered, after a series of follow-up observations and calculations, that most galaxies, with the exception of Andromeda, were receding from us. Hubble found that the ratio of a galaxy's recession speed to its distance is constant after analyzing and summing the recession speeds of all galaxies (Hubble 1929). To put it another way, galaxies are receding from us, and the further they are, the faster they are receding. We call this Hubble's law, also known as the Hubble-Lemaitre (Einstein, Die Grundlage der allgemeinen Relativitätstheorie 1923) law (see Fig. 1.11).

Aside from the accelerating expansion of the universe, no other explanation seemed viable. Hubble's findings caused an uproar in the astronomical community, akin to "raising up a hornet's nest." The theory of a continually

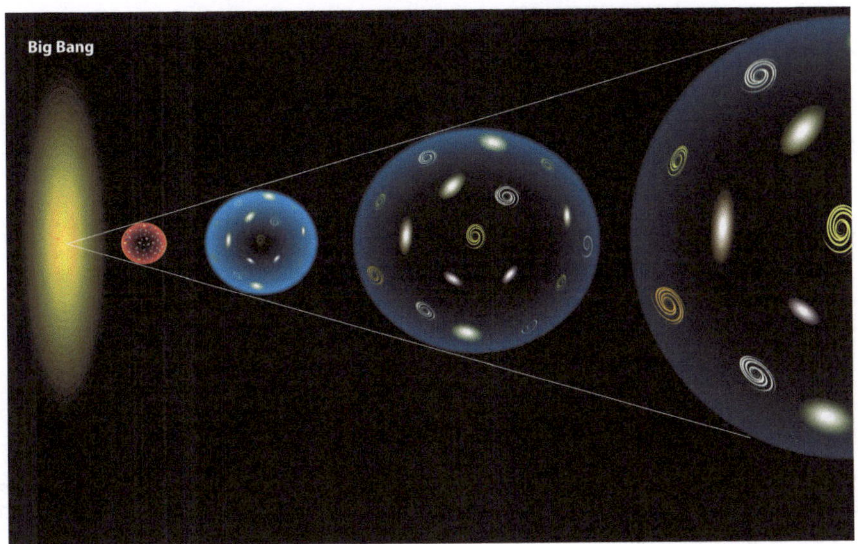

Fig. 1.11 Expanding universe (*Credit* Depositphotos/edesignua)

expanding universe is now widely recognized in the scientific community. But the concept of an expanding universe was revolutionary at the time. After receiving the news, Einstein, who was living on the opposite side of the planet, was also greatly astonished. He was unable to keep calm and felt compelled to travel to the United States in order to visit Hubble.

There was good cause for Einstein's surprise. Even before he proposed the general theory of relativity in 1916, Einstein discovered that its equation led to an inference that the universe was dynamic, either expanding or contracting. However, this conclusion ran counter to Einstein's intuition. Both his intuition and his naive cosmology led him to the conclusion that something must be missing from his calculations if the cosmos is not in a constant, static state. Then, after much trial and error, Einstein included a "cosmological constant" (Einstein, Kosmologische Betrachtungen zur allgemeinen Relativitätstheorie 2005) in the equations of general relativity to keep the cosmos from oscillating.

To confirm Hubble's startling discovery of an expanding universe, Einstein must travel to Hubble's observatory in person.

On the morning of January 29, 1931, Hubble and his wife drove Einstein to Mount Wilson Observatory as promised (Christianson 1996), the pleasant sea breeze in California having blown away the winter cold. On the way, Einstein told Hubble's wife, "Your husband's work is fantastic," from the depths of his heart. When Einstein saw the enormous telescope at the observatory, he became entranced like a child and forgot to go. After being shown

Fig. 1.12 Einstein (Left) visited Mount Wilson Observatory (*Source* Associated Press/ Caltech Archives). Einstein (second from left) with Hubble (first from right) (*Source* The Huntington Library, Art Museum, and Botanical Gardens)

the telescope by the crew, Albert Einstein took one glance at Hubble, flashed a grin, and remarked to everyone else there, "Hubble proved my prediction by using this telescope" (see Fig. 1.12).

By having Einstein accepted and acknowledged, Hubble and Mount Wilson Observatory gained widespread attention. Every major scientific prize there was, with the exception of the Nobel Prize, had been bestowed upon Hubble. Because there was no precedence for the field of astronomy to be given the Nobel Prize when Hubble was alive (Lee 2011), and the scientific community of the time did not consider astronomy and physics to be the same scientific discipline, hence Hubble did not receive the Nobel Prize in Physics that he deserved.

The discoveries that Hubble made at the Mount Wilson Observatory have had a profound impact on our understanding of the universe, and have ensured that both Hubble and the observatory will go down in astronomical history as landmarks. Mount Wilson Observatory is a lighthouse if Hubble is a beacon for mankind to enlighten the universe and guide humans to explore the cosmos in the proper way (see Fig. 1.13).

In 1986 (VT 100-Inch n.d.), the historic mission of the Hooker telescope was successfully fulfilled, and it was subsequently deactivated. With the advancement of science and technology, the Hooker Telescope was outfitted with an adaptive optics system in 1995. This reinvigorated the exceptional "Knight General" and led to its reactivation. The telescope continued to be a significant instrument for astronomers all around the world to use as they uncovered new parts of the cosmos in the years that followed.

Fig. 1.13 Hubble was examining a map of galaxies in 1931. (*Source* The Huntington Library, Art Museum, and Botanical Gardens)

Several thousands of people visit the Mount Wilson Observatory each year, almost as if they were making a pilgrimage to see the site where humanity's understanding of the cosmos was forever altered (see Fig. 1.14).

Humanity's ultimate inquiry, in my opinion, is "What is the universe?" In terms of magnitude, this issue is unrivaled. There is no word big enough to capture the magnitude of the disparity between humanity and the cosmos. It's incredible, though, that we humans, who are so insignificant compared to the cosmos, have figured out how to view the big picture. The origin of human civilization's comprehension of the universe can be traced back to Hubble's epic tale at the Mount Wilson Observatory, and through it we gain a profound appreciation of how science speaks with evidence. Even a genius like Einstein needs to see proof before he'll change his mind because all reasonable conclusions must be founded on evidence.

Fig. 1.14 Hooker telescope at Mount Wilson (*Credit* Ken Spencer)

References

Bartusiak, M.: The Day We Found the Universe. Vintage (2010)

Building the 100-inch Telescope. (n.d.). Mount Wilson Observatory: https://www.mtwilson.edu/building-the-100-inch-telescope/. Accessed 11 Apr 2023

Christianson, G.E.: Edwin Hubble: Mariner of the Nebulae. University of Chicago Press (1996)

Einstein, A.: Die Grundlage der allgemeinen Relativitätstheorie. In: Das Relativitätsprinzip, pp. 81–124. Vieweg+Teubner Verlag (1923). https://doi.org/10.1007/978-3-663-19510-8_7

Einstein, A.: Kosmologische Betrachtungen zur allgemeinen Relativitätstheorie. In: Albert Einstein: Akademie-Vorträge, pp. 119–130 (2005). https://doi.org/10.1002/3527608958.ch10

Florence, R.: The Perfect Machine: Building the Palomar Telescope. Harper Collins (2011)

Halliday, D., Resnick, R., Walker, J.: Fundamentals of Physics, Extended. Wiley (2013)

Hubble finds proof that the universe is expanding (n.d.). pbs.org: https://www.pbs.org/wgbh/aso/databank/entries/dp29hu.html. Accessed 11 Apr 2023

Hubble, E.: A relation between distance and radial velocity among extra-galactic nebulae. Proc. Natl. Acad. Sci. **15**(3), 168–173 (1929). https://doi.org/10.1073/pnas.15.3.168

Hubble, E.P.: 104. Cepheids in spiral nebulae. In: A Source Book in Astronomy and Astrophysics, 1900–1975, pp. 713–715. Harvard University Press (1979). https://doi.org/10.4159/harvard.9780674366688.c112

Hubble's Famous M31 VAR! plate (n.d.). Carnegie Science: https://obs.carnegiescience.edu/PAST/m31var. Accessed 11 Apr 2023

information@eso.org. (n.d.). Edwin Powell Hubble—The Man Who Discovered the Cosmos. ESA/Hubble: https://esahubble.org/about/history/the_scientist_behind_the_name/. Accessed 11 Apr 2023

Isaacson, W.: Einstein: His Life and Universe. Simon and Schuster (2007)

Leavitt, H.S.: 57. Periods of twenty-five variable stars in the small magellanic cloud. In: A Source Book in Astronomy and Astrophysics, 1900–1975, pp. 398–400. Harvard University Press (1979). https://doi.org/10.4159/harvard.9780674366688.c63

Lee, M.G.: Physics Nobel Prize in 2011: superstars and discovery of an accelerating universe. Phys. High Technol. **20**(12), 2 (2011). https://doi.org/10.3938/phit.20.054

Livio, M.: Brilliant Blunders: From Darwin to Einstein—Colossal Mistakes by Great Scientists that Changed Our Understanding of Life and the Universe. Simon and Schuster (2014)

NASA: Hubble Views the Star That Changed the Universe (n.d.). NASA.gov: https://www.nasa.gov/mission_pages/hubble/science/star-v1.html. Accessed 11 Apr 2023

Nasaw, D.: Andrew Carnegie. Penguin (2007)

Osterbrock, D.E.: Pauper & Prince: Ritchey, Hale & Big American Telescopes. University of Arizona Press (1993)

Our history (n.d.). Carnegie Science: https://carnegiescience.edu/about/our-history. Accessed 11 Apr 2023

Pais, A.: Einstein Lived Here. Oxford University Press, USA (1994)

Rottman, G.J.: The Solar Radiation and Climate Experiment (SORCE): Mission Description and Early Results. Springer Science & Business Media (2007)

Runge, M.S., Patterson, C.: Principles of Molecular Cardiology. Springer Science & Business Media (2007)

Smith, R.: Edwin Hubble, 20 Jul 1998. Encyclopedia Britannica: https://www.britannica.com/biography/Edwin-Hubble

The Observatory, by Alfred Noyes (n.d.). Mount Wilson Observatory: https://www.mtwilson.edu/about-mwo/the-observatory-by-alfred-noyes/. Accessed 11 Apr 2023

This Month in Physics History: Edwin Hubble Expands our View of the Universe (n.d.). Advancing Physics: https://www.aps.org/publications/apsnews/200801/physicshistory.cfm. Accessed 11 Apr 2023

VT 100-Inch (n.d.). Mount Wilson Observatory: https://www.mtwilson.edu/vt-100-inch/. Accessed 11 Apr 2023

Wikipedians. Albert Einstein. Wikipedia: https://en.wikipedia.org/wiki/Albert_Einstein. Accessed 10 Apr 2023

Wikipedians (27 Jan 2023). George Ellery Hale. Wikipedia: https://en.wikipedia.org/wiki/George_Ellery_Hale

Wikipedians (19 Mar 2023). Mount Wilson Observatory. Wikipedia: https://en.wikipedia.org/wiki/Mount_Wilson_Observatory

Wikipedians (10 Apr 2023) William Huggins. Wikipedia: https://en.wikipedia.org/wiki/William_Huggins

2

Parkes Radio Telescope: When a Star Shines with a More Intense Light than a Galaxy

Parkes Radio Telescope

Parkes, a small town in southeast Australia, was battered by an unusual storm on July 21, 1969. The last time it poured was a very long time ago. The dry dust on the ground was carried up by the 110 km/h storm, generating a sandstorm that obscured the Sun and made Parkes as dark as night in the middle of the day.

The tallest structure in the entire town of Parkes is a 64-m radio telescope named after the town (CSIRO, Parkes radio telescope, Murriyang 2023).

A dreadful cracking sound was coming from this telescope as its bowtie antenna swayed in the wind. Five astronomers were on standby in the control room at the massive antenna's base. They had to find a means to protect the radio telescope while getting ready to do something no one had ever done before: broadcast the Moon landing live (Bird 2019) (see Fig. 2.1).

"If the wind hasn't died down by the time the Moon rises, the live broadcast is doomed." Neil Fox Mason, the telescope's operator, didn't even squint as he went about his work.

East, where the Moon was rising, was where the wind was coming from. It's easy to envision what would happen if a 64-m-diameter "umbrella" were exposed to a force 12 gale. And yet, 380,000 km away on the Moon, astronaut Neil Armstrong had already donned his spacesuit and was getting ready to decompress the space capsule (NASA, July 20, 1969: One giant leap for mankind 2015b).

"In human history, this marks a significant step forward. Whatever the scenario may be, we shall use maximum effort to achieve success." Parkes

J. Wang, *Eye Beyond the Sky*, Astronomers' Universe,
https://doi.org/10.1007/978-981-99-9818-0_2

Fig. 2.1 Parkes' control room in charge of the 1969 moon landing live broadcasting (*Credit* CSIRO). © Copyright CSIRO Australia, (10 August 2020)

Observatory director and vocal supporter of the live lunar landing program John Bolton (1922–1993) (Sullivan 1994) gave the talk. Bolton had signed a contract with NASA (see Fig. 2.2).

The wind died down a little as the Moon rose slowly over the horizon, but it was still too gusty for the telescope to operate safely. As the massive antenna gently swung eastward, a historic live broadcast was initiated, reaching 600 million people around the world (Sarkissian 2001) (see Fig. 2.3).

When it came to this live streaming, Parkes wasn't the only radio telescope that got involved. The Honeysuckle Creek Radio Telescope in Canberra, Australia, and the Goldstone Deep Space Communication Complex in California, USA were also contributing to this endeavor (Harland 2011). In contrast to the other two tracking stations, the live signals provided by Parkes were of such high quality (Orchiston et al. 2006) that NASA only relied on them throughout (see Figs. 2.4 and 2.5).

During the Cold War era of the 1960s and 1970s, when the United States and the Soviet Union were vying for dominance, common folks frequently mistook this Australian space exploration instrument for Soviet or American. What is this mysterious Parkes radio telescope, anyway, that even in the worst of storms could send back significantly clearer signals than any other radio telescope? This chapter will detail its story for you.

Radar technology advanced quickly during World War II due to military demands. The Radiophysics Laboratory was the first Australian organization dedicated to radar research.

Fig. 2.2 Parkes radio telescope at daytime (*Credit* Depositphotos/NickRH)

The Council for Scientific and Industrial Research of Australia (CSIRO) founded this laboratory back in those days to perform covert radar studies (Robertson 1992). By the early 1950s, the Radiophysics Laboratory had become the largest, most diverse, and most successful research center of its kind (Haynes et al. 1996). Historically speaking, this is the only instance where Australian scientific research institutions have been on the cutting edge of developing fields (see Fig. 2.6).

Once upon a time, the terms radar and radio telescope were used interchangeably. The earliest radio telescopes frequently repurposed components

Fig. 2.3 Of the moonwalk on the monitor in the control room (*Credit* CSIRO). © Copyright CSIRO Australia, (1996)

Fig. 2.4 The 70-m antenna at Goldstone (*Credit* NASA JPL/DSN Status–Planetary Sciences Advisory Committee)

from decommissioned radars. Despite their lack of sophistication, these makeshift radio telescopes uncovered numerous extragalactic radio sources, which aided in the development of larger telescopes (Sullivan III 2009).

Fig. 2.5 Honeysuckle creek radio telescope (*Credit* National Museum of Australia/ Hamish Lindsay and Colin Mackellar)

In those years, Taffy Bowen (1911–1991) oversaw research and development at the Radiophysics Laboratory (Haynes et al. 1996). He rose to prominence in the field after developing a miniature radar that could be mounted on a fighter aircraft (Buderi 1998). On the other hand, progress in radio astronomy could be found in every region of the planet. Australia could not rely entirely on such antiquated machinery if it wished to keep its then position of preeminence. It's imperative that they procure the finest telescope available.

The primary distinction between radio astronomy research and the production of small radars is that the former not only does not generate profits but actually loses money. Australian politicians, who were notoriously

Fig. 2.6 Early simple antennas (*Credit* CSIRO). © Copyright CSIRO Australia, (November 2, 2021)

nearsighted, would rather invest in research on sheep shearing techniques than in telescopes.

As luck would have it, Taffy Bowen knew a lot of powerful people in the US scientific community. He convinced the Carnegie Corporation and the Rockefeller Foundation to put up half the money for the construction of Australian Super Telescope by leveraging his personal ties and Australia's surviving technological advantages. After receiving the funding from the United States, the then-Prime-Minister of Australia Robert Menzies (1894–1978) grudgingly gave his approval to the remaining monies (Robertson 1992).

The planning and building of the Parkes radio telescope took seven years, from 1954 to 1961 (Milne 1994). The practice of gifting scientific equipment was not a societal tradition in Australia. Anyway, Parkes still remains the country's most important scientific engineering endeavor. You could only imagine the amount of pushback and criticism this project had received over the seven years. Nonetheless, it quickly showed its worth with a significant finding (see Fig. 2.7).

Fig. 2.7 Construction site of Parkes (*Credit* CSIRO). © Copyright CSIRO Australia, (November 2nd, 2021)

A type of unknown radio source was detected long before the Parkes radio telescope was constructed. Some had even been spotted in the visible spectrum. These were not galaxies, despite appearances to the contrary, because they appeared as single points in the sky. Spectral detection, however, revealed that their spectra were radically distinct from those of regular stars. This class of objects is known as quasars. What this means is that they appear to be stars but are actually something else.

Exact astronomical source matching between optical and radio points is a crucial step in the investigation of unexplained quasars (Ackermann et al. 2015). For the most part, optical telescopes can observe more point sources and have higher resolution. Thus, there will be far too many optical matching alternatives for a direct correspondence inside the location area given by the radio telescope. As an optical source matching had not been discovered, no additional spectra could be taken for analysis.

Some brilliant astronomers at this time devised an ingenious technique. Lunar occultation of quasars allows for more precise position calculations, even while radio telescopes' spatial resolution is inadequate (Kastner and Bhatia 1990). A quasar's position on the celestial sphere is similar to rest, and the Moon's motion is quite regular. You may determine where these quasars are by using the Moon's position at the time of the occlusion and doing some simple math (Singal et al. 1979).

Using the Parkes radio telescope, astronomers Cyril Hazard, M.B. Mackey, and A.J. Shimmins were able to observe five lunar occultations of the quasar 3C 273 in the year 1962 (Hazard et al. n.d.). In practice, this resulted in a pinpoint precision of localization (see Fig. 2.8).

Using this coordinate, astronomer Maarten Schmidt was able to locate 3C 273's optical match and provided a spectral analysis of it Schmidt (1963) (see Fig. 2.9).

The spectra of such objects were distinct from stellar spectra, as had been previously discovered. Schmidt contemplated 3C 273's puzzling spectral features. The spectra of celestial bodies typically have emission lines from various elements, and the data contained in these lines can be thought of as

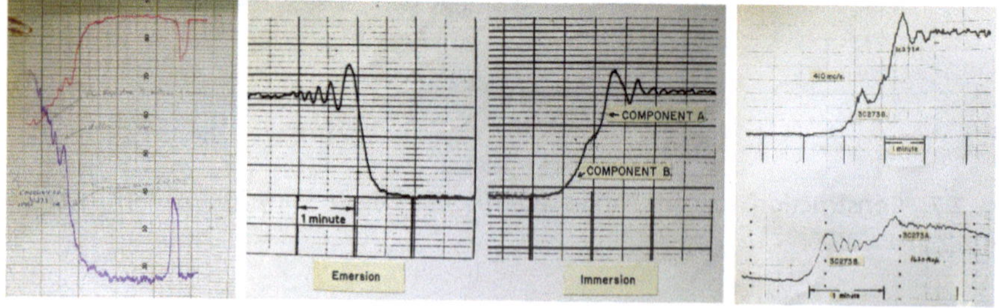

Fig. 2.8 Manuscript documenting five lunar occultations of quasar 3C 273 (*Credit* CSIRO/Parkes and 3C 273). © Copyright CSIRO Australia

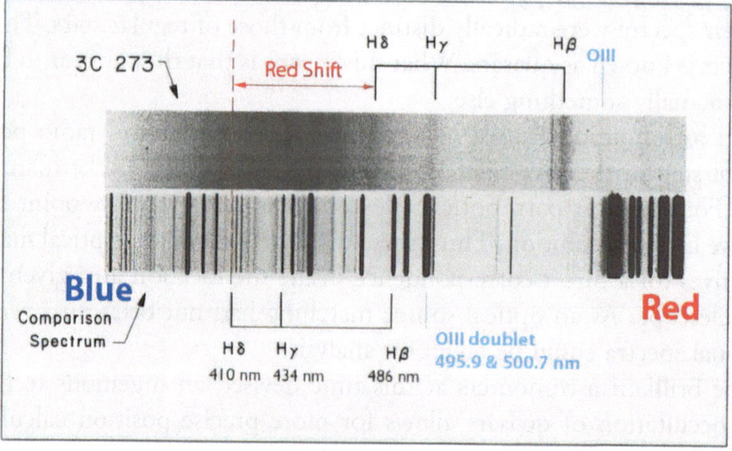

Fig. 2.9 The spectrum used by Maarten Schmidt to determine the redshift of 3C 273 (*Credit* CSIRO/Parkes and 3C 273). © Copyright CSIRO Australia

the celestial bodies' "ID card." On the other hand, it appeared to be challenging to determine what elements were present in the spectral lines of 3C 273 (Bolton 1969).

Schmidt, after considerable thought, ventured to propose that several of the apparent emission lines in the spectral range were from hydrogen. This meant that its spectrum had a redshift of 15.8% (Schmidt 1963), which was greater than anything that had been previously imagined. The redshift is primarily brought on by the expansion of the universe, and the greater the distance between the Earth and the star in question, the more pronounced the redshift will be. A redshift phenomenon of this magnitude placed 3C 273 at a distance of 2.4 billion light-years from Earth. Suddenly, quasars surpassed all other distant objects as the furthest ones spotted at the time.

Distance was not the most astonishing aspect. Even at such a great distance, there might be such a clear signal to the Earth due to the enormous amount of energy that was stored in quasars. This had left all scientists dumbfounded. As soon as the discovery was made, it sent shockwaves across the scientific world. Quasars are widely recognized as one of the four most significant scientific discoveries of astronomy made during the twentieth century. The accomplishment earned Maarten Schmidt a cover story in *Time* (TIME magazine cover: Maarten Schmidt—Mar. 11, 1966 n.d.). At the time of this significant discovery, the Parkes radio telescope, which played a prominent role in it, had just turned one year old, just in its infancy.

Parkes was so effective in its design that it became a model for later telescopes to emulate. While others have tried to replicate the success of Parkes, none have been able to match it.

On July 16, 1969, Apollo 11 carried three men to the Moon.

Four days later, on July 20, astronauts Neil Armstrong and Buzz Aldrin landed the Eagle lunar module on the Moon and broadcast the first-ever pictures of human beings on the Moon to the world (NASA, Apollo 11 mission overview 2015a).

The Goldstone Deep Space Communication Telescope (NASA, Goldstone Deep Space Communications Complex n.d.) was the primary instrument for transmitting data between the Earth and the Moon. This telescope in California, USA, was developed by NASA and, as its name suggests, was intended for interstellar communication.

The already gigantic Parkes' receiving area is outdone by 20% by Goldstone, which has an aperture of 70 m (Monaghan 2020). Despite sandstorms, Parkes, which served as a backup relay station, managed to outperform its California counterpart in terms of signal quality. This is the story at the beginning of this chapter.

In contrast to its larger rivals, Parkes was able to triumph thanks to its receivers' superior sensitivity. As the creator of compact high-precision radars, Taffy Bowen put an emphasis on receiver sensitivity, which has also been the key to creating a timeless mystique around the Parkes radio telescope.

In later years, the Parkes radio telescope was utilized as part of a wide variety of space programs, including the well-known Galileo expedition to Jupiter (Sarkissian 1997) and the Cassini mission to Saturn (Finley 2004).

Besides supporting space missions, the Parkes radio telescope studied pulsars (Manchester 2008).

While black holes may be the scariest thing out there, neutron stars likely come in a close second. A solar mass object will become a neutron star with a diameter of only 10 km when practically all the atoms inside the star collapse in a supernova explosion, the outer electrons and the protons in the nucleus synthesize neutrons, and the distant nuclei suddenly squeeze together. A teaspoon of this material would weigh as much as a mountain if it were located on the Earth. Certain neutron stars are dubbed "pulsars" because their compact size allows them to rotate at a fast speed per second and release constant pulses of radiation with a rotation period.

Everything about neutron stars, with their extremely harsh environments, is fascinating to astronomers. Parkes has excelled in this area. About 3,000 pulsars have been discovered by humanity so far, and Parkes was able to detect close to two-thirds of them (Lorimer et al. 2006).

After reading this, you'll likely agree that Parkes is a respected scientific "veteran" with a long list of accomplishments. Nonetheless, we are well aware that veterans have always had a hard time with the passage of time.

When 2001 rolled around, Parkes was approaching its fortieth year in operation. After being exposed to wind and rain for forty years, the mesh dish-shaped reflective surface had become rusty and had lost all of its shine. Like an old mill on the plain, the orange-red tower base that supported the enormous antenna had been baptized by time and had lost its sci-fi and futuristic aesthetic (see Fig. 2.10).

Nevertheless, this was really an optical illusion. Parkes was still accompanying astronomers and receiving data from the depths of the universe forty years after it was first constructed, and the surface rust had no discernible effect on its performance.

Payback is guaranteed in each situation. On July 24, 2001, an incredible signal from space was picked up and recorded by Parkes. The millisecond-long signal was too brief to cause the attention of the scientists using Parkes that day. Yet, six years later, this signal ushered in a new era of astronomy and helped us learn more about the universe.

Fig. 2.10 Parkes has rust on the surface. (*Credit* Benjamin Shaw)

An astronomer named Duncan Lorimer found an unexpected signal in 2007 while re-examining the data from 2001 (Lorimer et al. 2007) (see Fig. 2.11).

Being a millisecond pulse in the GHz band, this signal was quite similar to the pulsar signal. Still, the two weren't identical. As a result of interaction with plasmas in interstellar space, electromagnetic waves of varying frequencies can

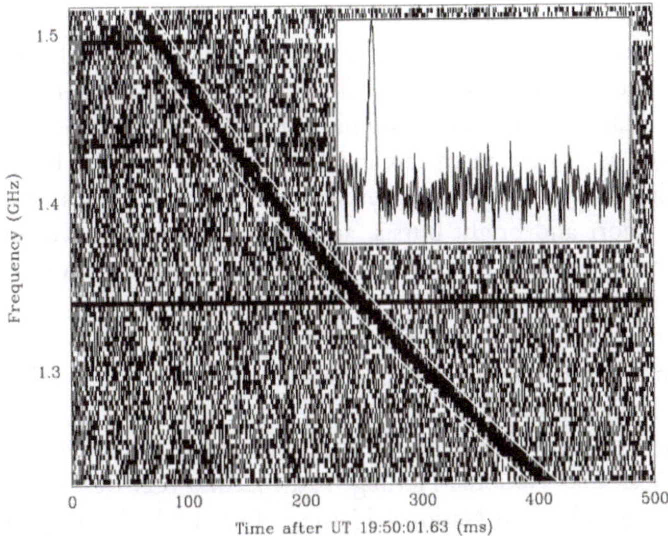

Fig. 2.11 Signal detected by Lorimer (*Credit* CSIRO). © Copyright CSIRO Australia

deviate very slightly from the speed of light in vacuum. This causes delays in the arrival of radio signals from faraway celestial bodies of varying frequencies. The distance traveled and the amount of plasma that these signals have had to travel through can be determined from their measured arrival time difference.

Different from past pulsars, the distance measured from this strange signal implied that this celestial object was not within our Milky Way Galaxy.

Due to the vastness of the universe, galaxies are typically located at great distances from the Earth. If the stars in our galaxy are the fish in a lake, then the galaxies themselves are the enormous bodies of water. When compared to the distance between fish in a lake, the distance between lakes is vastly greater. If these celestial objects are exceedingly distant and energetic like quasars, then astronomers will conclude that they are not in our galaxy. Astronomers have given this type of occurrence the catchy moniker "Fast Radio Burst" to better describe it.

By 2007, the Parkes Radio Telescope had lost its once-dominant position. Many new competitors had emerged on the stage in the more than forty years that have passed since it was constructed. The scientific community was slow to embrace Parkes' unexpected finding. Moreover, it is a standard practice in the scientific community that isolated evidence is not acknowledged. Yet, it appeared that some artificial signals could seem like fast radio bursts, thus compounding the problem. Further investigation revealed that these artificial signals originated from microwave ovens within the observatory (Petroff et al. 2015). The validity of Fast Radio Bursts (FRBs) was called into question for a period of time.

Six years later, in 2013, scientists using Parkes found four new FRBs after removing the microwave oven interference. The existence of FRBs has been verified at last (Petroff et al. 2014). The study of FRBs has come a long way since then. Many resources, both human and material, have been poured into this area, and as a result, many fascinating breakthroughs have been achieved.

Fast radio bursts are of great interest to astronomers because of their high event rate, short and bright bursts, and the vast distances over which they occur. Astronomers also have great expectations for them as a tool to use in the exploration of the universe.

There are a lot of unanswered questions about FRBs at the moment. Further unexpected results in their research are imminent. It all began with the Parkes radio telescope.

The Parkes Radio Telescope, which has surpassed sixty years old, was added to Australia's National Heritage List in 2020 (CSIRO, CSIRO Parkes radio telescope added to National Heritage List n.d.). Yet, its daily routine

Fig. 2.12 Parkes 1961 (*Credit* CSIRO). © Copyright CSIRO Australia, (10 August 2020) versus 2021 (*Credit* Depositphotos/Stokaji)

remains packed. It employs pulsars to seek for gravity waves and extraterrestrial civilization communications in addition to detecting fast radio bursts. The expectations for any given task are high (see Fig. 2.12).

Why, after sixty years of operation, is Parkes still holding its own in the vanguard of scientific inquiry?

The reason for this has been stated previously; the receiver is particularly sensitive.

The receiver is the radio telescope's retina, analogous to the photosensitive chip in an optical telescope. Although the large smooth reflector plays a crucial role, the signal it collects can only be retrieved by the receiver. With a radio telescope, the receiver's sensitivity is the limiting factor (Wilson et al. 2013).

Parkes has a three-legged focal point that extends out 27.4 m (CSIRO, Murriyang users guide n.d.) in front of the reflector, where the receivers are located. Several receivers can now be installed simultaneously in Parkes' focus cabin and switched swiftly depending on mission requirements. These sensitive receivers operate near absolute zero to minimize interference from thermal motion (see Fig. 2.13).

When a celestial body explodes, it often does so violently, releasing a great deal of energy, according to astronomers. To be more precise, the term "violent" only applies to the position of the heavenly body. Those enormous energies will continue to decrease as they traverse the vast universe, and by the time they reach the Earth, they will be extremely faint. If scientists want to "listen to thunder in silence," they need to keep improving telescopes' sensitivity.

Parkes is currently estimated to be 10,000 times more sensitive than it was in 1961 (Webster et al. 1999). This kind of adaptability has ensured that its legend will live on forever.

Fig. 2.13 The receiver of the Parkes radio telescope was replaced in 1996. (*Credit* Nick-D)

Because of the availability of radio observations at all hours, the Parkes radio telescope spends nearly all of its time staring into space, in addition to doing routine maintenance. More accurately, it resembles a penetrating "eye" in the fabric of human culture. It has survived the test of time and become more bright with age.

References

Ackermann, M., Ajello, M., Atwood, W.B., Baldini, L., Ballet, J., Barbiellini, G., Bregeon, J., et al.: The third catalog of active galactic nuclei detected by the fermi large area telescope. Astrophys. J. **810**, 14 (2015). https://doi.org/10.1088/0004-637X/810/1/14

Bird, T.S.: Role of the parkes radiotelescope in the first moon landing. In: 2019 IEEE International Symposium on Antennas and Propagation and USNC-URSI Radio Science Meeting, pp. 1377–1378 (Jul 2019). https://doi.org/10.1109/apusncursinrsm.2019.8888349

Bolton, J.G.: Extragalactic radio sources. Astron. J. **74**, 131 (1969). https://doi.org/10.1086/110786

Buderi, R.: The Invention That Changed the World: How a Small Group of Radar Pioneers Won the Second World War and Launched a Technological Revolution. Touchstone (1998)

CSIRO: Parkes radio telescope, Murriyang, 11 Apr 2023. CSIRO: https://www.csiro.au/en/about/facilities-collections/atnf/parkes-radio-telescope-murriyang

CSIRO: CSIRO Parkes radio telescope added to National Heritage List (n.d.). CSIRO: https://www.csiro.au/en/news/news-releases/2020/csiro-parkes-radio-telescope-added-to-national-heritage-list. Accessed 11 Apr 2023

CSIRO: Murriyang users guide (n.d.). CSIRO: https://www.parkes.atnf.csiro.au/observing/documentation/users_guide/pkug.html. Accessed 11 Apr 2023

Finley, D.: Radio Telescopes Will Add to Cassini-Huygens Discoveries, 22 Dec 2004. National Radio Astronomy Observatory: https://public.nrao.edu/news/radio-telescopes-will-add-to-cassini-huygens-discoveries/

Harland, D.: NASA's Moon Program: Paving the Way for Apollo 11. Springer (2011)

Haynes, R., Haynes, R.D., Malin, D., McGee, R.: Explorers of the Southern Sky: A History of Australian Astronomy. Cambridge University Press (1996)

Hazard, C., Mackey, M.B., Shimmins, A.J.: Investigation of the radio source 3C 273 by the method of lunar occultations. Nature **197**(4872), 1037 (n.d.). https://doi.org/10.1038/1971037a0

Kastner, S.O., Bhatia, A.K.: Explicit relations in Bowen fluorescence—applications to nebulae, the sun, Scorpius X-1, and laboratory plasmas. Astrophys. J. **362**, 745 (1990). https://doi.org/10.1086/169312

Lorimer, D.R., Bailes, M., McLaughlin, M.A., Narkevic, D.J., Crawford, J.: A bright millisecond radio burst of extragalactic origin. Science **318**(5851), 777–780 (2007). https://doi.org/10.1126/science.1147532

Lorimer, D.R., Faulkner, A.J., Lyne, A.G., Manchester, R.N., Kramer, M., McLaughlin, M.A., Crawford, F., et al.: The parkes multibeam pulsar survey–VI. Discovery and timing of 142 pulsars and a Galactic population analysis. Mon. Not. R. Astron. Soc. **372**(2), 777–800 (2006). https://doi.org/10.1111/j.1365-2966.2006.10887.x

Manchester, R.N.: The Parkes pulsar timing array project. In: AIP Conference Proceedings, vol. 983, pp. 584–592 (2008)https://doi.org/10.1063/1.2900303

Milne, D.: Parkes: Thirty Years of Radio Astronomy. CSIRO Publishing (1994)

Monaghan, H.: About the Deep Space Network, 30 Mar 2020. NASA: https://www.nasa.gov/directorates/heo/scan/services/networks/deep_space_network/about

NASAL: Apollo 11 mission overview, 17 Apr 2015. NASA: https://www.nasa.gov/mission_pages/apollo/missions/apollo11.html

NASA. (2015, February 19). July 20, 1969: One giant leap for mankind. NASA: https://www.nasa.gov/mission_pages/apollo/apollo11.html

NASA. (n.d.). Goldstone Deep Space Communications Complex. NASA: https://www.gdscc.nasa.gov/. Accessed 11 Apr 2023

Orchiston, W., Slee, B., Burman, R.: The genesis of solar radio astronomy in Australia. J. Astron. Hist. Herit. **9**(1), 35–56 (2006). https://doi.org/10.3724/SP.J.1440-2807.2006.01.03

Petroff, E., et al.: Identifying the source of perytons at the Parkes radio telescope. Mon. Not. R. Astron. Soc. **451**(4), 3933–3940 (2015).https://doi.org/10.1093/mnras/stv1242

Petroff, E., Bailes, M., Barr, E.D., Barsdell, B.R., Bhat, N.D., Bian, F., Keller, S., et al.: A real-time fast radio burst: polarization detection and multiwavelength follow-up. Mon. Not. R. Astron. Soc. **447**(1), 246–255 (2014).https://doi.org/10.1093/mnras/stu2419

Robertson, P.: Beyond Southern Skies: Radio Astronomy and the Parkes Telescope. Cambridge University Press (1992)

Sarkissian, J.M.: The parkes galileo tracks (Nov 1997). https://www.parkes.atnf.csiro.au/people/sar049/galileo/galileo_report.html

Sarkissian, J.M.: The Parkes observatory's support of the Apollo 11 Mission (2001)

Schmidt, M.: 3C 273: a star-like object with large red-shift. Nature **197**, 1040 (1963). https://doi.org/10.1038/1971040a0

Singal, A.K., Gopal-Krishna, Venugopal, V.R.: Lunar occultation observation of 65 radio sources at 327-MHZ list seven. Mem. Astron. Soc. India **1**, 14 (1979)

Sullivan III, W.T.: Cosmic Noise: A History of Early Radio Astronomy. Cambridge University Press (2009)

Sullivan, I.W.: John G. Bolton (1922–1993). BAAS (1994)

TIME magazine cover: Maarten Schmidt—Mar. 11, 1966 (n.d.). TIME.com: https://content.time.com/time/covers/0,16641,19660311,00.html. Accessed 11 Apr 2023

Webster, R., Kilborn, V., O'Brien, J., Staveley-Smith, L., Putman, M., Banks, G.: The Parkes multibeam blind HI survey. Int. Astron. Union Colloq. **171**, 291–298 (1999). https://doi.org/10.1017/s0252921100054464

Wilson, T.L., Rohlfs, K., Hüttemeister, S.: Tools of Radio Astronomy. Springer Science & Business Media (2013)

3

"Big Ear" Telescope: A Possible Alien Civilization's "Wow!" Signal

"Big Ear" Telescope

Although most Americans probably wouldn't remember anything about the 18th of August 1977 (Krulwich 2010), for Jerry R. Ehman, it was a momentous day. In the evening of that day, he finished his routine work, made himself a cup of coffee, and relaxed comfortably in a chair before picking up a lengthy paper tape that was full of data and looking at it with interest (Ehman 1997). He was looking through information collected by the "Big Ear" radio telescope at Ohio State University. When it came to displaying observational data, lengthy strips of paper were the norm in the 1970s (Ceruzzi 2003) rather than a screen. Ehman and other volunteers spent a significant amount of time per week going through paper tapes of data and analyzing them. Sadly, the tape's characters were usually as dense and useless as the substance they represent, amounting to little more than white noise (see Fig. 3.1).

Ehman had lost track of the number of tapes he had watched. His eyes were suddenly filled with a long string of characters. After reopening his eyes to take another look, he exclaimed, "Wow!" There were six characters at 10:16 on the tape from the telescope on August 15, 1977; these read 6EQUJ5 (Shostak 2012). This was three days before the tape was captured. The average person would not give this string of characters a second thought, but to Ehman, it was exceedingly odd. As he stayed there, his thoughts racing, he realized that among all the possibilities, the one that was least plausible also contained the most fascinating idea. Ehman was hesitant to expand on the idea, but it was churning up from the depths of his thoughts and would not

J. Wang, *Eye Beyond the Sky*, Astronomers' Universe,
https://doi.org/10.1007/978-981-99-9818-0_3

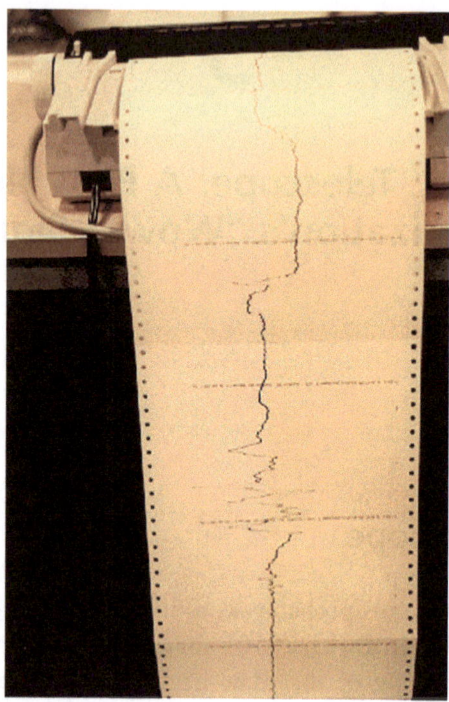

Fig. 3.1 The tape recorder is connected to the printer and generates the paper tape in real time. (*Credit* Ohio State University Radio Observatory and North American AstroPhysical Observatory)

be squelched. So, what does this character string actually mean? Let's start by deciphering the significance of these symbols (see Fig. 3.2).

There is just one interpretation of the symbols written on the paper tape: the signal intensity. Background noise is represented by a value of 1, 10 is indicated by the letter A, 11 by the letter B, and so on up to 35, which is indicated by the letter Z (Shuch 2011).

High-intensity signals (those with an amplitude of 5 or higher) were relatively uncommon on the signal tape. The string 6EQUJ5 emerged out of nowhere, rather "abrupt," in a sea of 1 s and 2 s. It's noteworthy that the signal strength U is 30 standard deviations above the background noise, the highest value ever recorded by the telescope (Gray and Marvel 2001) (see Fig. 3.3).

The signal might have lasted more than 72 s, despite the fact that the code 6EQUJ5 had only 6 characters, each of which indicated a length of 12 s. The telescope could only track an interplanetary signal for a maximum of 72 s due to its fixed position on the ground and the Earth's rotation. The

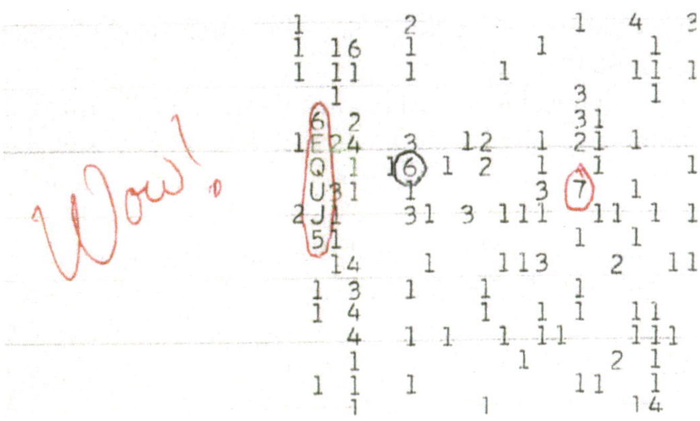

Fig. 3.2 "Wow!" signal (*Source* Big Ear Radio Observatory and North American AstroPhysical Observatory (NAAPO))

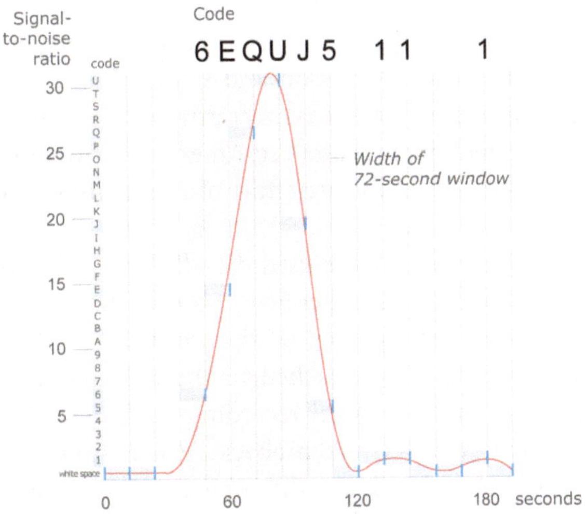

Fig. 3.3 "Wow!" signal's strength variation (*Credit* Maxrossomachin)

corresponding signal intensity varied in a highly conventional inverted bell form, suggesting it should originate from outside Earth (Kraus 1988).

The telescope was tuned to 1.42 GHz, which is also worth noting. Astronomers place a premium on this particular frequency. The name "hydrogen band" refers to a line in the electromagnetic spectrum that is generated by neutral hydrogen atoms as a result of variations in their energy levels. The wavelength corresponding to this frequency is about 21 cm, so hydrogen band is also called 21 cm line (Tarter 2001). The hydrogen band has many astronomical uses since hydrogen is the most common constituent of the universe. If an alien civilization ever tries to make contact with the

Earth, many scientists believe this is one of the most likely frequencies they will choose (Shostak 2012).

There were innumerable moments when Ehman had fantasized about what a signal from an alien civilization might look like, and the signal in front of him was consistent with his imagination. Ehman, overwhelmed with awe, drew a circle around the string and scribbled "Wow!" beside it in red with a trembling hand. The "Wow!" signal and the "Big Ear" radio telescope at Ohio State University Radio Observatory have been immortalized in SETI (Search for Extraterrestrial Intelligence) history (Shostak, Cosmic Company: The Search for Life in the Universe 2009) ever since Ehman marked that string with a red exclamation point.

Eavesdropping on cosmological electromagnetic waves for signals of advanced civilization is the most reliable technique to uncover extraterrestrial life.

In 1959, American physicist Philip Morrison and Italian physicist Giuseppe Cocconi released a paper in the journal *Nature* titled "*Search for Interstellar Communications*, (Morrison and Cocconi 1959)" in which they proposed using the hydrogen band to detect extraterrestrial life (Drake 1961).

Project Ozma, created by the renowned American astronomer Frank Drake (1930–2022), was the first deliberate effort of humanity to look for signs of extraterrestrial life.

One day in April of 1960, Drake used the US National Radio Astronomy Observatory's brand new 26-m telescope to observe the star Epsilon Eridani, ten light-years away, at a frequency of the exact 1.42 GHz. After only a few short minutes, the speaker attached to the telescope began emitting a succession of sounds, and the signal recording equipment began frantically scribbling on paper. Drake was taken aback. Was it really so easy to locate aliens if such a powerful signal was discovered at the outset of the search? That it could be so easy is, of course, absurd. After a few days of investigation, it was determined that the signal originated from an airplane. Further research eventually uncovered truth: this was a top-secret military experiment (Sagan and Shklovskii 1977).

During the course of 4 months and around 150 hours of intermittent monitoring, Drake found no valuable signals other than the airplane that led to the confusion (Tarter 2001). There were no results from the Project Ozma, although various SETI efforts had continued nonstop ever since.

To this end, NASA commissioned and published a report in 1971. A radio telescope array of 1,500 dishes, dubbed Project Cyclops (Oliver and Billingham 1971), was among the many helpful recommendations given in the report. This program persisted, albeit on a smaller scale, throughout the 1990s.

Ohio State University suggested a new SETI experiment in 1973, and the "Big Ear" radio telescope has become the hero of this tale.

Numerous large-aperture radio telescopes, such as the 76-m Lovell Radio Telescope in Britain, the 64-m Parkes Radio Telescope in Australia, and the 305-m Arecibo Telescope in Puerto Rico, sprang up one after another after World War II (Sullivan 2009) thanks to the civilian use of radar technology. John D. Kraus, a Professor of physics at Ohio State University (Kraus, Big Ear 1976), created a radio telescope that was quite "niche" in comparison to others. This telescope was not as elaborate or costly as others, but it used the principle of "simple but great, spending tiny money to do vast things" in its design (see Fig. 3.4).

Kraus put out an innovative plan for a radio telescope in 1955. It employed a fixed parabolic reflector as an antenna, which was both simple to produce and inexpensive, in contrast to conventional radio telescopes, which used a rotating paraboloid or spherical surface as a receiver and could point in different directions in the sky. Through another flat reflector that could be adjusted, Big Ear was able to see the sky between 63° north and 36° south with the rotation of the Earth, and the sensitivity is similar to a sphere with an aperture of 52.5 m for a very low price (Kraus et al. 1986), despite the immobility of the parabolic reflector. Kraus' students undertook a lot in the construction process for free, like making telescope models, designing steel beam designs, welding, and so on, which is why they were able to complete the project for under $250,000 (see Fig. 3.5).

By the time it was finished in 1962, the telescope had an area almost equal to three football fields (NASA 2020). This telescope, which consisted of a "bed surface" and two "bed heads," had an overall shape that resembled a high and low bed. Central bed dimensions were 152.4 m in length and

Fig. 3.4 From left to right: Lovell Radio Telescope (*Credit* Mike Peel; Jodrell Bank Centre for Astrophysics, University of Manchester), Parkes Telescope (*Credit* Diceman Stephen West), and Arecibo Telescope (*Credit* H. Schweiker/WIYN and NOAO/AURA/NSF)

Fig. 3.5 Staff member tested a corner reflector (*Source* NAAPO)

110.4 m in width. An aluminum sheet and concrete came together to form this plane. It's typically white in color. Its primary purpose was to protect weak signals from being swallowed up by the ground. Two "bed heads," of varying heights, sat at opposite ends of the "bed surface." The taller end was a flat reflector with 103.6 m in length and 30.68 m in height (Morrison et al., The Search for Extraterrestrial Intelligence, SETI 1977). Around 45° was the angle at which it was tilted with respect to the ground. A common feature of both high and low beds was an elevation at one end for support. The shorter end was a parabolic reflector measuring 110.4 m in length and 21.3 m in height. Cosmological radio transmissions were received by the plane reflector, reflected by the opposing parabolic reflector, focused by the horn feed adjacent to it, and finally processed by signal output. Many people lovingly refer to it as the "Big Ear" because, from a distance, the two massive reflecting surfaces resemble two human ears (see Fig. 3.6).

From the moment it was constructed, the "Big Ear" had been in constant use. Its primary mission at first was radio-band sky surveying. It issued a

Fig. 3.6 "Big Ear" radio telescope (*Source* NAAPO)

database of almost 20,000 radio sources in 1970 (Kraus, The Ohio Sky Survey and other radio surveys 1977). Of these, about half had never been discovered before, and two were the furthest objects ever spotted.

Since 1973, "Big Ear" had spent the vast majority of its time looking for transmissions coming from extraterrestrial civilizations (Kraus et al. 1986). When comparing electromagnetic signals produced by humans to those produced by nature, one finds that the former has a far narrower frequency range, while the latter has a much broader one. Collecting data from a wide variety of narrowband frequency bands simultaneously is a common requirement in the search for signals of extraterrestrial civilization. The earliest "Big Ear" had an 8-channel receiver in 1973; this was increased to 50 channels in the 1980s (Kraus, Big ear two: Listening for other-worlds 1995).

The data created by this telescope over several years now needed to be examined and interpreted visually. Many times, fresh information arrived before the old data had been thoroughly evaluated. Finding the right data amid a sea of information is like trying to find a needle in a haystack. The process of sifting through thick paper tapes in search of suspicious signals is time-consuming, labor-intensive, and prone to causing dissatisfaction with continued effort. Everyone was aware of the extremely low likelihood of the occurrence they anticipated. This meant there was no guarantee of success even if you put in your fullest effort. Regrettably, the National Science Foundation of the United States reduced funding for "Big Ear" since 1972, and data analysis had become the largest bottleneck. A large number of volunteers, however, provided their services to "Big Ear" for no charge (Zuckerman and Hart 1995).

Jerry R. Ehman was one of them. When he first saw the string of characters on August 18, 1977, he was so impressed that he scribbled a huge "Wow!" on the tape to express his amazement. This is why the signal is now known as "Wow!" signal.

The "Wow!" signal had a high intensity, a narrow frequency band, and an observation frequency of 1.42 GHz, which was consistent with scientists' hypothesis about the signal of extraterrestrial civilization, and it could be basically concluded that it did not originate from the Earth, from artificial satellites, or from natural phenomena. It originated in the direction of Sagittarius (Gray and Marvel 2001).

So, is the "Wow!" signal a message from another civilization? There is, sadly, no definite answer to this issue. Many observations have been made since then, including some by Ehman using the "Big Ear" in the months that followed, by scientists utilizing the META array at Oak Ridge Observatory from 1987 to 1989 (Horowitz and Sagan 1993), by researchers employing the 12-m radio telescope at the National Radio Astronomy Observatory of the United States in 1995, and by astronomers taking advantage of the powerful American Very Large Array Radio Telescope (VLA) in the twenty-first century. Countless scientists and amateur astronomers have tried repeatedly to tune in to the same frequency and point their telescopes in the same direction, but they have never managed to pick up the strong signal again (see Fig. 3.7).

Quite a few hypotheses have been proposed to explain the "Wow!" signal's origin. For example, Ehman once proposed that it originated on Earth and was then reflected back by debris in the planet's low Earth orbit. Because the hydrogen band is an astronomical protection frequency, the likelihood of artificial generation is low, and the calculation results do not back up the hypothesis that space debris can return such signals; establishing this premise is challenging. Two comets' neutral hydrogen clouds was proposed as the source of the signal by US astronomers in 2017, but this hypothesis was also disproved by other researchers (Paris and Davies 2015).

There hasn't been enough evidence to rule out the possibility of "Wow!" signal as a possible message from an advanced civilization outside our solar system, either. In astronomy, however, the possibility of extraterrestrial civilization is at the bottom of the list when trying to explain strange signals from space. Although it is every astronomer's fondest wish that humanity is not "orphaned" in the universe and that Earth's civilization is not the only

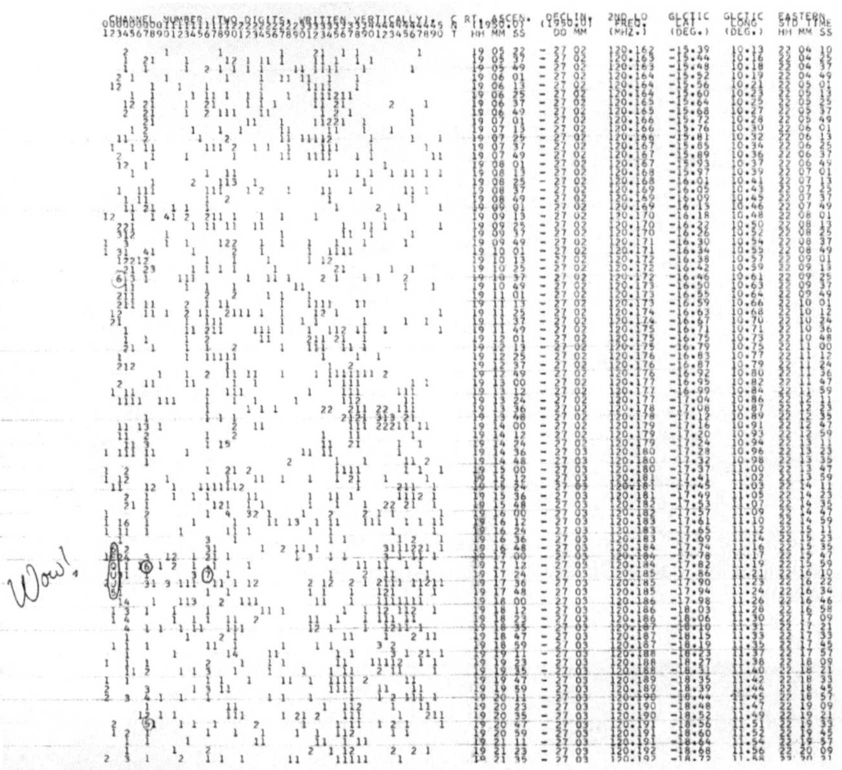

Fig. 3.7 Original paper tape of the "Wow!" signal (*Source* Ohio State University Radio Observatory and North American AstroPhysical Observatory.)

sentient one out there, science requires us to handle the most shocking results with the most rigorous evidence.

It's interesting that accompanied by the "Wow!" signal's fame, the string 6EQUJ5 experienced an unexpected surge in popularity. It has since become a personal mantra, the name of a popular song, and a password of mine (see Fig. 3.8).

Big Ear has gained widespread attention because of the "Wow!" signal, but we shall also bear in mind that throughout each peaceful night, it always diligently endeavored to prick its "ears" and listen closely to the sound from the universe. According to Guinness World Records, it has been the "longest-serving telescope in search of extraterrestrial intelligence," and it had spent decades trying to attain its crowning achievement. When the "Big Ear" radio telescope was taken down in 1998, the land it occupied was turned into a golf course (Kellermann et al. 2020) and homes. But humanity will never forget the work done by the "Big Ear" to discover an alien civilization. Even if we

Fig. 3.8 "Big Ear" before it was removed (*Source* Paul Wade (W1GHZ), Ohio State University Radio Observatory and North American AstroPhysical Observatory (NAAPO))

Fig. 3.9 Road sign after removal of "Big Ear" (*Source* Ohio State University Radio Observatory and North American AstroPhysical Observatory)

don't find the solution we're looking for, we'll learn a lot through trying, and that's worth a lot given the relative insignificance of the Earth in comparison to the rest of the cosmos (see Fig. 3.9).

References

Ceruzzi, P.E.: A History of Modern Computing, 2nd ed. MIT Press (2003)
Drake, F.: Project ozma. Phys. Today **14**(4), 40–46 (1961)

Ehman, J.R.: The Big Ear Wow! Signal What We Know and Don't Know About It After 20 Years, 1 Sep 1997. Ohio State University Radio Observatory: http://www.bigear.org/wow20th.htm

Gray, R.H., Marvel, K.B.: A VLA search for the Ohio state 'Wow.' Astrophys. J. **546**(2), 1171–1177 (2001). https://doi.org/10.1086/318272

Horowitz, P., Sagan, C.: Five years of Project META—an all-sky narrow-band radio search for extraterrestrial signals. Astrophys. J. **415**, 218 (1993). https://doi.org/10.1086/173157

Kellermann, K.I., Bouton, E.N., Brandt, S.S.: Is anyone out there? In: *Open Skies*, pp. 229–262. Springer, Cham (2020). https://doi.org/10.1007/978-3-030-32345-5_5

Kraus, J.D.: Big Ear. Cygnus-Quasar Books (1976)

Kraus, J.D.: The Ohio sky survey and other radio surveys. Vistas Astron. **20**, 445–474 (1977). https://doi.org/10.1016/0083-6656(77)90027-7

Kraus, J.D.: Big Ear Two: Listening for Other-Worlds. Cygnus-Quasar Books (1988)

Kraus, J.D.: Big Ear Two: Listening for Other-Worlds. Cygnus-Quasar Books (1995)

Kraus, J.D., Tiuri, M., Räisänen, A.V., Carr, T.D.: Radio Astronomy. Cygnus-Quasar (1986)

Krulwich, R.: Aliens found in Ohio? The "wow!" Signal, 28 May 2010. NPR: https://www.npr.org/sections/krulwich/2010/05/28/126510251/aliens-found-in-ohio-the-wow-signal

Morrison, P., Cocconi, G.: Searching for interstellar communications. Nature **184**(4690), 844–846 (1959)

Morrison, P., Billingham, J., Wolfe, J.: The Search for Extraterrestrial Intelligence, SETI. NASA - National Aeronautics and Space Administration (1977)

NASA: Radio, the big ear, and the wow! Signal, 2 May 2020. Science Mission Directorate: https://science.nasa.gov/radio-big-ear-and-wow-signal

Oliver, B.M., Billingham, J.: Project Cyclops: A Design Study of a System for Detecting Extraterrestrial Intelligent Life. NASA/Stanford University (1971)

Paris, A., Davies, E.: Hydrogen clouds from comets 266/P christensen and P/2008 Y2 (Gibbs) are candidates for the source of the 1977 "WOW" signal. J. Wash. Acad. Sci. **101**(4), 25–32 (2015). https://www.jstor.org/stable/jwashacadscie.101.4.25

Sagan, C., Shklovskii, I.S.: Intelligent Life in the Universe. Holden-Day, San Francisco (1977)

Shostak, S.: Cosmic Company: The Search for Life in the Universe. Cambridge University Press (2009)

Shostak, S.: Confessions of an Alien Hunter: A Scientist's Search for Extraterrestrial Intelligence. National Geographic Books (2012)

Shuch, H.P.: Searching for Extraterrestrial Intelligence: SETI Past, Present, and Future. Springer (2011)

Sullivan, W.T.: Cosmic Noise: A History of Early Radio Astronomy. Cambridge University Press (2009)

Tarter, J.: The search for extraterrestrial intelligence (SETI). Ann. Rev. Astron. Astrophys. **39**(1), 511–548 (2001)

Zuckerman, B., Hart, M.H.: Extraterrestrials: Where are They? CUP Archive (1995)

4

European Southern Observatory: The Reigning King of Optical Telescopes

European Southern Observatory

On the evening of May 25, 1998, a vehicle carrying several journalists raced along a straight road.

The Atacama Desert flanked the route in both directions. Without a cloud in sight, the afterglow of the sun gradually turned the western sky from deep blue to purple (see Fig. 4.1).

The Milky Way had finally shown, and the beautiful night sky shone brighter than the city's paved streets decorated by street lights.

After a long day of travel, the reporters finally made it to their destination in the dead of night. Meanwhile, the European Southern Observatory's Paranal Observatory control room was flooded with light. According to the engineers, there had never been so many people here since the construction was completed.

More than 1,300 km separated the control room from Santiago, Chile's capital. It had been ESO's first public test of its massive telescope—Very Large Telescope (VLT)—that attracted reporters to travel from far and wide (ESO, A Great Moment for Astronomy—VLT First Light Successfully Achieved 1998a). The control room was silent as everyone watched the central screen for the results of the maiden test.

Jason Spyromilio (ESO, VLT first light, 25 May 1998 1998b), then engineer in charge of debugging, leaped out of his chair as if possessed. Massimo Tarenghi (Catapano 2015), then project manager, controlled his emotions and yelled the number displayed on the screen "50"!

J. Wang, *Eye Beyond the Sky*, Astronomers' Universe,
https://doi.org/10.1007/978-981-99-9818-0_4

Fig. 4.1 The Atacama desert's winding mountain road (*Credit* ESO/J. Dommaget)

He was referring to VLT's observation resolution of 50 ms, which would make it one of the most powerful telescopes in the world (see Fig. 4.2).

Findings having been verified, the entire control room was filled with a feverish atmosphere (see Fig. 4.3).

Fig. 4.2 Paranal observatory's control room in 1998 (*Credit* ESO)

Fig. 4.3 ESO's construction site (*Credit* ESO)

In a joint news conference conducted by eight European member states two days later, the engineers unveiled the first images captured by the Very Large Telescope. For a moment, the news outlets around the globe were rushing to cover the breaking scoop (Wilkinson 1998), with many quoting Riccardo Giacconi's statements as then ESO director general, "This is a magnificent day in the history of international astronomy" (see Fig. 4.4).

After barely a day of being the center of spotlight, however, the focus shifted from VLT to another significant astronomical discovery. NASA made a prominent announcement that the Hubble Space Telescope might have captured the first image of an exoplanet (Nemiroff 1998) (see Fig. 4.5).

Despite this discovery's significance, it was not peer-reviewed. It didn't take long for NASA to acknowledge that the so-called exoplanets had been nothing more than the slightly dimmed stars. Regardless of whether NASA did it intentionally, the fact that two space agencies were competing for hot spots left astronomy fans the impression that a strong gunpowder smell pervaded the air.

So why exactly did the Very Large Telescope of ESO make the haughty folks at NASA green with envy? The establishment of the European Southern

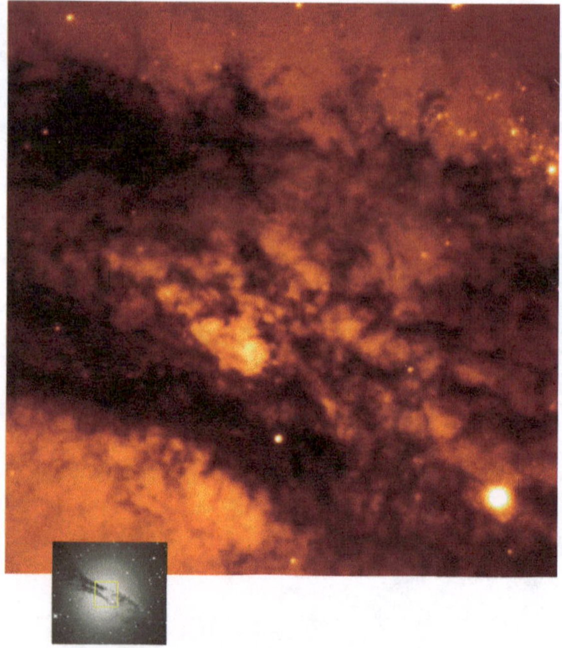

Fig. 4.4 Alpha Centauri A taken by VLT in May 1998 (*Credit* ESO)

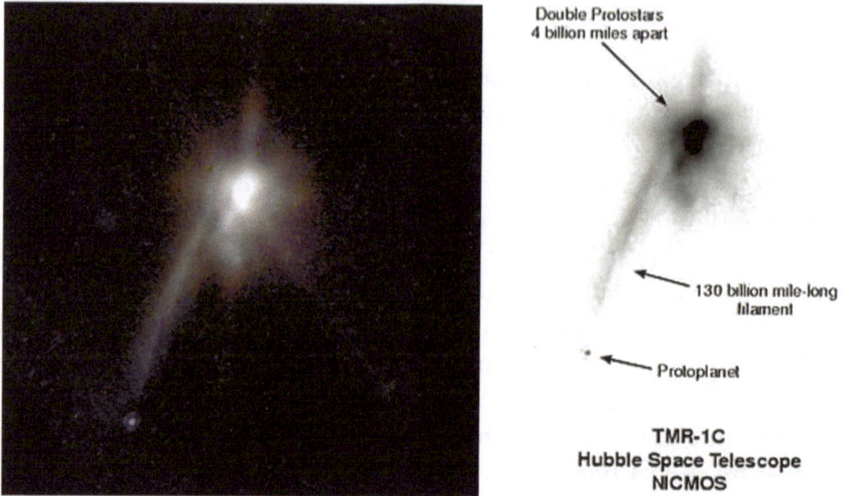

Fig. 4.5 Hubble's first direct look at a possible planet around another star (*Left Picture's Credit* Susan Terebey (Extrasolar Research Corp.), and NASA/ESA; *Right Picture's Credit* NASA/ESA)

Observatory in 1962 might be an ideal place to start looking for the answer to this question.

There is no doubt that Galileo's construction of the first astronomical telescope in 1609 helped Europe establish a strong identity as the world center for astronomy in the following three centuries. Taking advantage of the ever-advancing observational equipment, generations of outstanding astronomers turned Europe into the cradle of the greatest astronomical discoveries.

However, the United States became an upstart overnight on the other side of the ocean as a result of two world wars that followed one another, while Europe, which served as the primary battlefield, was riddled with holes. Successive and successful investments in the world's largest Hooker Telescope and Hale Telescope had propelled the United States past Europe to the throne of astronomy on a global scale.

European astronomers yearned to regain the splendor in the old days moment by moment.

Finally, a dozen European astronomers met at Leiden University in the Netherlands on January 26, 1954, and signed a declaration. Known to the world as The Leiden Declaration, this document has been historic in scope. It paved the way for the creation of the ESO (ESO, The Leiden Declaration 2009).

Large observatories had been primarily constructed in the northern hemisphere, and the sky monitoring of the southern hemisphere was limited to a few selective regions. Considering that the southern sky contains both the core of the Milky Way and the stars nearest to the solar system, the astronomers planned to pool the resources of European nations all together in order to construct a state-of-the-art observatory in the southern hemisphere—The European Observatory, abbreviated as ESO (Blaauw 1991).

Thus, ESO set its sights on becoming the world's premier observatory from the very outset of its founding, and the United States was unquestionably its main competitor.

However, the world's number one can never be achieved only by speaking aloud. If you're planning on constructing a state-of-the-art observatory, finding an appropriate location will be your first challenge. A high-altitude observatory places significant demands on the environment. First, the dense atmosphere and the water molecules in it can cause the attenuation of light. As a result, the greater the height and the drier the climate, the better the location (Leverington 2017).

To guarantee the adequate observation time, there must also be as many clear nights as feasible. Besides, the greater the distance from the city, the better it is for reducing light and air pollution interference. In addition, irregular turbulence in the air can also distort light, resulting in starlight distortion. Imagine looking at the moon through water. Only a calm body of water can reveal a distinct image. This is referred to as astronomical seeing, which is also a crucial requirement for the observatory's location (Burrows 2020).

Since 1955, scientists spent a total of nine years scouring the plateaus and deserts of the southern hemisphere in search of an optimal location. They frequently traveled by laden vehicles in pairs and endured hellish conditions under which they ate, slept, and worked.

A trip to the Atacama Desert in northern Chile in November of 1962 offered some hope at the end of the day. The Karoo Desert in South Africa had been selected as an alternate site, but the scientists were startled to discover that the climatic conditions of Atacama had been significantly superior (Blaauw 1991). It turned out to be a holy destination specifically designed for astronomical observation.

Over 300 clear nights a year characterizes the Atacama Desert, making it one of the driest places on Earth. It is difficult for even the most basic single-celled organisms to survive this harsh environment of sand and gravel that serves as the soil here (Schween et al. 2020). Ignoring the sky color, it looks a lot like Mars (see Fig. 4.6).

Fig. 4.6 The Atacama desert (*Credit* Depositphotos/xura)

Minerals are the only source of wealth in a place like this. In May of 1964, the Chilean government sold ESO 627 km^2 of land and a mountain named La Silla, which stood at an altitude of 2,400 m above sea level, after determining that the area contained no economically significant mineral deposits. A small cost of $8,000 was ultimately assessed (Lodewijk 2012). It's only worth around $100,000 now, after adjusting for inflation.

It wasn't until 1966 that the first telescope had been installed at ESO's La Silla Observatory. It was a reflective telescope with an aperture of 1.04 m, which continues to be utilized as a photometric telescope at the forefront of today's astronomy study (see Figs. 4.7 and 4.8).

After that, progress on the telescopes at the La Silla Observatory picked up speed dramatically. In its short history, the La Silla Observatory has earned the enviable moniker "Telescope Park" for its accommodation of approximately 30 astronomical telescopes (Saviane et al. 2020).

The most well-known are the 3.6-m Telescope on Cerro La Silla constructed in 1976 and the 3.58-m New Technology Telescope (NTT) constructed in 1989 (see Fig. 4.9).

The ESO's 3.6-m Telescope has been the first major instrument of its kind in an effort to compete for the best in the world. The 5.08-m-diameter Hale telescope, which was constructed in 1948 at the Palomar Observatory in the

Fig. 4.7 La Silla mountain in 1966 (*Credit* ESO/R.H.E.Holder)

Fig. 4.8 Staff installed a 1.04-m reflective telescope in 1966 (*Credit* ESO)

Fig. 4.9 The 3.6-m telescope (*Credit* ESO)

United States, was the largest optical telescope in the world at the time (ESO, Telescopes and Instrumentation n.d.) (see Fig. 4.10).

It finally took more than 20 years to develop and construct the 3.6-m Telescope. Even though it was still a long way from being the largest in the world, it made its way into the elite club of top telescopes. Overnight, scientists from all throughout Europe submitted applications for observation, and the 3.6-m Telescope quickly became a sought-after instrument.

The number of countries that made up the European Southern Observatory expanded from six to eight due to the attractiveness of the organization's high-quality instruments. Although the financing crisis was mitigated thanks to the increase in member states, tensions in employing the existing telescopes had increased. To ease the burden on the La Silla Observatory, several member states even transferred instruments from other observatories.

There had been calls from some astronomers for ESO to spend money on constructing a new, state-of-the-art telescope that could easily compete with or even surpass the performance of American Hale Telescope.

However, enormous telescopes cannot be constructed only with money, as it is not the lack of labor that stands in the way of their development, but rather the universal force of gravity.

The mirror of a small-aperture telescope will experience only minor deformation due to gravity. The effect of this deformation, however, will become

Fig. 4.10 Hale telescope (*Credit* Caltech/Palomar Observatory)

more noticeable as the lens size grows. Larger lenses can sometimes improve observational precision, but eventually they hit the limits of the telescope.

The cost of building a lens doubles every time engineers increase the thickness of the main mirror to counteract the effects of gravity. An optical telescope with a 5-m aperture is nearly the limit of what can be achieved within a reasonable budget (Rutten and Van Venrooij 1988).

In any case, somebody will always discover a solution if one is required. In 1982, Raymond Wilson, then an engineer at ESO, originally proposed "active optics" (Wilson et al. 1987). The premise is that a supporting mechanism is mounted under the mirror of the telescope, which can be detected and rectified in real time when the mirror is bent, thus counteracting the influence of gravity. That is to say, the telescope is able to monitor the lens's deformation and make necessary adjustments so that it performs optimally at all times.

In spite of this, many people were pessimistic about this technology since, at the time, under the level of automation technology, it was incredibly impossible to implement the real-time and autonomous detection and adjustment of the telescope. No surprise a respected American astronomer stated that Wilson was "talking about a pipe dream" (see Fig. 4.11).

Wilson, however, did not back down. After years of experiment, testing, and refinement, he ultimately helped ESO build the world's first telescope with active optical technology—The New Technology Telescope (NTT). It has a 3.58-m aperture, which is smaller than the previous 3.6-m telescope but results in a resolution that is three times as good. It also costs only a third as much and weighs half as much as its predecessor (Tarenghi and Wilson 1989).

Fig. 4.11 The New Technology Telescope (*Credit* ESO/José Francisco Salgado (jose-francisco.org))

Popular US astronomy publication *Sky & Telescope* declared the NTT "The Best Telescope Yet" and featured it on its cover that year (see Fig. 4.12).

Europe has finally made a comeback against the United States in telescope manufacturing technology (see Fig. 4.13).

The NTT's remarkable performance is a prime example of how crucial technological progress is to the field of astronomy.

Wilson was awarded the 2010 Kavli Award for Astrophysics for his pioneering work in developing active optical technology (ESO, ESO telescope designer Raymond Wilson wins prestigious Kavli Award for Astrophysics 2010), which is now a standard for all large telescopes around the world.

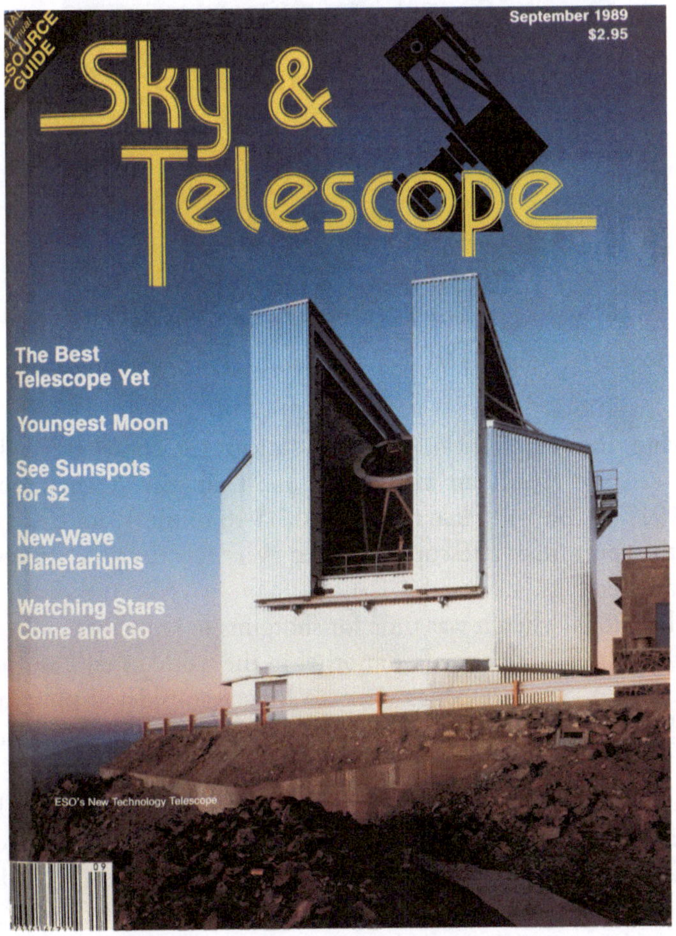

Fig. 4.12 NTT appeared on the cover of *Sky & Telescope* (*Credit* ESO/Sky & Telescope)

Fig. 4.13 Hooker telescope at Mt. Wilson (*Credit* Antony-22)

Observing the progression of astronomical telescopes throughout the twentieth century reveals an intriguing rule: from the 2.5-m Hooker telescope in 1917 to the 5-m Hale telescope in 1948 to the 10-m Keck telescope in 1993, the aperture of telescopes doubles every few decades, somewhat like Moore's law (see Figs. 4.14 and 4.15).

Since the 1980s, when it was time for shotguns to switch weapons, 8–10-m telescopes had been built one after another in the world's major industrialized countries. The 3–5-m telescopes were no longer considered premium products. Under these conditions, ESO obviously could not rest comfortably on its back.

ESO had planned to build a 10-m telescope—The Very Large Telescope (VLT) long before the completion of the New Technology Telescope. The design called for a location that was significantly larger than what had been available in order to house VLT, which would consist of four 8.2-m unit telescopes and four 1.8-m movable auxiliary telescopes (Bely 2006).

Fig. 4.14 Hale telescope dome (*Credit* Mike Peel)

Fig. 4.15 W. M. Keck observatory (*Credit* z2amiller)

Therefore, ESO re-signed a deal with the Chilean government to locate a new site for VLT: Mount Cerro Paranal, more than 600 km northwest of La Silla. At an altitude of over 2,600 m above sea level, this location offered superior viewing conditions to nearby La Silla.

In 1993, the VLT infrastructure project was in full swing when a Chilean court unexpectedly issued a subpoena. It turned out that a native family by the name of Santa Cruz claimed that their ancestors had served the state and so been awarded Mount Cerro Paranal 50 years ago. They would like to regain control of the mountain and demanded that the VLT project either compensate for their financial losses or be shut down.

Curiously, the Chilean court agreed to hear the complaint and ordered ESO to immediately cease all operations. Whether or not the Santa Cruz family had convincing evidence, ESO could not afford to wait for the trial to conclude.

After lengthy negotiations between the ESO's headquarters and the Chilean government, the latter caved and handed the Santa Cruz family $8.8 million. In return, astronomers in Chile would receive 10% of the VLT's total observation time. Despite the fact that the disagreement was resolved, the whole three years (ESO, ESO and Chile sign agreement (1996) 2011) have passed since it occurred.

The VLT's first unit telescope (UT1) was completed and put through its paces in May 1998 (Giacconi 1998). The news really spooked the Americans across the Atlantic. After their completion, the observational capabilities of the four 8.2-m telescopes would be comparable to that of America's largest Keck telescope. Consequently, even NASA was in a rush to provide images of exoplanets in an effort to gain prominence. This set the stage for the background story presented at the chapter's outset.

The VLT and its final auxiliary telescope (AT4) were finished in December 2006.

The VLT's single mirror may view stars 4 billion times fainter than the resolution limit of the human eye (Herbst 2003). Using all of its telescopes in tandem, VLT can achieve a resolution 25 times greater than its single mirror.

If you want to work in a first-class observatory, build one, as Massimo Tarenghi, then head of the VLT Project, once advised. His wish had been granted at last.

ESO's global standing has been significantly strengthened by the completion of VLT (see Fig. 4.16).

There are around 2,000 applications for observation time at ESO each year, which is four to six times as long as the observation time that can be scheduled. ESO's instruments have amassed 65 TB (Pasian 2013) of crucial astronomical data, including over 1.5 million images, on the basis of which, more than 2 papers have been published per day on average.

If you can't muster any emotion for these numbers, I'll list a few more well-known scientific findings.

Using the data of Type Ia supernova observed by ESO's 3.6-m Telescope, New Technology Telescope, and Very Large Telescope, scientists confirmed the accelerated expansion of the universe (Perlmutter et al. 1999). In 2011, this accomplishment was awarded the Nobel Prize in Physics (see Fig. 4.17).

Fig. 4.16 VLT after completion (*Credit* J.L. Dauvergne and G. Hüdepohl (atacamaphoto.com)/ESO)

Fig. 4.17 ESO's panoramic picture (*Credit* ESO/Y. Beletsky)

In 2004, scientists captured the first image of an exoplanet close to a brown dwarf star 230 light-years away using VLT (Chauvin et al. 2004). This discovery creates an entirely new field of astrophysics: spectral research and imaging of planetary systems (see Figs. 4.18 and 4.19).

In 2008, after a total of sixteen years, scientists confirmed the existence of a supermassive black hole in the center of the Milky Way by using ESO's New Technology Telescope and Very Large Telescope to track stars in the Milky Way's central region (Ghez et al. 2008). In 2020, this achievement was awarded the Nobel Prize in Physics.

In 2016, scientists using ESO's 3.6-m Telescope and Very Large Telescope discovered an Earth-like planet, Proxima b, in the habitable zone of the closest

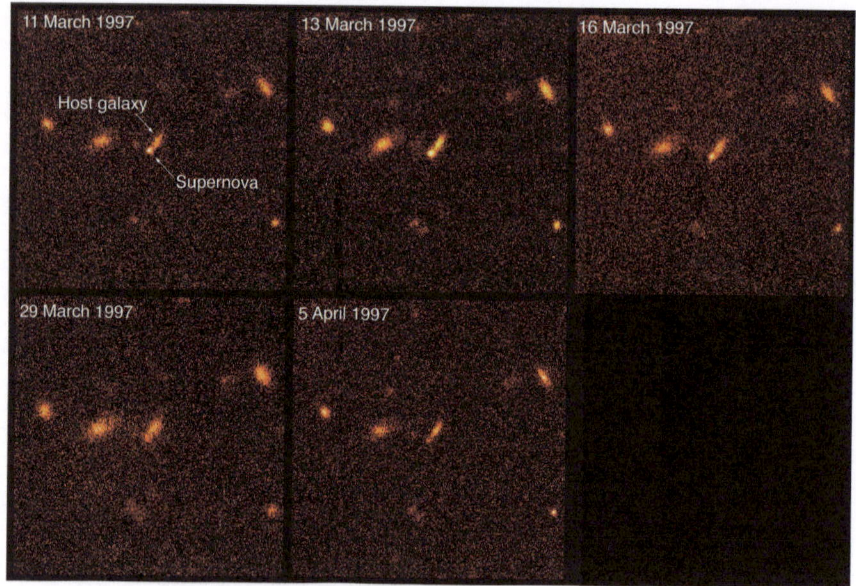

Fig. 4.18 ESO's observations and discoveries that won the 2011 Nobel Prize in Physics (*Credit* ESO)

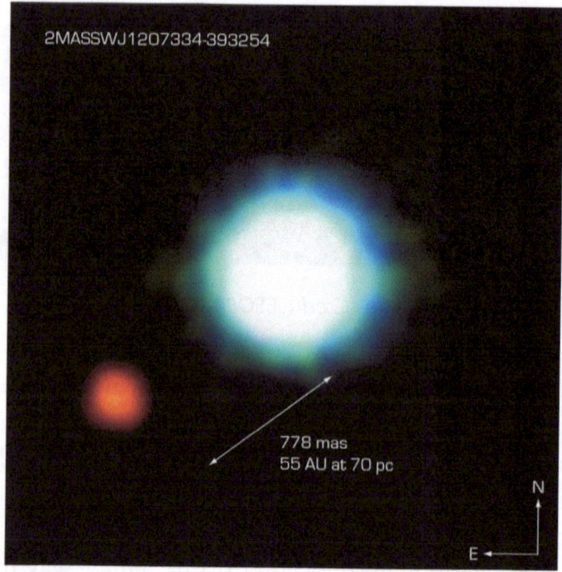

Fig. 4.19 ESO's discovery of an exoplanet (*Credit* ESO)

Fig. 4.20 ESO's observations and discoveries that won the 2020 Nobel Prize in Physics (*Credit* ESO/S. Gillessen et al.)

star, Proxima. Proxima b has been the closest exoplanet that may support life (Anglada-Escudé et al. 2016). To put it more science fictionally, the possibility of trisomy cannot be ruled out (see Fig. 4.20).

These scientific achievements, which have garnered worldwide attention, not only excite humans each and every time, but also arouse the insatiable thirst for knowledge possessed by the human race (see Fig. 4.21).

ESO's Council gave their stamp of approval to the European Extremely Large Telescope (E-ELT) in 2012. The extraordinary aperture of the telescope, at 39 m, is matched only by its high standard of design (ESO, Green Light for E-ELT Construction 2014). The finished product will be the largest optical telescope ever built (see Fig. 4.22).

It's incredible that in the desert with the fewest signs of life on Earth, there erect massive devices that are most likely to discover extraterrestrial life. Every clear night is utilized to its utmost. In this inaccessible desert, celestial artifacts are observing the stars in the southern hemisphere, and a group of astronomers, far from the bustle of urban centers, are dedicated to uncovering the secrets of the universe. Nothing, in my opinion, could be more emblematic of human's enduring curiosity than such a picture.

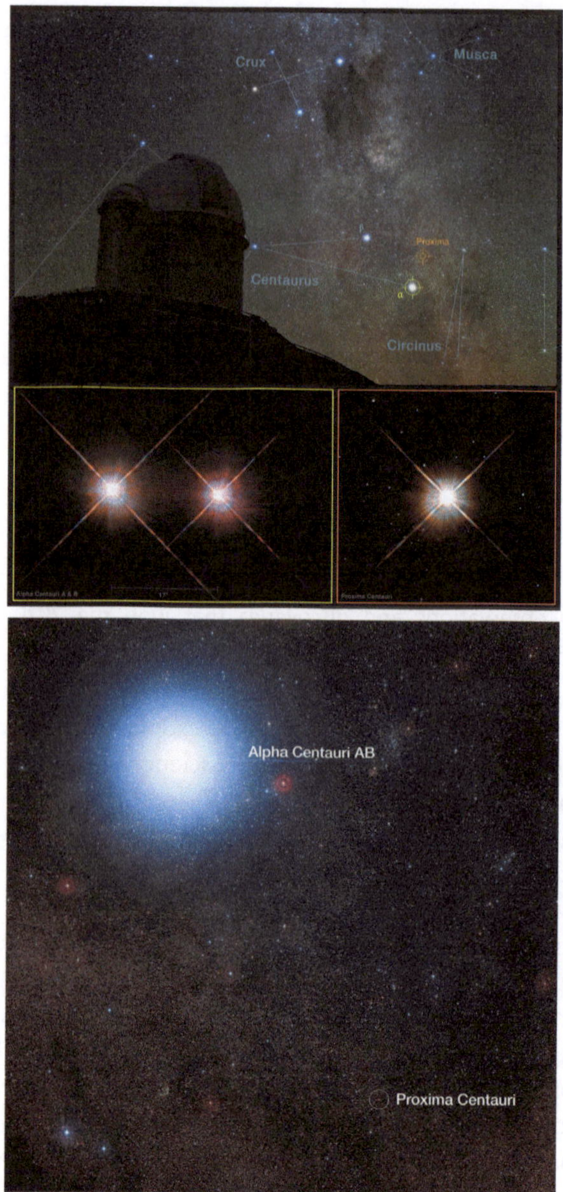

Fig. 4.21 ESO's discovery of the Earth-like planet Proxima b (*Upper Picture's Credit* Y. Beletsky (LCO)/ESO/ESA/NASA/M. Zamani, *Lower Picture's Credit* Digitized Sky Survey 2)

Fig. 4.22 E-ELT under construction (*Credit* ESO)

References

Anglada-Escudé, G., Amado, P.J., Barnes, J., Berdiñas, Z.M., Butler, R.P., Coleman, G.A., Kiraga, M., et al.: A terrestrial planet candidate in a temperate orbit around Proxima Centauri. Nature **536**, 437–440 (2016). https://doi.org/10.1038/nature19106

Bely, P.: The Design and Construction of Large Optical Telescopes. Springer Science & Business Media (2006)

Blaauw, A.: ESO's Early History: The European Southern Observatory from Concept to Reality (1991)

Burrows, D.N.: The WSPC Handbook of Astronomical Instrumentation (In 5 Volumes). World Scientific Publishing Co. (2020)

Catapano, P.: Massimo Tarenghi: a lifetime in the stars, 26 Aug 2015. CERN Courier: https://cerncourier.com/a/massimo-tarenghi-a-lifetime-in-the-stars/

Chauvin, G., Lagrange, A.-M., Dumas, C., Zuckerman, B., Mouillet, D., Song, I., Lowrance, P., et al.: A giant planet candidate near a young brown dwarf. Astron. Astrophys. **425**(2), 29–32 (2004)https://doi.org/10.1051/0004-6361:200400056

ESO: A Great Moment for Astronomy—VLT First Light Successfully Achieved, 27 May 1998. Eso.org: https://www.eso.org/public/news/eso9820/. Accessed 11 Apr 2023

ESO: VLT first light, 25 May 1998. Eso.org: https://www.eso.org/public/images/im15f6_1-cc/. Accessed 11 Apr 2023

ESO: The Leiden Declaration, 3 Dec 2009. Eso.org: https://www.eso.org/public/images/eso_statement_26-01-54/. Accessed 11 Apr 2023

ESO: ESO telescope designer raymond wilson wins prestigious kavli award for astrophysics, 4 Jun 2010. Eso.org: https://www.eso.org/public/news/eso1022/. Accessed 11 Apr 2023

ESO: ESO and Chile sign agreement (1996), 18 Feb 2011. Eso.org: https://www.eso.org/public/images/chile-sign-agreement-vlt/. Accessed 11 Apr 2023

ESO: Green Light for E-ELT Construction, 4 Dec 2014. Eso.org: https://www.eso.org/public/news/eso1440/. Accessed 11 Apr 2023

ESO: Telescopes and Instrumentation, (n.d.). Eso.org: https://www.eso.org/public/teles-instr/. Accessed 11 Apr 2023

Ghez, A.M., Salim, S., Weinberg, N.N., Lu, J.R., Do, T., Dunn, J.K., Naiman, J., et al.: Measuring distance and properties of the milky way's central supermassive black hole with stellar orbits. Astrophys. J. **689**(2), 1044–1062 (2008). https://doi.org/10.1086/592738

Giacconi, R.: First light of the VLT unit telescope 1. Mess. **92**, 2–3 (1998)

Herbst, T.: Interferometry with the large binocular telescope. In: The Very Large Telescope Interferometer Challenges for the Future, pp. 45–53. Springer Netherlands (2003). https://doi.org/10.1007/978-94-017-0157-0_4

Leverington, D.: Observatories and Telescopes of Modern Times. Cambridge University Press (2017)

Lodewijk, W.: Europe's Quest for the Universe. EDP Sciences (2012)

Nemiroff, R.: Astronomy Picture of the Day: May 29, 1998—An Extrasolar Planet?, 29 May 1998. APOD: https://apod.nasa.gov/apod/ap980529.html. Accessed 11 Apr 2023

Pasian, F.: Data Management and Archiving in Astronomy: Status and Challenges for the Future (2013)

Perlmutter, S., Aldering, G., Goldhaber, G., Knop, R.A., Nugent, P., Castro, P.G., Nunes, N.J.: Measurements of Ω and Λ from 42 high-redshift supernovae. Astrophys. J. **517**, 565(1999)https://doi.org/10.1086/307221

Rutten, H.G., Van Venrooij, M.A.: Telescope Optics : A Comprehensive Manual for Amateur Astronomers. Willmann-Bell (1988)

Saviane, I., Leibundgut, B., Schmidtobreick, L.: Report on the ESO workshop "The La Silla Observatory—From Inauguration to the Future." Mess. **179**, 36–40 (2020)

Schween, J.H., Hoffmeister, D., Löhnert, U.: Filling the observational gap in the Atacama Desert with a new network of climate stations. Global Planet. Change **184** (2020).https://doi.org/10.1016/j.gloplacha.2019.103034

Tarenghi, M., Wilson, R.N.: The ESO NTT (new technology telescope): the first active optics telescope. In: SPIE Proceedings (1989). https://doi.org/10.1117/12.960835

Wilkinson, J.: BBC News—BBC One O'Clock News, 27/05/1998, First pictures from the Very Large Telescope, 27 May 1998. BBC: https://webcache.googleusercontent.com/search?q=cache:tqCBGzdeX34J:https://www.bbc.co.uk/programmes/p00bc9yk&hl=zh-CN&gl=us. Accessed 11 Apr 2023

Wilson, R.N., Franza, F., Noethe, L.: Active optics. J. Mod. Opt. **34**(4), 485–509 (1987). https://doi.org/10.1080/09500348714550501

5

Arecibo Telescope: On the Radio Telescope Throne for Four Decades

Arecibo Telescope

Puerto Rico, situated at the crossroads of North and South America, is known as the "Pearl of the Caribbean Sea" because of its beautiful karst topography and tropical flora (Pitzer 2009). Formerly the largest single-aperture radio telescope in the world, Arecibo is nestled amid the lush mountains of Puerto Rico (see Fig. 5.1).

Cloudless blue skies greeted people on the afternoon of November 16, 1974. Around the Arecibo radio telescope, tents doted the mountainside.

Time had come. Small groups of people emerged from the tent and gathered around the enormous 305-m-diameter spherical antenna, as though awaiting the start of a ceremony. The atmosphere on the spot reached a boiling point when a ticking sound, reminiscent of a radio transmission, emanated from the speaker.

Around three minutes were occupied by the tickling sounds. The Arecibo radio telescope in front of the audience used a transmitter with a peak power of up to 1 megawatt to send a binary number with a length of 1,679 bits into deep space during this time period. Since the information travels through space at the speed of light, it might eventually be intercepted. The ticking sounds that listeners picked up on were the result of this data being converted to audible frequencies (Wolfram 2017).

The building blocks of life, the structure of the solar system, and the DNA double helix are all included in this binary number, along with countless other pieces of information crucial to human civilization. The M13 cluster,

J. Wang, *Eye Beyond the Sky*, Astronomers' Universe,
https://doi.org/10.1007/978-981-99-9818-0_5

Fig. 5.1 Arecibo in the lush mountains (*Credit* Depositphotos/brianirwin.aol.com)

which is located at a distance of 25,000 light-years and contains 300,000 stars, was the intended recipient of the data (The Staff at the NAIC 1975).

This is reportedly the first time that humans have formally communicated with an alien culture through radio waves. These signals will, over the course of tens of thousands of years, reach distant star clusters with the assistance of the immense transmission power of the Arecibo radio telescope (see Fig. 5.2).

To those who have read *The Three-Body Problem*, the question "How can human beings do such a ridiculous thing?" may have occurred to them unintentionally. Doesn't this amount to a self-murder? Remember, though; the year was 1974. It is necessary to begin with the space race between the United States and the Soviet Union in order to make sense of the American actions, which appear to be somewhat irrational (see Fig. 5.3).

The first artificial satellite around the Earth was launched into orbit by the Soviet Union on October 4, 1957. The National Aeronautics and Space Administration (NASA) and the Advanced Research Projects Agency (ARPA) were both founded in the following year (Divine and Divine 1993). With that, the curtain was thrown open on the space race between the United States and the Soviet Union.

On the surface, it appeared that the space rivalry was about sending humans into space, but the real motivation was to develop effective intercontinental ballistic missiles (ICBMs) capable of worldwide deployment. The

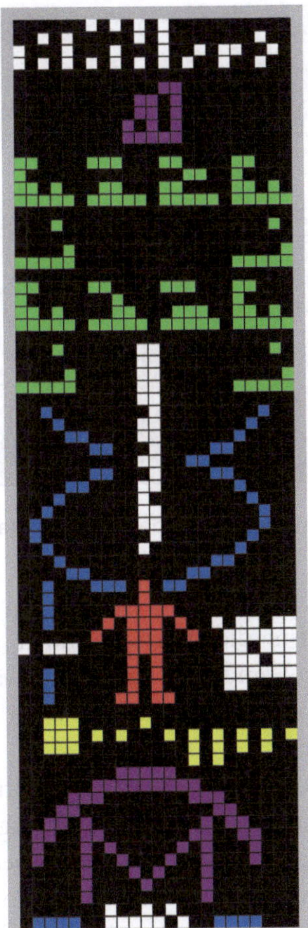

Fig. 5.2 Arecibo message (*Credit* Arne Nordmann (norro))

threat posed by ICBMs to the United States increased with each new Soviet aeronautical achievement. For this reason alone, it was evidently insufficient to compete in space (McDougall 1997). In order to counter intercontinental missiles, the United States desperately needed cutting-edge defense systems.

As a branch of the Department of Defense, ARPA was naturally interested in exploring ballistic missile defense technologies, while NASA's responsibility was to develop space technology.

Radar monitoring of the ionosphere was one of the many missile defense initiatives receiving a great deal of attention. Several scientific investigations had demonstrated that re-entering ICBMs from space resulted in distinctive radar echoes. Once the missile had taken off, its location could be tracked if a radar with enough sensitivity was constructed (Swords 1986).

Fig. 5.3 Sputnik 1, the first human-made satellite placed in orbit around Earth (*Source* NSSDC, NASA)

Yet back then, scientists didn't know much about the upper atmosphere (i.e., the ionosphere), and that knowledge was crucial for deciphering the radar echo of intercontinental missiles. Both initiatives needed to be conducted at the same time.

Dr. William E. Gordon (1918–2010) of Cornell University sensed a window of opportunity here. Gordon had a background in electrical engineering and military aviation, and he had always been interested in radar. By doing the math, he determined that the needs of ionospheric research might be addressed by a massive radar antenna with a diameter of 1,000 feet (or 305 m) (Altschuler and Salter 2013) (see Fig. 5.4).

For the *IRE Transactions on Antennas and Propagation*, Gordon penned an article detailing the huge radar's idea and design (Gordon and LaLonde 1961). Hence, Gordon's institution Cornell University proposed constructing a massive radar to the Ministry of Defense. The notion fitted well with ARPA's quest for a new and improved missile defense system.

Gordon determined that the giant radar's diameter was required to be at least 305 m, but at the time there was no engineering precedent for constructing such a radar. Even the well-endowed ARPA would be unable to cover the enormous construction cost if it had to be built in the old-fashioned way.

Gordon contemplated the ways in which he could lower the expense of constructing a massive radar for a long time during the clear, starry nights. Cornell University Civil Engineering Professor William McGuire (1920–2013) (Mathews 2013) had a brilliant idea one day. He informed Gordon

Fig. 5.4 Gordon (left) and McGuire discussed about Arecibo's model (*Source* ResearchGate/John David Mathews)

that many funnel-shaped limestone pits could be found on the outskirts of Arecibo, Puerto Rico. The massive cost of constructing the support frame for the gigantic radar could be saved if the structure was developed in accordance with the mountain form (Murdin 2001).

As a result of this brilliant suggestion, the issue of excessive building costs was resolved (see Fig. 5.5).

The original parabolic design of the telescope was discarded in favor of a more practical spherical one at ARPA's request. In order to concentrate all of the incoming radio waves at a single point, the receiving surface of a radio telescope was typically shaped like a paraboloid. But, Arecibo was stuck in the valley and could not change its course. If it was still a paraboloid, then it meant that Arecibo could only study the ionosphere and could not perform radio astronomy research in an effective manner. It also meant that it could only look at a tiny area of the sky in order to receive signals.

As the spherical reflector's focal point was a line rather than a point, the telescope's field of view was effectively expanded. The revised plan called for constructing three towers, each 100 m tall, surrounding the telescope, and then suspending a column-shaped receiver above the antenna by means of 18 steel cables. The signal reflected by the spherical mirror was collected by the columnar receiver, which was pointed in the true direction of the antenna

Fig. 5.5 The area of the Arecibo Observatory (*Credit* USGS/NASA)

(Breakall 2021). This meant the telescope could be used for both national security and scientific inquiry.

Arecibo Ionospheric Observatory, the name given to the telescope after it had been under construction for over 2 years, opened to the public on November 1, 1963. Considering his role as Arecibo's primary architect, Gordon was appointed the facility's first director (see Fig. 5.6).

Less than half a year after its operation, Arecibo had already achieved remarkable results.

Mercury's rotation and revolution periods had both been theorized to be 88 days due to the planet's seeming lock by the Sun's powerful gravity. Full observation, however, led the Arecibo Observatory to determine on April 7, 1964 that Mercury's rotation period was not 88 days, but rather substantially shorter, 58.646 days. In contrast to the Moon, Mercury was not held in place by gravitational forces. It fulfills three complete rotations for every two solar revolutions. The rotation period of Venus was later determined to be exactly 243.16 days after Arecibo turned its "sight" toward the planet (Taylor 2016).

The Arecibo radio telescope's ability to transmit and receive high-power radar waves made it ideal for this kind of precise monitoring and study of planetary rotation. As the Earth spins, distant planets will draw nearer or further away from both sides, and the time it takes for radar waves to return

Fig. 5.6 Arecibo Ionospheric Observatory officially opened on November 1, 1963. (*Source* ResearchGate/John David Mathews)

will vary slightly. From this slight variation, Arecibo determined the precise period by which the planet rotates (see Fig. 5.7).

Arecibo didn't stop. The Crab Nebula, about 6,500 light-years away, became its target (Hewish et al. 1968). Ancient Chinese astronomers made the discovery that there were bright stars visible during the daytime around the area of the Crab Nebula as early as the year 1054. The Crab Nebula was thought by scientists to be the supernova's remnant.

The question then became how to demonstrate this hypothesis. What happened to the supernova after the explosion? Did it become a neutron star? If there were no evidence based on observations, then these hypotheses could never amount to more than guesswork.

Using data from the Arecibo radio telescope's observations, physicist Richard Lovelace of the Arecibo Observatory discovered a series of pulses near the center of the Crab Nebula with a duration of 33 ms on November 10, 1968 (Lovelace and Tyler 2012). A rapidly revolving neutron star was proven to be the source of these signals.

As a result, the hypothesis that the Crab Nebula was the remnant of a supernova explosion had been verified. When the Arecibo radio telescope

Fig. 5.7 Arecibo radio telescope's feed cabin system (*Source* ResearchGate/John David Mathews)

made this discovery, it sparked a new era in the hunt for pulsars (all pulsars are neutron stars). It was quickly discovered that the first generation of stars had burned out and exploded, leaving behind their remnants all over the universe.

In light of the Arecibo radio telescope's string of successes in astronomical exploration (NSF, Arecibo: Facts And Figures n.d.), the United States Department of Defense had decided that this super radar first constructed to counteract intercontinental missiles was better suited to scientific purposes.

Consequently, on October 1, 1969, the United States Department of Defense officially handed over control of Arecibo to the National Science Foundation (NSF) so that the observatory could become more involved in astronomical research.

The Arecibo Ionosphere Observatory was officially renamed the National Astronomy and Ionosphere Center (NAIC) by NSF in September 1971. To keep Arecibo running, NASA subsequently partnered with NSF to contribute funds.

The years in which Arecibo reaped the benefits of fruitful results and benefited from adequate funding were the telescope's most successful period.

Cornell University, which oversaw the Arecibo Observatory, made significant improvements to the already-excellent telescope in 1973. They used 38,778 aluminum plates (each measuring 1 m by 2 m and could be adjusted and replaced individually) in lieu of the original reflective antenna, which was made of barbed wire that had been galvanized. The upgrade allowed Arecibo's radar transmission frequency to rise from 500 to 5,000 MHz (Altschuler and Salter 2013) (see Fig. 5.8).

In the presence of this super telescope, the scientists' confidence levels skyrocketed to unprecedented heights: This radio telescope would usher in a new era of radio astronomy research, boasting the world's largest aperture, most cutting-edge equipment, and most sensitive "hearing." Cornell University Astronomy Professor Donald Campbell (1921–1967) recalled: "At that time, people wanted a really symbolic event to illustrate our competence."

Fig. 5.8 Arecibo with upgraded transmitting antenna in 1973 (*Source* ResearchGate/John David Mathews)

The event commemorating Arecibo's upgrade was planned in such a way that radio waves would be sent to extraterrestrials. Announcing Earth's existence to an alien society located 25,000 light-years away was the pinnacle of human pride and technological prowess.

As I explained at the outset of this chapter, on November 16, 1974, a radio wave message was transmitted from the site of the celebration to the faraway globular cluster M13 in the constellation Hercules.

The "Arecibo Message" is the name given to this signal that gained Arecibo worldwide recognition. It has become a connection between the Earth and other civilizations, a beloved Hollywood filming site, and a lasting symbol in the minds of numerous enthusiasts of science fiction and astronomy.

The upgraded Arecibo did not let astronomers down, as new discoveries were unveiled to the world one by one.

In 1974, Russell Hulse, then 24 years old and a PhD student in physics, and Joseph Taylor, then his tutor, discovered a binary system in the massive trove of pulsar data. Humanity has made history with the discovery of PSR1913 + 16, the first binary pulsar (Hulse and Taylor 1975). In 1993, they were awarded the Nobel Prize in Physics for their finding, which indirectly confirmed a prediction of general relativity.

Arecibo's enhanced capabilities allowed for the emission of stronger electromagnetic waves. These electromagnetic waves were powerful enough to break through Venus's thick atmosphere, revealing the planet's surface for the first time (Carter 2004).

Astronomer John Harmon turned Arecibo's focus back to Mercury in 1994 after discovering a thick layer of ice at the planet's north pole (Harmon 2001). Harmon was able to finish charting the polar ice on Mercury thanks to the precise radar echoes he received (see Fig. 5.9).

In addition, NASA had tasked Arecibo with a unique duty to safeguard the Earth. Arecibo was upgraded to the level of a planetary-class radar system, allowing for more accurate tracking of potentially hazardous asteroids passing close to Earth and earlier warning of their impending arrival. Arecibo was regarded as a heroic "figure." Numerous people were encouraged to pursue their space adventures after seeing its appearance in movies, TV shows, and other media.

The larger the accomplishment, the greater the subsequent level of anticipation. Arecibo was promoted once again at the height of its career in 1997. This time, scientists upgraded Arecibo's sensitivity by replacing the square column feeder with a Gregorian reflector system that doubled the radio frequency to 10,000 MHz (Kildal and Davis 2016).

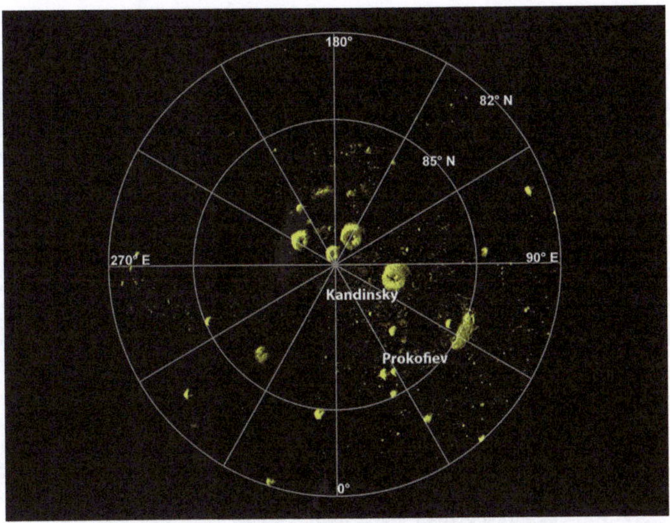

Fig. 5.9 A radar image of Mercury's north polar region acquired by the Arecibo Observatory (*Credit* National Astronomy and Ionosphere Center, Arecibo Observatory)

Unfortunately, the weight of the already substantial suspended feed cabin had nearly doubled as a result of this improvement. Nine steel cables held the 900-ton apparatus in place above the massive reflecting surface. In addition, Arecibo's future had been put in jeopardy by the frequent tropical hurricanes that struck the island of Puerto Rico (see Fig. 5.10).

The now-famous "SETI @ home" project (Anderson et al. 2002) was launched in the month of May 1999. This project used distributed computing technologies to allow any astronomy enthusiast interested in the age-old issue of "Are we alone in the universe?" to take part in the hunt for extraterrestrial intelligence.

Since Arecibo was the pioneering radio telescope to communicate with aliens, it was charged with the primary mission of detecting transmissions from distant civilizations. Almost 290,000 computers all across the world were involved in the examination of the data collected. With the use of a piece of software that could be downloaded to a home computer, interested volunteers could receive a radio wave that was picked up by Arecibo and then analyze it in an effort to detect signs of intelligent life among the massive amounts of data.

But compared to the vastness of the radio wave transmissions in the universe, the data that Arecibo could collect was, unfortunately, a drop in the bucket. The "SETI @ home" project was more of a symbol than a scientific endeavor.

Fig. 5.10 After Arecibo's upgrade in 1997 (*Credit* Alessandro Cai)

With a new visitor center that can accommodate more than 100,000 people each year, Arecibo was starting to appear like a cultural ambassador for space. In the face of this breathtaking super project, people were stunned by the wonders of science and the cosmos (see Fig. 5.11).

With the turn of the new millennium, several researchers had focused their attention on developing enormous radio telescope arrays that could be scaled up or down as needed. Even the title of "the largest in the world" would not be able to stop a "rookie" with a 500-m aperture dubbed FAST from taking it.

Arecibo had no longer been indispensable to the conduct of scientific study as it once was. In order to encourage the exploration of more innovative technologies, NASA and NSF were forced to reduce financing for it. Arecibo had to eke out a meager existence due to the lack of precious funding.

NSF made the announcement in 2006 that the Arecibo Observatory might be shut down if other funding sources were not discovered. The statement caused an outcry, with people from all walks of life declaring that Arecibo had made humanity's voice heard throughout the cosmos and that the observatory was therefore impregnable. Thousands of astronomy fans showed their

Fig. 5.11 Visitors sightseeing in Arecibo (*Credit* Arecibo Observatory)

support for Arecibo by signing a petition to raise money for the observatory. This gave Arecibo a new lease on life for a short while.

As the National Science Foundation was pressed for funding in 2011, it had to take control of the Arecibo Observatory away from Cornell. A group of researchers from Stanford University and the University of Puerto Rico took its place. The only benefit of this new group was the reduced management charge and cheaper overall cost (Scoles 2016).

Due to a lack of available funds, the Arecibo radio telescope had continued to operate without the necessary maintenance. Damage to more than 30 aluminum panels on the telescope's reflective surface was caused by Hurricane Maria on September 21, 2017. Without upkeep money, Arecibo could only drag his ailing body and run with all his might.

It was the 10th of August, 2020, when another hurricane hit. The roots have yanked off a steel cable from the 900-ton feed cabin. The steel cable was dragged down from a height of 150 m like a giant whip, ripping a gaping hole in the sphere-shaped antenna that was more than 30 m in length (Clery 2020) (see Fig. 5.12).

Managers were caught off guard by the crisis and had to reevaluate how tough it would be to restore the Arecibo radio telescope. Using precise stress distribution measurements, the technical team evaluated the likelihood of repairing the cables.

Fig. 5.12 August 2020 hurricane knocked down a steel cable of Arecibo (*Courtesy* National Science Foundation)

The evaluation's final verdict is disheartening. It was determined by engineers that the remaining cable had a lower load-bearing capability than originally estimated, and that the only safe option for repairing Arecibo was to attempt dismantling.

On November 19, 2020 (NSF, Arecibo Observatory's 305-m telescope suffers collapse 2020), NSF made the announcement on Twitter that it would stop working on Arecibo's repairs and instead take precautions to dismantle the potentially hazardous equipment.

Sadly, though, the accident occurred. The demolition plan was still in the works as of December 1, 2020, when the engineering team took a job on it. At 7:53, the remaining steel cable snapped under the strain. The 900-ton feed cabin plummeted from a height of 150 m to a giant mirror that had been staring at it for the previous 57 years, much like a ripe fruit returning to its roots. The ancient hero, like a stalwart soldier, picked a courageous method to end his life abruptly; even in death, Arecibo kept his fighting stance (see Fig. 5.13).

Despite the fact that the Arecibo radio telescope has officially "died," the tremendous amount of data it collected is still a priceless gift for humanity. Whatever your reluctance, don't look back with regret. Immediately prior to

Fig. 5.13 Fallen Arecibo (*Credit* Michelle Negron, National Science Foundation, *Courtesy* National Science Foundation) after cleanup (*Credit* Tedder)

Arecibo's departure, the Chinese "Heavenly Eye" radio telescope, which is 500 m in aperture, successfully took over the duties previously performed by Arecibo. Civilization on Earth, with its "eyes" still open, has been surveying the vast sea of stars. My heart is full of hope and expectation that Arecibo will one day rise again on this land of magic treasures.

References

Altschuler, D.R., Salter, C.J.: The Arecibo observatory: fifty astronomical years. Phys. Today **66**(11), 43–49 (2013). https://doi.org/10.1063/pt.3.2179

Anderson, D.P., Cobb, J., Korpela, F., Lebofsky, M., Werthimer, D.: SETI@home: an experiment in public-resource computing. Commun. ACM **45**(11), 56–61 (2002). https://doi.org/10.1145/581571.581573

Breakall, J.K.: History of antenna technology at the Arecibo observatory in Arecibo, Puerto Rico [historically speaking]. IEEE Antennas Propag. Mag. **63**(2), 103–105 (2021). https://doi.org/10.1109/map.2021.3054028

Carter, L.M.: Impact crater related surficial deposits on venus: multipolarization radar observations with Arecibo. J. Geophys. Res. **109**(E6) (2004). https://doi.org/10.1029/2003je002227

Clery, D.: Famed Arecibo telescope, on the brink of collapse, will be dismantled, 19 Nov 2020. Science.org: https://www.science.org/content/article/famed-arecibo-telescope-brink-collapse-will-be-dismantled

Divine, R.A., Divine, R.A.: The Sputnik Challenge. Oxford University Press (1993)

Gordon, W., LaLonde, L.: The design and capabilities of an ionospheric radar probe. IRE Trans. Antennas Propag. **9**(1), 17–22 (1961). https://doi.org/10.1109/tap.1961.1144946

Harmon, J.: High-resolution radar imaging of Mercury's north pole. Icarus **149**(1), 1–15 (2001). https://doi.org/10.1006/icar.2000.6544

Hewish, A., Bell, S.J., Pilkington, J.D., Scott, P.F., Collins, R.A.: Observation of a rapidly pulsating radio source. Nature **217**(5130), 709–713 (1968). https://doi.org/10.1038/217709a0

Hulse, R.A., Taylor, J.H.: Discovery of a pulsar in a binary system. Astrophys. J. **195**, 51–53 (1975). https://doi.org/10.1086/181708

Kildal, P.S., Davis, M.M.: Modern antenna technologies stimulated by the Arecibo radio telescope upgrade project 1984–1997. In: 2016 IEEE International Symposium on Antennas and Propagation (APSURSI) (2016). https://doi.org/10.1109/aps.2016.7695898

Lovelace, R., Tyler, G.: On the discovery of the period of the Crab Nebular pulsar. Obs. **132**, 186–188 (2012)

Mathews, J.D.: A short history of geophysical radar at Arecibo observatory. Hist. Geo- Space Sci. **4**(1), 19–33 (2013). https://doi.org/10.5194/hgss-4-19-2013

McDougall, W.A.: The Heavens and the Earth: A Political History of the Space Age. JHU Press (1997)

Murdin, P.: Encyclopedia of Astronomy & Astrophysics. CRC Press, Boca Raton (2001). https://doi.org/10.1201/9781003220435

NSF: Arecibo Observatory's 305-m telescope suffers collapse, 1 Dec 2020. NSF News: https://new.nsf.gov/news/arecibo-observatorys-305-meter-telescope-suffers

NSF: Arecibo: Facts And Figures (n.d.). National Science Foundation: https://www.nsf.gov/news/special_reports/arecibo/Arecibo_Fact_Sheet_11_20.pdf

Pitzer, K.: Puerto Rico's Eastern Coast & El Yunque Rainforest. Hunter Publishing, Inc. (2009)

Scoles, S.: World's largest radio telescope faces retirement due to stagnant funding, 6 Jun 2016. PBS News Hour: https://www.pbs.org/newshour/science/worlds-largest-radio-telescope-faces-retirement

Swords, S.S.: Technical History of the Beginnings of Radar. The Institution of Engineering and Technology, London (1986)

Taylor, P.A.: The arecibo observatory planetary radar system. In: 47th Lunar and Planetary Science Conference, held March 21–25, 2016 at The Woodlands, Texas. LPI Contribution No. 1903, p. 2534 (2016)

The Staff at the NAIC: The Arecibo message of November, 1974. Icarus **26**(4), 462–466 (1975). https://doi.org/10.1016/0019-1035(75)90116-5

Wolfram, S.: Arecibo Telescope 1974 Transmitted SETI Message, 2 Jun 2017. Wolfram Data Repository: https://doi.org/10.24097/wolfram.73460.data

6

Mauna Kea Observatories: Creator of Space–Time Magic

Mauna Kea Observatories

Tycho Brahe (1546–1601), as he made his way home on the evening of November 11, 1572, cast his gaze upward toward the night sky. Tycho was already quite familiar with the road under his feet; in fact, he was so familiar with it that he could not take a stride in the wrong direction even if he looked up at the sky (Brahe 1573).

> A brilliant star in the sky's uppermost tier attracted his attention.
> "No! That has no place there!"

In an instant, Tycho had stopped and was staring fixedly at the star. Cassiopeia is the constellation with the W-shaped pattern of five brilliant stars. No one, not even a child, could ever misidentify this constellation. Only Ursa Major, the Big Dipper, and Polaris are more easily identifiable than Cassiopeia throughout the entirety of the sky. However, there were now six stars instead of five, and the sixth was brighter than the others (Norton and Kitchin 1973).

Putting his head down, Tycho scratched his eyes. Soothing himself and dispelling the delusion were both necessities. Nonetheless, the star was still visible in the sky when he looked up again. Tycho was astounded by what he saw. He located the others and directed their attention to the star. His inability to believe that he could actually see the star persisted until it was established beyond reasonable question by everyone else.

J. Wang, *Eye Beyond the Sky*, Astronomers' Universe,
https://doi.org/10.1007/978-981-99-9818-0_6

From that point forward, Tycho logged daily observations of the star's evolution. In just two weeks, he observed, the star's luminosity increased dramatically. It soon eclipsed Venus as the brightest nighttime star, and its light was visible during the day (Fig. 6.1).

Almost a year later, the formerly brilliant star dimmed and turned from white to dark red. After 16 months of diligent observation, Tycho finally lost the sight of this star in March 1574.

Although Tycho referred to the star's unexpected appearance and then gradual disappearance as the "birth" of a new star, he was actually seeing the star's "death" (Hellman 1959). Later generations recognized Tycho's contributions by giving the supernova the name "Tycho Supernova" in his honor.

No supernovas were observed for hundreds of years. Tycho Supernova was only seen by humans again in 2008 through a telescope named Subaru. Incredible as it may be, what we observe through this telescope is not the remnant of the supernova explosion. It is more like a kind of space–time magic, which transports us back through history to the time when Tycho lived and helps us see the true picture of a supernova explosion being created.

This chapter's heroes are the Mauna Kea Observatories in Hawaii, which show humanity the wonders of space and time. Permit me to begin with the enchanting state of Hawaii.

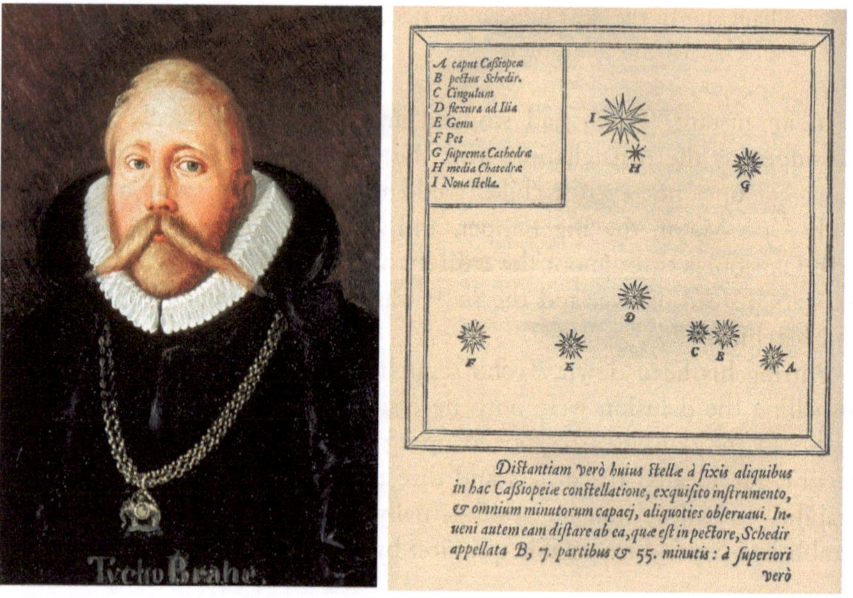

Distantiam verò huius stellæ à fixis aliquibus in hac Caßiopeiæ constellatione, exquisito instrumento, & omnium minutorum capacj, aliquoties obseruaui. Inueni autem eam distare ab ea, quæ est in pectore, Schedir appellata B, 7. partibus & 55. minutis: à superiori verò

Fig. 6.1 Tycho Brahe and his manuscript of Tycho Supernova (Public Domain)

The Hawaiian Islands are a chain of submerged volcanoes that form a component of the mid-ocean ridge in the Pacific Ocean's north-central region. Mauna Kea is the name of the largest volcano on the island. Beginning at a depth of 5,998 m, the mountain rises to an incredible 10,203 m at its peak. It is the tallest mountain from base to peak on our planet (Porter 1979). Only Olympus Mons on Mars is taller in the entire solar system (Fig. 6.2).

Despite being more than halfway submerged, Mauna Kea still towers at a height of 4,205 m above sea level. It is at this elevation that the inversion layer of the atmosphere sets in, preventing the rise of moisture from the Pacific Ocean and so ensuring that Mauna Kea enjoys more than 300 clear nights annually. Because of Mauna Kea's low latitude, the Northern and Southern Crosses can be viewed from here as well.

Hawaii's pleasant climate and stunning scenery made it a popular tourist destination, and Mauna Kea quickly earned a reputation as the best spot in the world to view the stars.

After reaching the summit of Mauna Kea, a stargazer once penned the following words: "Nowhere is the Milky Way as bright as it is on Mauna Kea. That evening, I felt like I was a part of the universe, and the dreamlike experience is something I will never be able to forget" (Fig. 6.3).

A world-class observatory on the summit of Mauna Kea had always been a dream for astronomers, who would not give up their hold on this spectacular viewing location anyway. Yet, there was a problem in that this was not only a stargazing shrine in the eyes of astronomers but also a sacred mountain (Lang and Byrne 2013) in the hearts of native Hawaiians. This created a conflict between the two groups.

Fig. 6.2 Panorama of Mauna Kea (*Credit* depositphotos/MN studio)

Fig. 6.3 Night sky at Mauna Kea (*Credit* depositphotos/cavan)

When so-called modern humans first set foot on the Hawaiian Islands more than a thousand years ago, they were astounded by the height of Mauna Kea, which left a lasting impression on both their hearts and minds. Native Hawaiians considered the peak of Mauna Kea to be the "Land of the Gods" due to their strong religious beliefs. Heaven is the final destination of the souls of the righteous, and it is also the location where God resides. None but the highest-ranking priests were allowed to ascend the holy mountaintop before the advent of modern civilization.

The Apollo moon landing program was initiated by the United States in the year 1960. The construction of huge telescopes has taken on a greater amount of significance as an astronomical observation project supporting the lunar landing program. In this scenario, NASA gave Gerard Kuiper (1905–1973), who was the director of the Lunar and Planetary Laboratory at the time, a blueprint for the construction of an optical telescope (Fig. 6.4).

Kuiper should be a household name for any reader with even a passing interest in the universe. Kuiper was an influential American astronomer. Along with finding the moons of Uranus and Neptune, he also established the presence of an area rich in asteroids beyond the orbit of Neptune. Kuiper has earned the title of "Father of modern planetary science" for his contributions (Sears 2019). The region beyond Neptune where numerous asteroids have been found is officially known as the "Kuiper Belt," in honor of Kuiper.

Kuiper had already surveyed numerous places that might be adequate for astronomical observation, but the holy mountain Mauna Kea was his

Fig. 6.4 Gerard Kuiper in 1964 (*Source* Dutch National Archives, The Hague, Fotocollectie Algemeen Nederlands Persbureau (ANEFO), 1945–1989, Nummer toegang 2.24.01.05 Bestanddeelnummer 916–8171)

ultimate goal. Conditions for both observing and traveling were perfect at the time. Being a volcano, the summit of Mauna Kea is very wide, with many mild peaks strewn out about the crater. Sizable observatories with multiple stations for gathering data would thrive in such a setting (Fig. 6.5).

However, a cautious approach was required. Kuiper was well aware that in order to fulfill his dream of constructing observatories atop Mauna Kea, he would need to first make an effort to overcome the cultural taboo that the Hawaiian people had regarding the mountain's sacred status. Therefore, he convinced Hawaii's governor, John Burns (1909–1975), to install a 30-cm-aperture telescope atop the island's second-highest peak for testing. This was justified on the grounds that it would aid in the construction of roads

Fig. 6.5 Crater of Mauna Kea (*Credit* depositphotos/kamchatka)

on the mountain. Kuiper deftly sidestepped a potential cultural crisis by not installing it on the summit of Mauna Kea's first peak, the mountain's most revered peak.

Further testing by Kuiper revealed that despite perpetual snowfall, the air at the peak of Mauna Kea was unusually dry, making it ideal for the collection of infrared signals. As a result, he promptly submitted a proposal to convince NASA to construct a huge telescope capable of observing in the infrared (West 2015).

Kuiper's persistent efforts quickly bore results. Both good and bad news came his way. The proposal had been approved by NASA, and funding was not an issue. Unfortunately, NASA did not hand over the project to him personally. They decided to call for a bid open to the public.

Fortunately, Kuiper did not have any competitors with comparable competency. The University of Hawaii was represented in the bid by local physics professor John Jefferies, who proposed allocating $3 million toward the construction of a 2-m telescope on Sacramento Peak. Typically, a physicist wouldn't be the first choice to lead the team responsible for building a telescope of this magnitude.

Nonetheless, events could turn out unexpectedly, and geopolitics often ended up winning. After a final evaluation of the benefits and drawbacks,

NASA determined that Kuiper was not the best option and awarded the contract to the University of Hawaii, despite the latter's inexperience.

When Kuiper heard the news, he got enraged. NASA had "stolen" the holy mountain in his heart, so there was no reason for him to stay. Kuiper had moved over his feelings of nostalgia and decided to depart in a decisive manner.

Despite this, NASA was pleased with the results that the University of Hawaii produced. The university might be lacking in intellectual accomplishments, but it unquestionably possessed first-rate operational competence and vision (Marichalar 2021).

In 1967, the Hawaiian government constructed its own all-weather road to the peak of the mountain and other essential infrastructure. Subsequently, they were able to openly solicit funding from telescope initiatives in a number of countries (Fig. 6.6).

The 3.6-m Canada-France-Hawaii telescope (CFHT) was the next to be constructed after the US Air Force's 0.6-m telescope primarily for satellite tracking and the Lowell Observatory's 0.6-m telescope mainly for planetary observations (Leverington 2017).

The University of Hawaii's astronomy major has made itself from amateur to professional status thanks to these cutting-edge tools. Not only has it produced more than a hundred exemplary doctorates in astronomy but has also attracted quite a number of well-known astronomers (Murdin 2001). Hawaiians' most revered peak is now a pilgrimage site for astronomers.

Fig. 6.6 Keck, NASA and CFHT observatories on Mauna Kea (*Credit* Bob Linsdell)

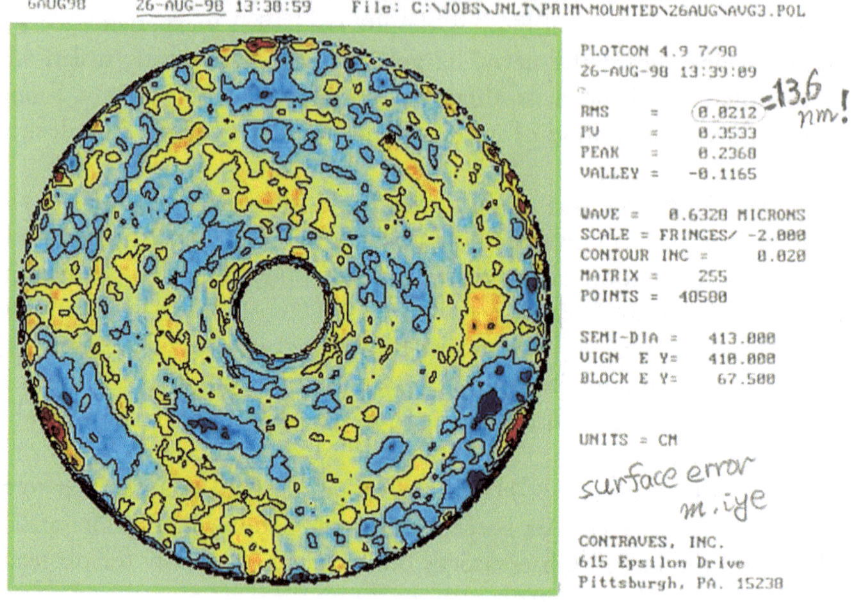

Fig. 6.7 Error figure of the surface shape of the primary mirror in error (*Credit* NAOJ)

The Japanese-built Subaru, which measures 8.2 m in aperture, is one of the most impressive telescopes at Mauna Kea Observatories.

The construction of the Subaru Telescope did not begin on Mauna Kea until 1991, seven years after the project's commencement in 1985 when it was first proposed. The majority of it was held up because the mirror plate grinding took so long.

To fix the lens distortion, scientists specifically utilized the active optics technology that had been proposed by the European Southern Observatory. As a result of the complexity of its sensors and automatic control systems, active optics technology was initially met with skepticism from many countries. In contrast, Japan, a world power in electronics, had enthusiastically embraced active optics and was employing it to overcome the limitations of telescope lens production (Murdin 2001).

At a thickness of only 20 cm, the Subaru telescope's lens was the thinnest of its time. The new glass utilized by Japanese engineers for the lens material significantly reduced the lens's deformation due to changes in temperature (IYE 2021).

Although engineers had made great strides in those years, they continued to run across problems with the lens grinding process. When it came to the

8.2-m telescope's lens, surface errors of no more than 0.1 microns (about one ten-thousandth of the diameter of a hair) are required. Japanese engineers had learned the hard way that mechanical grinding techniques could not provide such a high level of precision. At some point, they gave up on the machines and started polishing the lens by hand. One might ask how grinding by hand could possibly result in better precision. As a matter of fact, it was difficult to collect mistakes in a single direction when grinding lenses by hand because the direction of manual force was more erratic than mechanical force (Kaifu 1997). Before grinding with their bare hands, engineers applied a specific abrasive to their hands.

When some time had passed, they would check the precision of the lens, pinpoint the flaw's location, and continue polishing by hand (Shelton 2002). After seven years and a lot of hard work, this massive mirror was created.

Engineers built 261 computer-controlled supports behind the ultra-thin giant lens to keep the primary mirror in pristine condition regardless of the telescope's position or temperature (Fig. 6.8).

The Subaru telescope has a peculiarly shaped dome instead of a standard hemisphere. Using a cylindrical design with improved air circulation, engineers were able to reduce the turbulence caused by the temperature differences near the lens and improve the atmospheric seeing in its dome (Fig. 6.9).

Fig. 6.8 The finished primary mirror (*Credit* NAOJ)

Fig. 6.9 Subaru Telescope under night sky (*Credit* NAOJ)

From its commencement in 1984 until its completion in 1999, the construction of the Subaru Telescope took a whole 15 years. Despite the power of modern telescopes, distant objects are still barely visible. In most cases, the telescopes function more like a digital camera, requiring lengthy exposures in order to capture images of faraway celestial bodies. Because of this limitation, huge telescopes, in general, no longer have the capability of doing direct observations using the human eye alone. Subaru, however, is unique in that it allows observers to examine the night sky in front of the eyepiece.

Princess Sayako, on behalf of the Japanese royal family, was invited to attend Subaru's dedication ceremony (NAOJ 2000). Elegantly, she positioned herself in front of the telescope's eyepiece and made her first historic astronomical observation.

Seeing the night sky through a telescope's eyepiece is essentially the same as looking at images on a screen. But, considering that humans are such a peculiar creature, "seeing with one's own eyes through an eyepiece" is likely to result in a much more profound emotional response than viewing images would.

After looking at the night sky through the eyepiece, one staff member claimed that everything that can be seen in the photographs taken by the Hubble Space Telescope, including the colors and shapes of the nebulae, can be seen with their own eyes, and that every star displays unexpected colors (Fig. 6.10).

From 1999 until 2004, Subaru held the title of the world's largest optical telescope.

Since its completion, Subaru has provided scientists with numerous unexpected findings (NAOJ, The Subaru Telescope, n.d.). Its observation of Tycho Supernova is a spectacular example of space–time magic, as we indicated in the opening of this chapter.

Fig. 6.10 Mauna Kea observatories under night sky (*Credit* NAOJ)

Supernovae are the explosive deaths of massive stars. Extreme events like explosions occur when a star's nuclear fusion reactions accelerate out of control (NASA 2015). Supernovae are unusual because stars have such long lifespans. There are hundreds of billions of stars in the Milky Way, but it still takes an average of 30 years to spot a supernova explosion.

Due to the enormous violence of the energy outburst that occurs during a supernova explosion, the entire process lasts for only a matter of weeks or months at most. Nonetheless, it took the Tycho Supernova 16 months to exhaust its energy and burn out completely (Arnett 2020).

However, the story does not end here.

Just the direct light from the supernova was visible to Tycho when he looked at it with his bare eyes. Nevertheless, the power of supernova explosions was simply too great. Interstellar dust clouds have the potential to scatter light from distant sources in all directions (Dwek 2006), including towards the Earth. While the scattered light will reach the Earth at a later period, it can still be bright enough to be noticed. The dust clouds act as enormous mirrors in the universe and scatter light from supernovae.

Tycho Supernova's light finally reached the Earth in 2008, nine years after the Subaru Telescope was constructed, thanks to a reflection of a dust cloud (Bai 2008). They were a staggering 436 years behind the direct light that reached the Earth!

Several 4-m telescopes were initially used to spot the dust cloud, but they lacked the capabilities to capture the necessary spectra. The spectral information of the reflected light was then acquired by astronomers using the powerful Subaru Telescope. At a glance, astronomers who specialized in studying supernovae could tell that it must be the light from the supernova instead of from the dust cloud itself. When multiple sources of reflected light were analyzed together, their directions of origin could be identified as Tycho Supernova (Ruiz-Lapuente 2004).

Based on the ensuing spectra, astronomers have been able to classify Tycho Supernova as typical type Ia (NAOJ 2008). This put an end to the decades-long controversy about Tycho Supernova's classification and cleared the path for new research.

Astronomers living in the contemporary day can, with the assistance of the Subaru Telescope, experience the same process of a supernova explosion that their patriarch Tycho did. It is not an exaggeration to call this a form of space–time magic, and it would not diminish the awe-inspiring nature of the findings even if we did comprehend the principle.

The observation of the Tycho Supernova by Subaru is only one of the Mauna Kea observatories' abundant achievements.

Using the Gemini North Telescope on Mauna Kea, astronomers in 2014 uncovered the incredible sight of a millisecond pulsar and a white dwarf in mutual orbit around each other (Kilic et al. 2014) (Fig. 6.11).

In 2018, utilizing the Subaru Telescope, researchers were able to locate a previously unknown minor celestial object on the outskirts of the solar system. Its distance from the Sun, measured in astronomical units, is 120, or almost 2.5 times that of Pluto (Sheppard 2018) (Fig. 6.12).

A pair of faraway brown dwarfs in orbit around each other was discovered in 2022 by researchers from Arizona State University using the Keck Telescope on Mauna Kea. The distance between the brown dwarfs is almost three times that of Pluto to the Sun, according to the data (Softich et al. 2022) (Fig. 6.13).

Within this area that is smaller than 2 square kilometers in size, there are currently 13 (Hollier 2016) astronomical observatories that have been constructed. Mauna Kea is worthy of being called the "Holy Mountain of Astronomy" since it has the most dense distribution of heavyweight astronomical telescopes on the planet (Fig. 6.14).

Construction is now underway at Mauna Kea on the TMT (Thirty Meter Telescope) (Zastrow 2023). The projected completion date of construction is 2027. When that day comes, the brightest pearl that can be found on Mauna

Fig. 6.11 Illustration of a millisecond pulsar and a white dwarf in mutual orbit around each other (*Credit* ESO/L. Calçada)

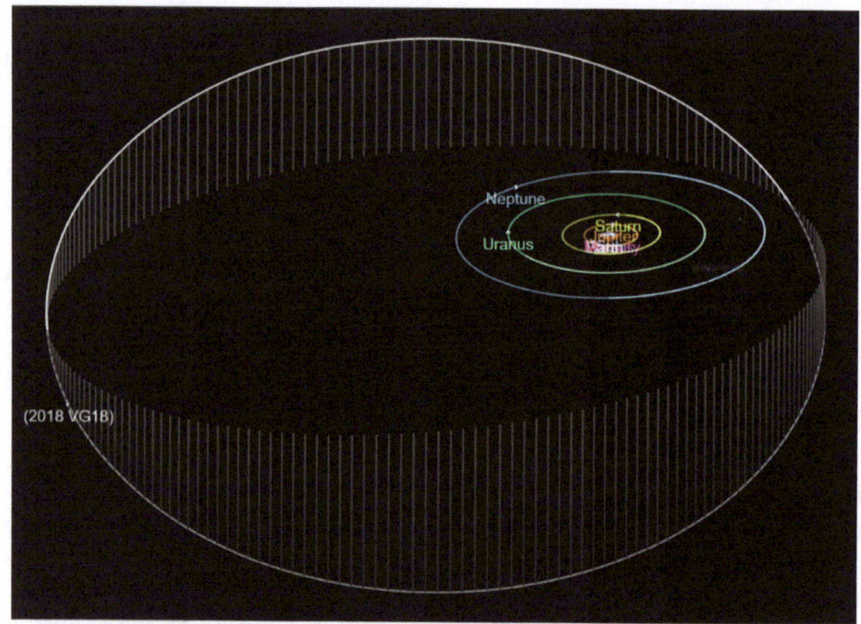

Fig. 6.12 Orbit of this previously unknown minor celestial object (*Credit* NASA/JPL)

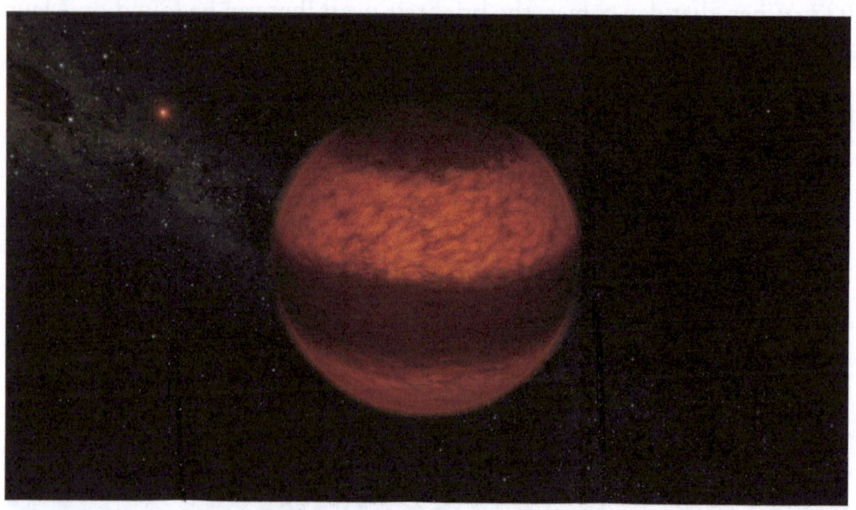

Fig. 6.13 A faraway pair of brown dwarfs in orbit around each other (*Credit* Caltech/R. Hurt (IPAC))

Fig. 6.14 Road leading to the "Holy Mountain of Astronomy" (*Credit* depositphotos/ MN studio)

Kea will also be the one to wear the crown of the largest optical telescope in the world (Fig. 6.15).

Those standing atop Mauna Kea are able to gaze out at infinity, thanks to the mountain's pristine summits, red volcanic rocks, and gorgeous stars.

Fig. 6.15 Illustration of TMT (*Credit* TMT Observatory Corporation)

As long as we continue to gaze at the night sky, the vastness of the universe, which is also exclusive to human beings and represents their own style of romance, will get entwined within our hearts.

References

Arnett, D.: Supernovae and Nucleosynthesis: an Investigation of the History of Matter, from the Big Bang to the Present. Princeton University Press (2020)

Bai, N.: Detoured Light From Tycho's Supernova Finally Makes it to Earth (2008). Retrieved from Discover Magazine: https://www.discovermagazine.com/the-sciences/detoured-light-from-tychos-supernova-finally-makes-it-to-earth

Brahe, T.: De nova stella anno a nato christo (1573)

Dwek, E.: The supernova origin of interstellar dust. Science **313**(5784), 178–180 (2006). https://doi.org/10.1126/science.1130423

Hellman, C.D.: The New Star of 1572: Its place in the history of astronomy (1959)

Hollier, D.: Big Glass and the Age of New Astronomy (2016). Retrieved from Smithsonian Magazine: https://www.smithsonianmag.com/air-space-magazine/big-astronomy-how-built-180960364/

IYE, M.: Subaru Telescope—History, active/adaptive optics, instruments, and scientific achievements. Proc. Jpn. Acad. Ser. B, Phys. Biol. Sci. **97**(7), 337–370 (2021). https://doi.org/10.2183/pjab.97.019

Kaifu, N.: Status and perspective of Subaru Telescope Project. SPIE Proc. (1997). https://doi.org/10.1117/12.269058

Kilic, M., Hermes, J.J., Gianninas, A., Brown, W.R.: PSR J1738+0333: the first millisecond pulsar + pulsating white dwarf binary. Monthly Notices Roy. Astron. Soc.: Lett. **446**(1), L26–L30 (2014). https://doi.org/10.1093/mnrasl/slu152

Lang, L., Byrne, D.A.: Mauna Kea: A Guide to Hawaii's Sacred Mountain. Watermark Publishing (2013)

Leverington, D.: Observatories and Telescopes of Modern Times. Cambridge University Press (2017)

Marichalar, P.: 'This Mountain Is It': how Hawai'i's Mauna Kea was 'Discovered' for astronomy (1959–79). J. Pac. Hist. **56**(2), 119–143 (2021). https://doi.org/10.1080/00223344.2021.1913402

Murdin, P.: Encyclopedia of Astronomy & Astrophysics. CRC Press, Boca Raton (2001). https://doi.org/10.1201/9781003220435

NAOJ: The Island of Hawaii, Hilo, and Subaru's Base Facility (2000). Retrieved from Subaru Telescope: https://subarutelescope.org/old/Topics/2000/09/11/index.html

NAOJ: Supernova Time Travel at Subaru Telescope: Super Explosion in 16th Century Caught by Subaru in 21st Century (2008). Retrieved from Subaru Telescope: https://subarutelescope.org/old/Pressrelease/2008/12/03/index.html

NAOJ. (n.d.). The Subaru Telescope. Retrieved April 12, 2023, from NAOJ: https://www.nao.ac.jp/en/research/telescope/subaru.html

NASA: What is a Supernova? (2015). Retrieved from NASA: https://www.nasa.gov/audience/forstudents/5-8/features/nasa-knows/what-is-a-supernova.html

Norton, A.P., Kitchin, C.R.: Norton's Star Atlas and Reference Handbook (epoch 1950.0) (1973)

Porter, S.C.: Geologic Map of Mauna Kea Volcano. Geological Society of America, Hawaii (1979)

Ruiz-Lapuente, P.: Tycho Brahe's Supernova: light from centuries past. Astrophys J **612**(1), 357–363 (2004). https://doi.org/10.1086/422419

Sears, D.W.: Gerard P. Kuiper and the Rise of Modern Planetary Science. University of Arizona Press (2019)

Shelton, I.: Subaru telescope. AccessScience (2002). https://doi.org/10.1036/1097-8542.YB021275

Sheppard, S.S.: Discovered: The Most-Distant Solar System Object Ever Observed (2018). Retrieved from Carnegie Science: https://epl.carnegiescience.edu/news/discovered-most-distant-solar-system-object-ever-observed

Softich, E., Schneider, A.C., Patience, J., Burgasser, A.J., Shkolnik, E., Faherty, J.K., Kuchner, M.J.: CWISE J014611.20–050850.0AB: the widest known brown dwarf binary in the field. Astrophys. J. Lett. **926**(2), L12 (2022). https://doi.org/10.3847/2041-8213/ac51d8

West, M.J.: A Sky WONDERFUL With Stars: 50 Years of Modern Astronomy on Maunakea. University of Hawaii Press (2015)

Zastrow, M.: Path Forward for Thirty Meter Telescope, Mauna Kea Begins to Emerge (2023). Retrieved from Astronomy.com: https://astronomy.com/news/2023/03/path-forward-for-mauna-kea--and-maybe-the-thirty-meter-telescope--begins-to-emerge

7

RATAN-600: The Lord Ring Staring at Sky

RATAN-600

On the morning of August 29, 2016, Steve Croft (Steve Croft—Astronomy Department n.d.) of the University of California, Berkeley rushed into the Berkeley SETI Research Center located in the North Building of the College of Physics, as the bells of the Sather Tower, which had been over a century old, rang in the background (Fig. 7.1).

As he reached the research center, he found it crowded with people who all seemed to have arrived at the same moment. Andrew Siemion (Andrew Siemion Director of the BSRC n.d.), one of his coworkers, was sitting in front of a computer, which displayed what appeared to be some mega project with a large ring of structure on a field. This ring structure, which seemed like a circular arena, had drawn the attention of the entire community of astronomers and even the media.

As Croft came in, he immediately turned to Siemion and said, "Sure enough, Andrew, you've witnessed it as well. HD 164595 Hercules. We've never paid much attention to this direction before (Croft et al. 2016)."

Siemion swiveled in his seat and grinned at Croft: "Have you been up all night like me? The news just made me awake with anticipation, and by the time morning came around, I couldn't wait for it to start."

"IAA SETI Permanent Committee sent us the relevant information this morning, and I am examining it," Siemion said as he sipped his coffee. "The most pressing concern, in my opinion, is to direct our efforts there and see what results we can get."

J. Wang, *Eye Beyond the Sky*, Astronomers' Universe,
https://doi.org/10.1007/978-981-99-9818-0_7

Fig. 7.1 UC Berkeley's Sather Tower (*Credit* depositphotos/cavan)

"Just what I was thinking." Croft gestured for everyone to begin working and then snapped his fingers. "Green Bank, David, tilt in the direction of HD 164595 (Shostak 2016). Dan, get ready to examine everything that Green Bank receives in real time. I really hope the Russians didn't make a mistake this time around."

During the early hours of the day, the Green Bank telescope slowly rotated its massive body to point toward of Hercules. From this direction came the most peculiar signal ever picked up by a Russian radio telescope (Fig. 7.2).

It's not the Green Bank telescope that takes center stage in this chapter, though; rather, it's the massive Russian radio telescope constructed in the 1970s, which has lived through some truly incredible things.

The United States and the Soviet Union launched their long space race in the late 1950s, during the height of the Cold War. Both countries had made

Fig. 7.2 Green Bank telescope (*Credit* cyberbaud)

significant attempts to develop huge telescope projects one after another due to the obvious benefits that building large radio telescopes would have for space exploration and national defense (History 2018).

The construction of the Radio Telescope of the Academy of Sciences 600 m in the Soviet Union began on August 18, 1965. It's officially known in English by its acronym "RATAN-600," where 600 m is the diameter of this telescope. As a whole, the telescope occupies a space roughly 260,000 m^2 in size, or nearly 40 standard football fields (Fig. 7.3).

When you consider that the "Chinese Heavenly Eye" was constructed 50 years later and had an aperture of only 500 m, RATAN's 600-m diameter seems terrifying. But take note that diameter and aperture are two different words with significant meanings.

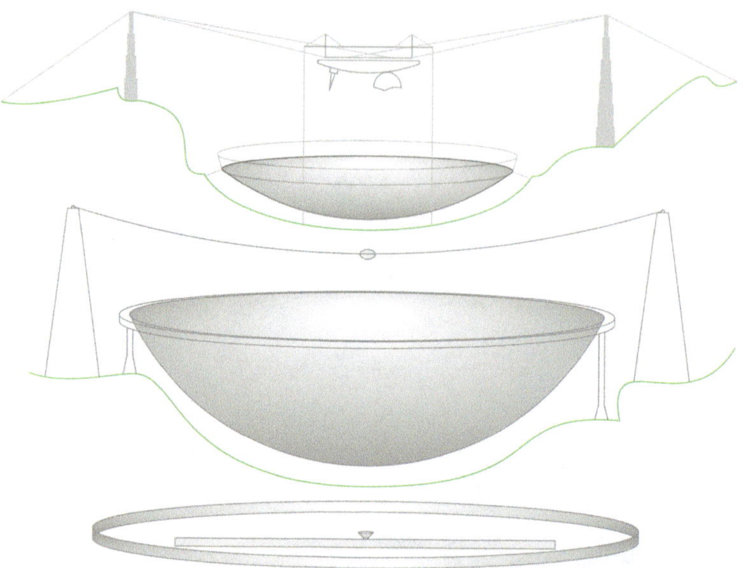

Fig. 7.3 From top to bottom: Arecibo, FAST, and RATAN-600. (*Credit* Cmglee)

Let's start with this telescope's unusual design. At a great height, it appears like a ring because it is a giant circle surrounded by 895 reflective components. Each reflecting unit measures 2 m in width and 7.4 m in height, or about the size of a standard football goal when rotated through 90° (Pariiskii and Korol'kov 1989). This telescope may have a frighteningly big diameter, but its effective receiving area is just 16.5% as large as the Arecibo in Puerto Rico and 6% as large as the Chinese Heavenly Eye (Fig. 7.4).

Weird-looking as it is, RATAN-600 can function in one of three ways (Parijskij and Korol'kov 1998) (Fig. 7.5).

Fig. 7.4 RATAN-600 close-up (*Credit* александр с кавказа)

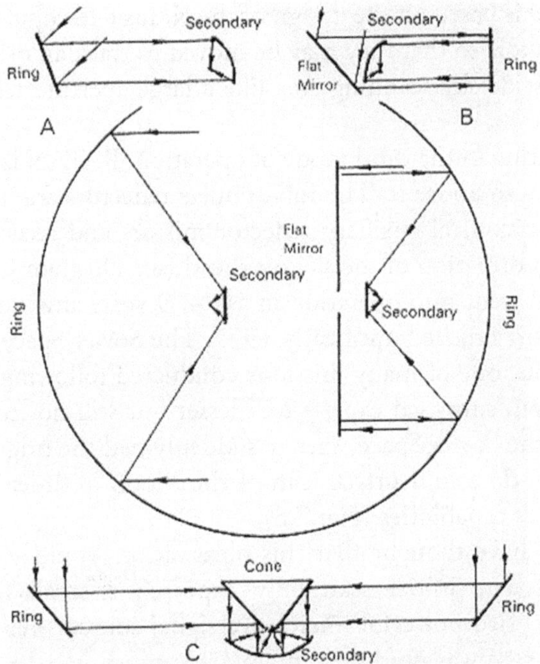

Fig. 7.5 Observation mode diagram of RATAN-600 (*Credit* SAO)

One is the "double-mirror mode," in which a sector made of multiple-ring reflection plates directs microwaves onto a cylindrical secondary mirror before sending them onto a receiver cabin. Each of RATAN's three primary mirror sectors is paired with a corresponding secondary mirror, all of which are cylinders. These secondary mirrors are movable in orbit, allowing the signal to be focused in accordance with the primary mirror's viewing angle (see Diagram A).

The three cylindrical secondary mirrors may monitor not only various objects simultaneously but also different azimuths of the same object, creating a synthetic azimuth map in two dimensions.

Three-mirror mode is the second configuration. On the southern side of RATAN-600, there is a flat reflector plate. Along with the primary mirror in the south, this reflector plate directs microwaves reflected from the flat reflector toward the cylindrical primary mirror reflector and finally the cylindrical secondary mirror (see Diagram B).

The Ohio State University Telescope, better known as "Big Ear," was the first to use this design structure proposed by John Kraus (1910–2004), a respected US astronomer.

RATAN-600 is based on the design of the Kraus telescope, but it adds rails to the flat reflectors so that they may be moved to track an observation target and acquire a wide sky scanning area like a large aperture telescope (Letrou et al. 2012).

When operating in the third mode of operation, RATAN is able to directly monitor the zenith above it. The full annular reflectors are used here, along with the central conical auxiliary reflector mirrors and receivers, to acquire the scanning information of the sky overhead (see Diagram C).

RATAN-600 went into operation in 1974, 9 years after construction had begun, but wasn't finished until early 1977. The Soviet Space Agency's lunar observation plan, one of many missions conducted following RATAN-600's official operation, surprised experts to a lesser but still noticeable extent. At that moment, the Soviet Space Agency suddenly had the bright idea of using RATAN-600 to do a full-surface scan of the Moon in order to evaluate the new instrument's capabilities (Fig. 7.6).

Who would have thought that this observation would yield such unexpected results? The initial scan only captured a small fraction of the Moon's surface. Two powerful microwave signal sources were discovered by astronomers. There was no doubt in anyone's mind that the Moon was not the source of these microwaves. The Soviet experts were taken aback by this. Some even speculated that aliens were responsible. They continued to probe

Fig. 7.6 RATAN-600 under night sky (*Credit* depositphotos/mulderphoto)

the rest of the Moon's surface and were even more taken aback when they discovered three more microwave signal sources.

These unexpected findings were brought to the Moscow Space Research Institute by specialists. After consulting with numerous specialists and poring over a mountain of data, they concluded that the lunar coordinates of these signal sources were remarkably consistent with the five landing locations of the Apollo spacecraft, proving that the Americans must have left these signal transmitters behind (Naugolnaia et al. 1978).

The Apollo Lunar Surface Experiments Package (Murtaugh 1966), of which no specifics were previously known at the time, was later revealed to be the source of the detected transmitters. From Apollo 11 in 1969 until Apollo 17 in 1972, the program ran its course. A control center, radioisotope thermoelectric generator, laser ranging reflector, passive seismic experiment package, lunar surface magnetometer, passive seismic experiment, and solar wind spectrometer were among the more than a dozen scientific instruments sent to the Moon's surface.

These instruments on the Moon served as a formidable research station. For at least a year after astronauts returned to the Earth, astronomers would be able to monitor the lunar environment remotely from the Earth. Throughout the course of 8 years, these instruments did not fail to fulfill their intended purpose. What we know about the Moon now is largely determined by the findings from this experiment.

On September 30, 1977, NASA officially ended the project due to funding concerns, but the transmitters of five stations were not immediately shut off, leading to some strange signals picked up by RATAN-600.

Yet, the Soviet's discovery became solid evidence of the Apollo lunar landing program and a hot topic of discussion among the RATAN-600 staff members.

Moon scanning was merely a pilot project. RATAN-600's actual scientific research responsibilities were quite extensive, and they included but were not limited to, radio research of solar activity; measurement of the extreme angular resolution of binary, multiple, and single stars using speckle interferometry; radio and spectral detection of nebulae in galaxies; photometric determination of pulsars, black holes, and other objects; development of local galaxy group space and kinematic scenarios; establishment of cosmic microwave background maps, etc. (Tsybulev et al. 2022).

Of course, the vast majority of these scientific research initiatives were carried out in collaboration with a number of other countries, which perfectly exemplified the spirit of RATAN-600's open and collaborative approach to astronomical study.

The search for extraterrestrial civilization was an essential part of RATAN-600's scientific research mission.

The star system HD 164595 in the constellation Hercules emitted a powerful microwave signal at 11 GHz in the 2.7 cm band on May 15, 2015, at 18:01:15 UTC, which was picked up by RATAN-600. A study authored by Alexander Panov detailing this discovery appeared in 2016's 67th International Astronautical Congress pre-conference materials (IAF 2016) (Fig. 7.7).

Paul Gilster, a well-known popular science author in the field of astronomy, came across this information and found it to be quite significant. Overnight, he tweeted about the possibility of an alien civilization in Hercules and posted an article to his website discussing the topic (Gilster 2016).

When Paul's post appeared online, it immediately sparked a worldwide hunt for alien life. The process was reported by The Times with a headline "Is that aliens calling, or just a wrong number?" (Moody 2016) Concurrently, CNN conducted a follow-up interview with the RATAN-600 research team, and one member told reporters there was a chance of an alien civilization near any strong signal detected (Griffiths 2016).

HD 164595 is a star about 95 light-years from the Earth that may be found near the waist region of the constellation Hercules. This star's surface

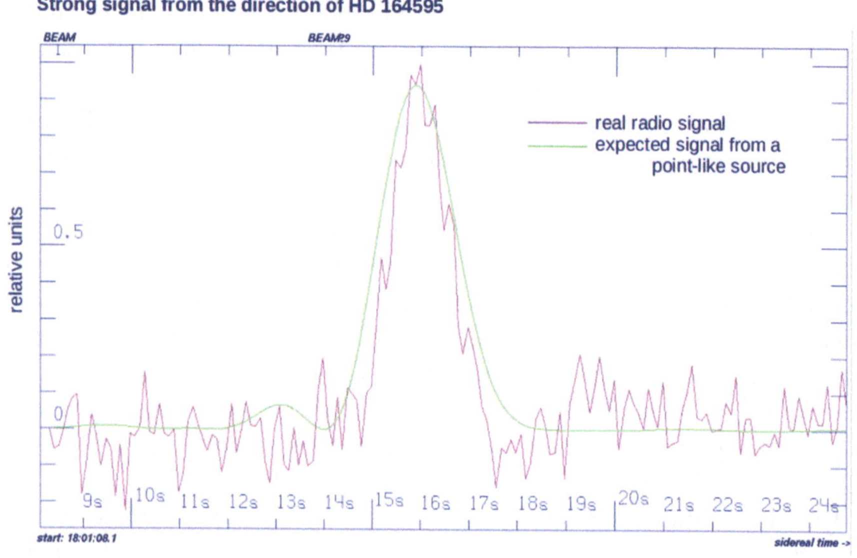

Fig. 7.7 A strong signal from the star system HD 164595 Hercules recorded by RATAN-600 (*Credit* Bursov et al. (2016))

temperature is comparable to that of the Sun. HD 164595 b, the star's companion, orbits the primary every 40 days (Is there anybody out there? SETI researchers look to explain mysterious signal 2016), suggesting the presence of planets in the system. The planet orbits its parent star once every 40 days, which is too close for liquid water on the surface to exist, but we cannot rule out the possibility of other Earth-like planets in the system.

Consequently, a large number of astronomical telescopes were pointed at HD 164595, including the Green Bank telescope, the Allen Telescope Array (Is there anybody out there? SETI researchers look to explain mysterious signal 2016) donated by Paul Allen (1953–2018), co-founder of Microsoft, and the Optical SETI Telescopes (Bruce 2007), all of whose primary missions were to search for signs of intelligent extraterrestrial life. It's as though then everybody was prepared to answer the extraterrestrial call at any moment (Fig. 7.8).

Regrettably, these telescopes could not detect any unusual signals after many days of monitoring. The crew at the Green Bank Telescope, led by Steve Croft, produced a comprehensive report (Isaacson et al. 2019).

According to the report, if this was indeed a signal from an alien civilization, it would most certainly be picked up by other telescopes as well. This signal, however, was likely caused by instrument interference or other human signals suddenly entering, as his team consulted the historical data of the Very Large Array at the National Radio Astronomy Observatory and compared it with the data continuously observed by Green Bank Telescope and found no sign of any signal in this region of the sky.

Fig. 7.8 From left to right: Allen Telescope Array (*Credit* Seth Shostak/SETI Institute) and SETI bucket optical telescope (*Credit* Harvard/SETI Institute)

The Green Bank Telescope was unable to provide compelling evidence for the RATAN-600 event since it did not identify any follow-up signals coming from the direction of HD 164595.

Another senior astronomer at SETI Seth Shostak agreed that the signal could be genuine but was more likely the result of ground interference or gravitational lensing at a longer distance than it had been from an alien civilization (Cool Worlds 2016).

Astronomer Nick Suntzeff from Texas A&M University argued the 11-GHz radiation was unlikely to originate from space. Throughout the universe, there is a high-energy astrophysical event called a "fast radio burst," which can generate thousands of megahertz microwaves but often lasts for only a millisecond, and the signal source of which is typically hundreds of millions of light-years from the Earth. There was little chance that the signal had been a fast radio burst. It's probable that the source of the signal was a top-secret military experiment, given that the signal's frequency band coincided with the frequency band in use by the military (McLaughlin 2016).

Despite all these, the media's spotlight on the incident persisted, and the Special Astrophysical Observatory of the Russian Academy of Sciences eventually issued a statement proposing that Earth was most likely the source of the signal (Sotnikova 2016) (Fig. 7.9).

Fig. 7.9 Russian Academy of Sciences (*Credit* depositphotos/demerzel21)

This unanticipated signal incident was frequently compared to 1977 "Wow!" signal event. We desperately want to find out whether our species is the only advanced one out there. The whole incident may seem like a prank God is playing on humanity, but, in the grand scheme of things, it's just a cup of mojito that adds a bit of exotic, romantic flavor to otherwise mundane exploration labor.

Humanity has been wondering if there might be other intelligent species in the universe for thousands of years (Cofield 2022), whenever we look up at the stars and try to count them. More than five thousand exoplanets have now been confirmed, and many more are likely waiting to be found. Can we tell if any of these planets meet the requirements for life? In the evolution of planetary systems, under what circumstances do terrestrial planets tend to form? These are questions that astronomers devote their careers to answering (Fig. 7.10).

It's incredible that somewhere in the expanse of space, on a tiny planet, there exists a community of intelligent beings, among whom one of you is reading this book about space.

Fig. 7.10 RATAN-600 keeps on its exploration. (*Credit* depositphotos/mulderphoto)

References

Andrew Siemion Director of the BSRC. (n.d.): Retrieved April 12, 2023, from Berkeley SETI: https://seti.berkeley.edu/people/Andrew.html

Bruce, B.: Planetary society's optical SETI telescope offers online view of night sky (2007). Retrieved from The Planetary Society: https://www.planetary.org/articles/oseti_103007

Cofield, C.: Cosmic Milestone: NASA Confirms 5,000 Exoplanets (2022). Retrieved from NASA Jet Propulsion Laboratory (JPL): https://www.jpl.nasa.gov/news/cosmic-milestone-nasa-confirms-5000-exoplanets

Cool Worlds: Alien signal received from HD 164595? [Video] (2016). Retrieved from YouTube: https://www.youtube.com/watch?v=wCW5I9UQkZY

Croft, S., Siemion, A., Macmahon, D., Lebofsky, M., Isaacson, H.T., Hickish, J., Deboer, D.R.: Breakthrough listen follow-up of a transient signal from the RATAN-600 telescope in the direction of HD 164595. Res. Gate (2016). https://doi.org/10.13140/RG.2.2.10561.20326

Gilster, P.: An Interesting SETI Candidate in Hercules (2016). Retrieved from Centauri Dreams: https://www.centauri-dreams.org/2016/08/27/an-interesting-seti-candidate-in-hercules/

Griffiths, J.: Hear Me Now? 'Strong signal' from Sun-Like Star Sparks Alien Speculation (2016). Retrieved from CNN: https://www.cnn.com/2016/08/30/health/seti-signal-hd-164595-alien-civilization/index.html

History, H.: The Space Race: A History from Beginning to End. Createspace Independent Publishing Platform (2018)

IAF: 67th International Astronautical (2016). Retrieved from proceedings.com: http://toc.proceedings.com/33478webtoc.pdf

Is there anybody out there? SETI researchers look to explain mysterious signal. (2016). Retrieved from Room The Space Journal of Asgardia: https://room.eu.com/news/is-there-anybody-out-there-seti-researchers-look-to-explain-myster ious-signal

Isaacson, H., Siemion, A.P., Marcy, G.W., Hickish, J., Price, D.C., Enriquez, J.E., Gizani, N.: The breakthrough listen search for intelligent life: no evidence of claimed periodic spectral modulations in high-resolution optical spectra of nearby stars. Publ. Astron. Soc. Pac. **131**(995), 014201 (2019). https://doi.org/10.1088/1538-3873/aaeae0

Letrou, C., Khaikin, V., Boag, A.: Analysis of the RATAN-600 radiotelescope antenna with a multilevel physical optics algorithm. C. r. Phys. **13**(1), 38–45 (2012). https://doi.org/10.1016/j.crhy.2011.10.011

McLaughlin, E.C.: Signal from Sun-like Star Likely of Earthling Variety (2016). Retrieved from CNN: https://www.cnn.com/2016/08/31/europe/seti-signal-hd-164595-not-alien-civilization/index.html

Moody, O.: Is that Aliens Calling, or Just a Wrong Number? (2016). Retrieved from The Times: https://www.thetimes.co.uk/article/is-that-aliens-calling-or-just-a-wrong-number-jbscwfrzf

Murtaugh, C.: Apollo Lunar Surface Experiments Package. 3rd Annual Meeting (1966). https://doi.org/10.2514/6.1966-919

Naugolnaia, M.N., Spangenberg, E.E., Soboleva, N.S., Fomin, V.A.: Selenographic coordinate measurements with the RATAN-600 radio telescope. Sov. Astron. Lett. **4**, 562–565 (1978)

Pariiskii, Y.N., Korol'kov, D.V.: Experiment cold: the first deep sky survey with the Ratan-600 Radio Telescope. Soviet Sci. Rev. (1989)

Parijskij, Y.N., Korol'kov, D.V.: 2. The RATAN-600 Radio Telescope (1998). Retrieved from Experiment Cold: https://www.sao.ru/hq/CG/cold/part2.htm

Shostak, S.: A SETI signal? (2016). Retrieved from SETI Institute: https://www.seti.org/seti-institute/a-seti-signal

Sotnikova, Y.V.: Monitoring of the Continuum of SETI Candidates with RATAN-600 (2016). Retrieved from SAO RAS official comment: https://www.sao.ru/Doc-en/SciNews/2016/Sotnikova/

Steve Croft—Astronomy Department. (n.d.): Retrieved April 12, 2023, from University of California Berkeley Department of Astronomy: https://astro.berkeley.edu/people/steve-croft/

Tsybulev, P., Nizhelskij, N., Dugin, M., Borisov, A., Udovitskiy, R., Kratov, D.: Continuum radiometers of the RATAN-600 radio telescope: Current state and directions of development. Astron. Instrum. Methods **425** (2022). https://doi.org/10.22323/1.425.0015

8

The Karl G. Jansky Very Large Array (VLA): A Veteran Super Array

The Karl G. Jansky Very Large Array

This signal was first picked up by NASA's Deep Space Network at 3:40 a.m. on June 5, 1989. The moment marked the beginning of Voyager 2's journey to Neptune, the furthest planet in our solar system (NASA, Voyager 2 Enters Final Planetary Encounter 1989) (Fig. 8.1).

Voyager 2 had been the only spacecraft to approach Neptune (NASA, Voyager 2 n.d.). Fifty images would be taken each day by Voyager 2 and transmitted to the Earth, as per the plan of the mission's scientists (Fig. 8.2).

Although the Voyager team had already completed five planetary exploration missions and was already experienced with the workflow before Voyager 2 arrived at Neptune, the task of capturing 50 photos a day was still considered as an enormous challenge by the entire team.

The fact that Neptune was so much further from the Earth than any other planet ever studied presented an additional hurdle for the Voyager mission team.

Be aware that the strength of a radio wave signal decreases in a manner that is inversely proportional to the square of the distance between the source and the receiver. Voyager 2 had a large 3.7-m antenna, but to conserve energy, scientists had to limit the radio transmitter's output to just 23 W (Frenzel 2017), which was almost the power required to charge a mobile phone. These radio waves, emitted with a transmission power of 13 W, were almost too feeble to be detected once they reached the Earth. Furthermore, the considerable distance resulted in a communication time gap of over four hours,

J. Wang, *Eye Beyond the Sky*, Astronomers' Universe,
https://doi.org/10.1007/978-981-99-9818-0_8

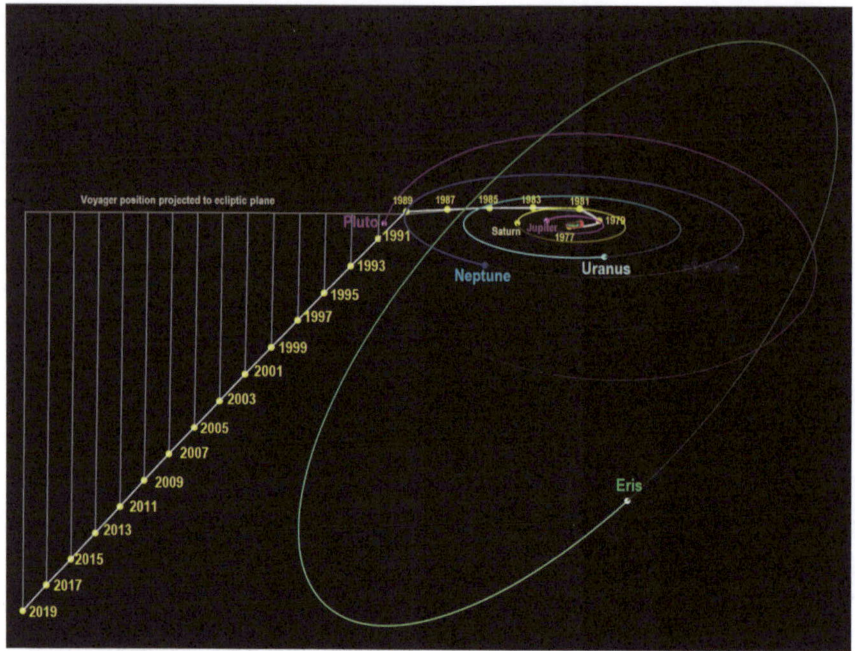

Fig. 8.1 Voyager 2 post-launch exploration trajectory (*Credit* Tomruen)

Fig. 8.2 Voyager 2 took these two images of Neptune's rings on August 26, 1989. (*Credit* Voyager 2, NASA/JPL)

rendering data verification an impossibility. If there were even one incorrect piece of information, the entire priceless picture would be ruined.

Early in the mission, around the time Voyager 2 passed through Uranus in 1986 (Mars 2021), the Voyager team began making preparations to improve the ground-based deep space communication network. Engineers had increased the aperture of the three largest radio telescopes in the network from 64 to 70 m in order to improve signal reception. The very sensitive Parkes radio telescope in Australia was also incorporated to receive data as part of the project so that the reception effect might be improved (Lazio and Deutsch 2014) (Fig. 8.3).

Still, the Voyager 2 team was unable to extend the daily photo transmission limit to 50. The contribution of the Very Large Array (VLA), which consisted of dozens of parabolic dishes that were each 25 m in diameter (Brown et al. 2013), was crucial in finally solving this issue.

But why could an array of radio telescopes each with a 25-m dish achieve what a 70-m antenna cannot? (Brown et al. 2013) To provide you with an answer to this question, our story will open in the year 1931.

1931 was the year when Bell Laboratories made the discovery that when transatlantic telephone operations were carried out utilizing radio waves of 10–20 m in length, there was always some unavoidable static interference. A

Fig. 8.3 Canberra deep space communication complex could send commands to Voyager 2. (*Credit* Robert Kerton, CSIRO)

young engineer named Karl G. Jansky (1905–1950) (Jansky n.d.) working at Bell Laboratories agreed to take on the study.

According to Jansky, tracing the origin of radio wave interference required a specially designed antenna. In addition, in order to pick up weak signals, the size of the antenna needed to be increased to its maximum capacity (Fig. 8.4).

In order to achieve this, Jansky constructed a peculiar antenna 30 m in length out of metal tubes and timber strips shaped like the wings of a biplane. He additionally built a circular chassis and four wheels below to enable the antenna to be freely steered in any direction. With just a little bit of help from a single person, the 30-m giant could be pushed around without any obstructions (Kraus 1981). Jansky utilized the volumes of the hissing noise produced in his headphone to locate the sources of the radio wave signal that the antenna had picked up. Over time, Jansky noticed that the strongest noise signal occurred when the antenna was pointing in a certain direction inside the Milky Way.

Jansky thoroughly ruled out any potential source of interference from power lines or electrical equipment, and after additional investigation, he came to the audacious conclusion that the noise originated from beyond the Earth (Jansky 1933).

Fig. 8.4 The "antenna" that Jansky invented in the 1930s (*Credit* NRAO/AUI/NSF)

Fig. 8.5 Karl G. Jansky (*Credit* NRAO/AUI/NSF)

The radio waves detected by Jansky have since been traced back to the Galactic Core. His work paved the way for fresh scientific inquiry and reshaped the way astronomers thought about the cosmos.

As a result of Jansky's work, astronomers have been able to use radio telescopes to investigate the emission of radio waves in the cosmos.

The wavelength of radio waves varies from 1 mm to 100 km. Radio waves with wavelengths longer than 10 m are absorbed and reflected by the atmosphere of the Earth. Radio waves with wavelengths shorter than 1 cm are likewise easily absorbed by the atmosphere of the Earth. For this reason, radio telescopes' primary foci of study are radio waves in the 1 cm to 10 m range (Lowe n.d.).

Yet, a telescope's angular resolution is inversely proportional to the ratio of wavelength to aperture width (Born and Wolf 2000). In other words, the greater the aperture of the telescope, the greater the resolution, as long as the frequency of the electromagnetic wave under study has been calculated.

Radio waves' wavelengths with a centimeter scale are still eight orders of magnitude longer than the those of visible light (Rybicki and Lightman

2008). In spite of their great size (often measured in the tens of meters), radio telescopes had just a few pixels of resolution while observing radio waves. A key issue in radio astronomy had been the limited resolution and challenging imaging of radio telescopes (Fig. 8.6).

A British astronomer named Martin Ryle (1918–1984) performed an experiment in 1946 in which he pointed two antennas in the direction of the Sun at the same time. He then superimposed the signals that were acquired independently in order to cause the signals to interfere with each other. The interference signal enhanced not only the strength but also the precision, he discovered (Ryle and Vonberg 1946).

Distance between the two antennas increases the aperture of the resulting virtual telescope. This is an example of the methodology known as interferometry.

An interferometric array is a collection of telescopes that work together to produce a single image. It is preferable to have a larger number of telescopes participating in the interferometric array, to have the telescopes spread out across a greater distance, and to have a more diverse frequency coverage. It might appear as though every radio telescope in the globe could be connected to one another to construct the most advanced radio telescope array, but in reality, this was not the case. The time calibration of atomic clocks was

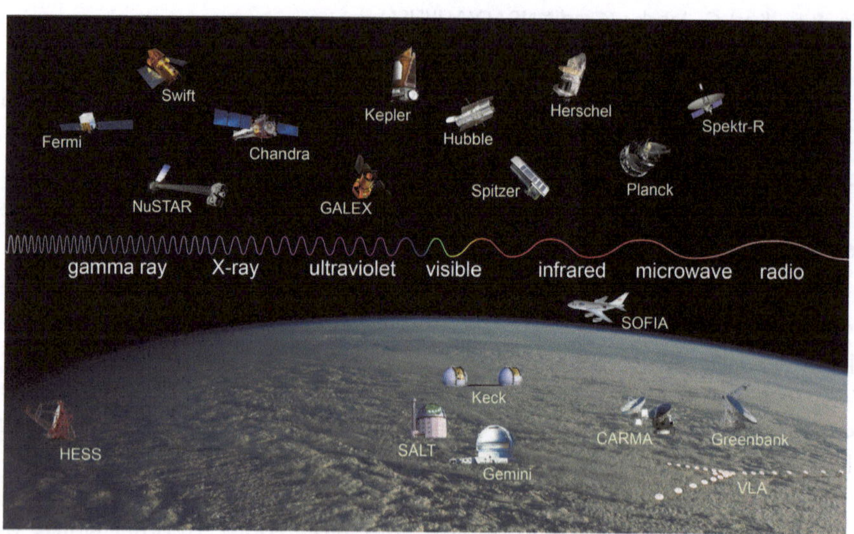

Fig. 8.6 Telescopes and radio waves (*Creidit* Observatory images from NASA, ESA (Herschel and Planck), Lavochkin Association (Specktr-R), HESS Collaboration (HESS), Salt Foundation (SALT), Rick Peterson/WMKO (Keck), Germini Observatory/AURA (Gemini), CARMA team (CARMA), and NRAO/AUI (Greenbank and VLA); background image from NASA)

the second step in getting a large number of radio telescopes to function together after making sure they could all operate in the same frequency range. The construction of a specially designed radio telescope array was required (Thompson et al. 2017).

To complement the massive single telescope, the National Radio Astronomy Observatory (NRAO) proposed the construction of a dedicated radio telescope array in the early 1960s.

Selecting an ideal location to put up radio telescopes is the first step before commencing the project. In comparison to terrestrial broadcast radio signals, the power of cosmic radio waves is on the order of a few billionths. For radio telescopes, even the smallest amount of human activity constitutes a significant disruption that just cannot be ignored.

The Plains of San Agustin in New Mexico, northwest of Socorro, is a flat stretch of desert far from major cities. Water molecules cause much less disruption to radio waves in dry desert environments. The desert is surrounded by rugged mountains, which act as a natural radio barrier from the cities located hundreds of kilometers away.

This remote location is perfect for radio astronomy because of the absence of interference from other radio waves. To waste such a prime building location would be a shame.

The National Radio Astronomy Observatory completed a ground-breaking design in 1967. They settled on erecting 27 antennas, each with a 25-m aperture with radio receivers for eight different frequency ranges. The 230-ton dish antenna was mounted on a movable tripod that allowed for 360-degree rotation (National Radio Astronomy Observatory 1967).

One unusual feature of the tripod was that it was not anchored to the ground. In the plan developed by the engineers, the maximum distance between the two telescopes was 36 km, and each of the three orbits had a length of 21 km and an inclination of 120 degrees (Finley 1985). A powerful tractor was another invention of the engineers. The tractors could be employed to transport the massive antennas when it was time to relocate the telescopes.

Because of its configuration, this radio telescope array was referred to as the Very Large Array (VLA). It would not only have the highest resolution humans had ever seen but also be ground-breaking in terms of adaptability (Perley et al. 2011).

All that was needed to get started on this planned endeavor had already been assembled.

The Very Large Array telescope Project, funded by the National Science Foundation, was authorized by the US Congress in August 1972. The project began without a hitch in April of 1973 (Kellermann et al. 2020) (Fig. 8.7).

In September 1975, the VLA's first antenna entered service and was used to study the Virgo Galaxy located 50 million light-years away. Successful interferometric astronomical observations were made for the first time in 1976 when construction of the second antenna was completed (Sullivan 2005).

After the VLA's successful commissioning, construction ramped up. Within the next year alone, six antennas had been deployed. Routine astronomical observations were initiated afterward.

The year 1980 witnessed the activation of the 28th antenna. To this point, the Very Large Array telescope had reached an official state of completion. In the end, they constructed 28 antennas rather than the 27 that had been planned initially. This was done so that all of the antennas could be rotated for maintenance, essentially ensuring that the VLA would keep functioning for a long period (NRAO 2016) (Fig. 8.8).

It had been the most costly radio telescope project at the time of its 1980 inauguration, costing a total of $78 million (about $485 million in today's US dollars) (Mann 2020).

Fig. 8.7 The first antenna that entered service in 1975 (*Credit* NRAO/AUI/NSF)

Fig. 8.8 Newly completed Very Large Array telescope (*Credit* NRAO/AUI/NSF)

A telescope's effectiveness is determined mostly by its sensitivity and resolution. Astronomers use the term "sensitivity" to describe how faint electromagnetic waves a telescope can pick up, and "resolution" to describe how clear the images it creates are.

It is difficult to simultaneously improve sensitivity and resolution, however. Telescopes require extensive periods of time to collect signals in order to detect faint electromagnetic signals.

The antennas in the VLA can be arranged in a variety of shapes to achieve a right balance between sensitivity and resolution. Distances between the antennas of the VLA can be adjusted by astronomers to optimize the observation of signals from sources of varying intensities (Fig. 8.9).

An extremely detailed picture was generated after computers combined the information from all 27 dishes. The VLA could produce images with the same resolution as a single, massive telescope with an aperture of 36 km (Thompson et al. 2017).

VLA has been the first astronomical instrument capable of producing color images using radio waves and the most widely used radio telescope in the world (NRAO, VLA Science 2023).

More than 14,000 observation programs from more than 5,000 astronomers had been approved before its formal opening in 1980. The

Fig. 8.9 Very Large Array telescope with adjustable antenna spacing (*Credit* depositphotos/jose1983)

scientific findings from these initiatives spanned virtually every branch of astronomy.

In 1983, utilizing the Very Large Array telescope, researchers were able to capture an image of the Milky Way's galactic core. The core seemed obscured by interstellar dust when viewed through optical telescopes, but this image revealed stars whirling around the central black hole as well as enormous swirls of gas and dust. Here was where radio astronomy pioneer Karl G. Jansky pointed his 30-m directional antenna (Fig. 8.10).

In 1987, astronomers were able to take pictures of a vast number of radio sources across the sky with the help of VLA. Multiple mirror images of a distant quasar were discovered by astronomers in a galaxy's central region, all contained in a single picture. This was the effect of gravitational lensing that Einstein had predicted back in 1936 (Eigenbrod 2011) (Fig. 8.11).

In 1989, the VLA telescope was given the opportunity to take part in receiving data for the flyby of Neptune, which was performed by Voyager 2. This marked the end of the Voyager mission team's exploration of the four planets of the solar system and the first time humans had seen the eighth planet up close. Since then, no other spacecraft has ever made the journey to Neptune, making this historic event both the first and only time a spacecraft has ever made the journey.

The Voyager 2 mission was successful in conveying a wealth of information and photos back to the Earth, much of which was made possible by the

Fig. 8.10 Image of the galactic core from the Very Large Array telescope (*Credit* A. Angelich (NRAO/AUI/NSF); NASA/JPL-Caltech/ESA/CXC/STScI)

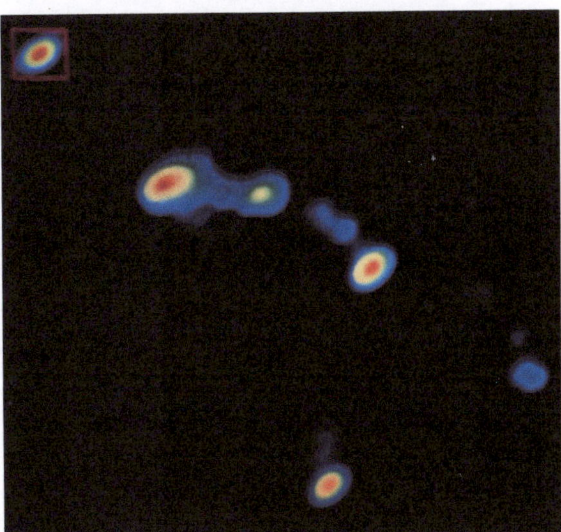

Fig. 8.11 This image from the Very Large Array telescope verified the presence of gravitational lensing. (*Credit* NRAO/AUI/NSF)

contribution of the VLA telescope. This is the story that was told at the start of this chapter (Warwick 1991).

In 1991, planetary scientists intended to send a detection signal to Mercury at a frequency of 8.5 GHz using the 70-m Goldstone Deep Space Communications Telescope and a 500,000-W transmitter. Mercury, however, is not only too far from the Earth but also too near to the Sun. After the calculation, it became clear that the solitary radio telescope available back then wouldn't be able to finish the job of receiving signal echoes. In the end, not only was the VLA telescope able to successfully accomplish its job, but it also managed to capture a distinct radar reflection image of Mercury (Butrica 1996) (Fig. 8.12).

VLA has been undergoing a comprehensive upgrading of its equipment since 2001. In the succeeding decade, virtually all of its electronic machinery was updated. The new and improved VLA had a sensitivity increase of a factor of ten. Antennas could produce 100 times more data than they did in the past. To say the least, this was a major shift (Perley et al. 2011).

"After 10 years of upgrading, the Very Large Array can identify practically all radio sources from outside the Milky Way," stated Jim Condon, a staff member of the US Radio Astronomical Observatory. "Previously, it had been

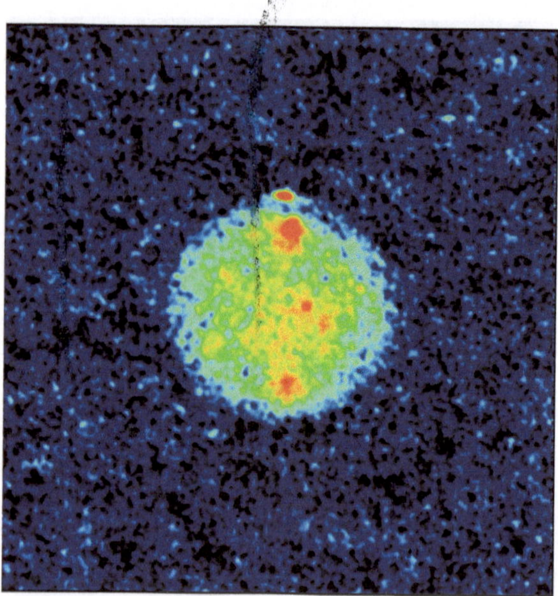

Fig. 8.12 This image of Mercury was the result of the radar experiment using the 70-m antenna of NASA JPL/DSN in Goldstone, California, as the transmitter and the Very Large Array (VLA) as the receiver. (*Credit* Duane O. Muhleman (Caltech); NRAO/AUI/NSF)

impossible to pick out such a faint signal from the noise of the background radiation."

Staff members agreed that a new, more fitting name for the Very Large Array telescope was needed to highlight the significance of this transformation. They subsequently collected suggestions for new names from the general public.

VLA was officially renamed as the Karl G. Jansky Very Large Array on January 10, 2012 by the observatory's leadership (Finley 2012). The "handshake" between the forefather and forerunner of radio astronomy was accomplished after 80 years (Fig. 8.13).

The Karl G. Jansky Very Large Array sky survey was initiated in the month of September 2017.

The survey would observe for over 5,500 h over the course of seven years (Lacy et al. 2020). Meanwhile, VLA would conduct three scans of the sky, each 32 months apart. This mission has the potential to uncover 10 million new celestial objects, which is four times the current count.

Upon completion of the observations, the National Radio Astronomy Observatory had pledged to make all survey data freely accessible to astronomers around the world. The second sky scan's data was made public as of March 10, 2022. While the study is ongoing, it has already produced

Fig. 8.13 The greatly-improved Karl G. Jansky Very Large Array (*Credit* Depositphotos/MaRoPictures)

Fig. 8.14 Driving along US Route 60 to the Very Large Array telescope (*Credit* Nan Janney/NRAO/AUI/NSF)

the most comprehensive and detailed map of radio emission sources in the heavens (NRAO, The Very Large Array Sky Survey (VLASS) 2017).

This map allows scientists to see through the dust clouds that have obscured the truth in visible light instruments. Research into supernovae, neutron stars, and supermassive black holes can benefit for a long time from the information contained within.

As one travels west from Socorro along the US Route 60, the lengthy stretch of mountains embracing the plains serves as a lone beacon along the barren highway. Deer, antelope, and roaming cattle are common sights on either side of the road, yet no people are in sight. There is no question that this is one of the most enchanted locations on the entire planet (Fig. 8.14).

Along with the clear skies and beautiful landscape, there are also 28 enormous antennae, which shift their configurations occasionally like loyal sentinels of all humanity, standing silently in the vast desert and listening to the whispers from the depths of the universe.

References

Born, M., Wolf, E.: Principles of optics: electromagnetic theory of propagation, interference and diffraction of light. CUP Archive (2000)

Brown, D.W., Brundage, W.D., Ulvestad, J.S., Kent, S.S., Bartos, K.P.: Interagency Telemetry Arraying for Voyager-Neptune Encounter (2013). Retrieved from NASA Technical Reports Server (NTRS): https://ntrs.nasa.gov/citations/19910002669

Butrica, A.J.: To see the unseen: a history of planetary radar astronomy (1996)

Eigenbrod, A.: Gravitational lensing. In: Gravitational Lensing of Quasars, pp. 49–66. EPFL Press, New York (2011). https://doi.org/10.1007/978-4-431-68162-5_31

Finley, D.G.: Very long baseline array (VLBA). Access Sci. (1985). https://doi.org/10.1036/1097-8542.yb980660

Finley, D.: Iconic Telescope Renamed to Honor Founder of Radio Astronomy (2012). Retrieved from NRAO: https://www.nrao.edu/pr/2012/jansky/

Frenzel, L.: The Speed of Light is Not Fast Enough (2017). Retrieved from Microwaves & RF: https://www.mwrf.com/community/article/21848722/the-speed-of-light-is-not-fast-enough

Jansky, K.G.: Electrical disturbances apparently of extraterrestrial origin. Proc. IRE **21**(10), 1387–1398 (1933). https://doi.org/10.1109/jrproc.1933.227458

Jansky, K. (n.d.): Retrieved April 12, 2023, from The National MagLab: https://nationalmaglab.org/magnet-academy/history-of-electricity-magnetism/pioneers/karl-jansky/

Kellermann, K.I., Bouton, E.N., Brandt, S.S.: The very large array. In: Open Skies, pp. 319–390. Springer International Publishing (2020)

Kraus, J.: The first 50 years of radio astronomy, part 1: Karl Jansky and his discovery of radio waves from our galaxy. Cosmic Search **3**(12), 8 (1981). Retrieved from http://www.bigear.org/CSMO/HTML/CS12/cs12p08.htm

Lacy, M., Baum, S.A., Chandler, C.J., Chatterjee, S., Clarke, T.E., Deustua, S., English, J., Farnes, J., Gaensler, B.M., Gugliucci, N., Hallinan, G., Murphy, E.J.: The Karl G. Jansky very large array sky survey (VLASS). science case and survey design. Publ. Astron. Soc. Pacific **132**(1009), 1–29 (2020). Retrieved from https://www.jstor.org/stable/26973251

Lazio, J., Deutsch, L.: The deep space network at 50. Phys. Today **67**(12), 31–37 (2014). https://doi.org/10.1063/PT.3.2619

Lowe, S. (n.d.).: Radio telescopes. Retrieved April 12, 2023, from Las Cumbres Observatory: https://lco.global/spacebook/telescopes/radio-telescopes/

Mann, A.: The VERY large Array: 40 Years of Groundbreaking Radio Astronomy (2020). Retrieved from Space.com: https://www.space.com/very-large-array.html

Mars, K.: 35 Years Ago: Voyager 2 Explores Uranus (2021). Retrieved from NASA: https://www.nasa.gov/feature/35-years-ago-voyager-2-explores-uranus

NASA: Voyager 2 Enters Final Planetary Encounter (1989). Retrieved from NASA: https://www.jpl.nasa.gov/news/voyager-2-enters-final-planetary-encounter

NASA. (n.d.): Voyager 2. Retrieved April 12, 2023, from NASA Solar System Exploration: https://solarsystem.nasa.gov/missions/voyager-2/in-depth/

NRAO: Very Large Array (2016). Retrieved from National Radio Astronomy Observatory: https://public.nrao.edu/telescopes/vla/

NRAO: The Very Large Array Sky Survey (VLASS) (2017). Retrieved from National Radio Astronomy Observatory: https://public.nrao.edu/vlass/

NRAO: VLA Science (2023). Retrieved from National Radio Astronomy Observatory: https://public.nrao.edu/telescopes/vla/science/

National Radio Astronomy Observatory: A proposal for a very large array radio telescope. West Virginia (1967)

Perley, R.A., Chandler, C.J., Butler, B.J., Wrobel, J.M.: The expanded very large array: a new telescope for new science. Astrophys J **739**(1), L1 (2011). https://doi.org/10.1088/2041-8205/739/1/l1

Rybicki, G.B., Lightman, A.P.: Radiative Processes in Astrophysics. John Wiley & Sons (2008)

Ryle, M., Vonberg, D.D.: Solar radiation on 175 Mc./s. In: W.T. Sullivan (ed.) Classics in Radio Astronomy, pp. 184–187. Springer Netherlands (1946). https://doi.org/10.1007/978-94-009-7752-5_18

Sullivan, W.T.: The Early Years of Radio Astronomy: Reflections Fifty Years After Jansky's Discovery. Cambridge University Press (2005)

Thompson, A.R., Moran, J.M., Swenson, G.W., Jr.: Interferometry and Synthesis in Radio Astronomy. Springer (2017)

Warwick, J.W.: Voyager Neptune encounter. In: H. Kikuchi (ed.) Environmental and Space Electromagnetics, pp. 327–330. Springer Japan (1991). https://doi.org/10.1007/978-4-431-68162-5_31

9

LAMOST: Numerous Stars at a Glance

LAMOST

A press conference was held at the National Astronomical Observatory of China on November 28, 2019, at 9:30 a.m. Beijing time (Lu 2019). As a researcher on the platform, Liu Jifeng stood out in his bright red tie against his black suit and white shirt. Heavyweight media correspondents from outlets including People's Daily, Xinhua News Agency, Guangming Daily, China Media Group, the Associated Press, Reuters, and Agence France-Presse were also present at the press conference (Fig. 9.1).

Researcher Liu Jifeng commented with great satisfaction once the conference had begun:

> "Using LAMOST's significant breakthrough, we have gained the 'holy grail' in this field. Therefore, Mr. David Reitze, Director of the Laser Interferometer Gravitational-Wave Observatory in the US, sent us a special letter to congratulate us on this extraordinary achievement, which, along with the merger of two black holes detected by his observatory over the past four years, will promote the revival of black hole astrophysics research."

Has there been any other scientific breakthrough that can be matched to the 2016 gravitational wave event? A common misconception is that black holes have been there for a very long time and that astronomers have located a great deal of them around the cosmos. Actually, only more than 20 black holes have been proven to exist in the Milky Way. Even now, in this day and age, when astronomy has reached such a high level of development, the finding of black holes, or more precisely, the discovery of conclusive

J. Wang, *Eye Beyond the Sky*, Astronomers' Universe,
https://doi.org/10.1007/978-981-99-9818-0_9

Fig. 9.1 Two black holes merging (*Credit* Depositphotos/Pike-28)

observational evidence of a black hole, is still considered to be a momentous achievement. This is in addition to the fact that we have discovered a black hole beyond human comprehension (Tatum 2020).

Undoubtedly, we can confidently claim that this "holy grail" discovery made by Chinese astronomers is a significant contribution to the field of astronomy worldwide, and one that will almost certainly force a complete rewrite of the current theory of stellar evolution. Our main character, the Guoshoujing Telescope in Xinglong County, Hebei Province, also known as the Large Sky Area Multi-Object Optical Fiber Spectroscopic Telescope (LAMOST) (Zhao 2012), is the one responsible for this groundbreaking finding. In honor of Guo Shoujing, an astronomer from the Yuan Dynasty who developed multiple observational instruments and other tools for watching the heavens, the telescope was formally named "Guoshoujing Telescope" in 2010 (Fig. 9.2).

So, why is this newly discovered black hole going to force a rewrite of textbooks? Let me regale you with the tale of LAMOST.

Prior to the 1980s, China had almost no major professional astronomical telescopes that were solely our own intellectual property. The dream of the previous generations of astronomers had always been to construct our own massive astronomical telescope.

Large aperture and wide field of vision seem to be at odds with each other, like fish and a bear's paw, in the field of astronomical telescope design. As any

Fig. 9.2 The LAMOST Telescope at Xinglong Station (*Credit* Sheliak)

of your photography-savvy friends will tell you, the more powerfully a "bird-shooting" lens focuses its light, the narrower its field of view. Yet, these ultra-wide-angle lenses can be made quite tiny. This is the contradiction between having a huge aperture and having a vast field of view, or, to put it more succinctly, being able to "see far" and "see wide" at the same time is not possible (Janesick 2001).

The construction concept of the Guoshoujing Telescope was originally proposed in the 1990s by a Chinese astronomical engineering team led by academicians Wang Shouguan (1923–2021) (IAU n.d.) and Su Dingqiang (NIAOT n.d.). It sought to address the conflict between large aperture and wide vision field, an issue that had long plagued the astronomical community, and potentially pave the way for China's independent production of big aperture telescopes.

Established in 1997, LAMOST is China's pioneering scientific project in astronomy.

Surely, you're curious how our telescope R&D team reconciled the seeming incompatibility between a huge aperture and a wide field of view. Optical fiber is the key to the answer. In its entirety, LAMOST is a Schmidt reflecting telescope. First, light from above is reflected by a flat primary mirror (MA) onto a concave primary mirror (MB). Four thousand optical fibers receive the light at the focal plane of the MB mirror. Light coming from all directions will be accurately directed into the spectrometer by these fibers. This is a brilliantly creative design. We may theoretically examine the spectra of up to 4,000 distinct stars at the same time because of the presence of these 4,000 optical fibers. It's the same as if you suddenly had a massively enlarged field of vision (CAS n.d.). Hence, LAMOST, in contrast to conventional optical telescopes, may capture spectra of astronomical objects rather than picturesque images of them.

It's simple to state this theory but much harder to implement.

To begin, LAMOST's two primary mirrors are composed of 24 and 37 small hexagonal mirrors, respectively, to cut down on building costs. For the purpose of accurate tracking of celestial objects, we must be able to modify the precise location of each mirror at any time, as the influence of the Earth's gravity will be slightly different depending on the direction the telescope is pointed. The R&D team put actuators behind each lens to make the spliced mirrors function as a single, integrated optic. These actuators are responsible for both supporting the weight of the mirror and modifying its form. Computer algorithms, which are part of a technique known as active optics technique (NAOC n.d.), make it possible to make adjustments in real time that are just one thousandth of a millimeter in magnitude. The completion of the LAMOST project is due in large part to the creative efforts of the Chinese research team working in the field of active optics technology.

In addition, correctly aligning each of the 4,000 optical fibers with a celestial body presents a formidable challenge for the engineering team. The "optical fiber positioning system with parallel control of partition operations" (Hu et al. 2006) is the innovative approach that LAMOST has taken to solve this issue. The system segments a focal plane that is 1.75 m in diameter into 4,000 circles, each of which is 33 mm in diameter. A rotatable fiber positioning unit is then positioned on each circle, and a fiber is attached to each unit. After the star catalogue's location is known, the telescope can be programmed to precisely locate 4,000 optical fibers in the focal plane in a matter of minutes.

As crucial as the technology itself is, the accompanying software is also essential. The LAMOST team not only developed a robust hardware device but also an extremely sophisticated software system for analyzing light (Guo et al. 2012) (Figs. 9.3 and 9.4).

In order to form the final spectrum, the light from distant celestial bodies must travel through the atmosphere, telescopes, optical fibers, and spectrometers, where it is mixed with the interference of stray light. This means that the original data cannot be used directly, and a complex algorithm is required

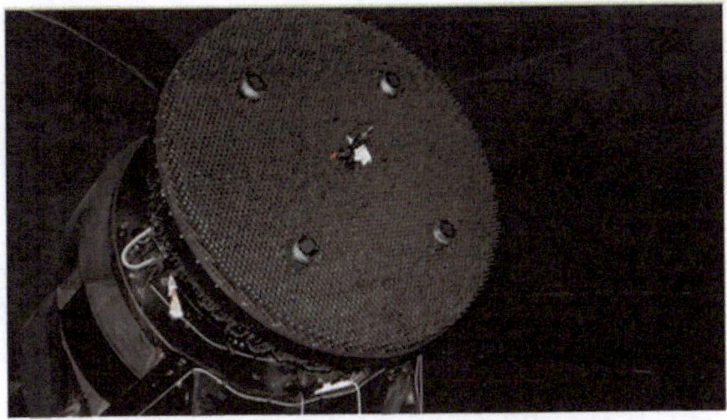

Fig. 9.3 1.75 m diameter focal plane (*Credit* LAMOST)

Fig. 9.4 Optical fiber positioning unit (*Credit* LAMOST)

for data processing. Since 2004, this piece of software, which was developed independently by a group of Chinese scientists, has not ceased the process of going through many iterations. More than 80,000 lines of code have been added to the software system over the course of more than 3,000 upgrades, guaranteeing that LAMOST's data correctness has achieved an internationally advanced level.

LAMOST was accepted for use in June 2009 after 5 years of construction.

In the popular imagination, astronomers spend their time idly seated in front of telescopes. This is actually a long-held misconception. Most modern astronomers rely on telescopes for gathering information about the universe, which they then use to draw conclusions about the celestial bodies they study. LAMOST is capable of recording the entire visible-light spectrum of a heavenly body. Each spectrum is unique in the same way that each individual's fingerprint is. A star's unique spectral characteristics come from the variety of elements it contains. Furthermore, celestial bodies in motion will exhibit

Fig. 9.5 LAMOST under the Milky Way (*Credit* Yingwei Chen, LAMOST)

varying spectral properties. So, if astronomers can obtain a celestial body's spectrum, they can use it to infer its composition, velocity, distance, and a host of other properties. An analogy between the spectrum and a celestial object's identification card is not too far-fetched (Carroll and Ostlie 2007) (Fig. 9.5).

The first significant accomplishment of LAMOST was that the number of sky spectra it had surveyed surpassed 10 million. LAMOST's 7-year sky survey spectral data was made publicly available in March 2019. This data set included 11.25 million spectra (Xinhua 2019), which was roughly twice as much as the total number of spectra made public by other international sky survey studies. The survey conducted by the Guoshoujing Telescope has been the first-ever spectral survey to collect more than 10 million spectra. These spectra are the most comprehensive astronomical data set ever collected, boasting the highest survey volume and sampling density, the most samples, and the best statistical consistency. Hundreds of academic institutions both at home and abroad are making use of these records for their research projects (Fig. 9.6).

The LAMOST team was proud of its massive spectrum data. And the members were ecstatic to report yet another surprise discovery in November of the same year. The National Astronomical Observatory called an unprecedented press conference to announce this discovery. This scene was described when you first started reading this chapter.

Before I reveal the primary finding discovered by the LAMOST team, I'll give you a brief introduction to how black holes are formed.

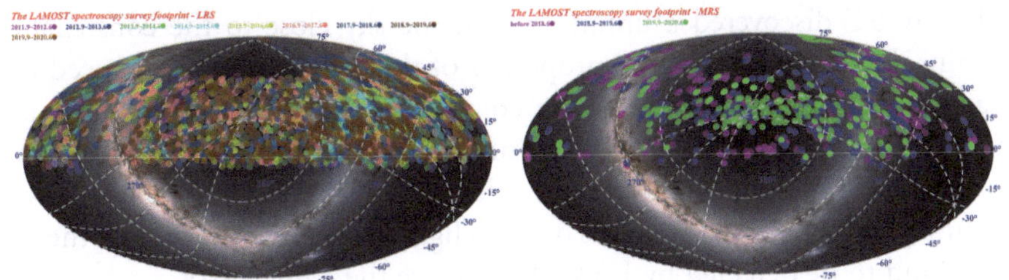

Fig. 9.6 LAMOST's "footprint" of sky survey (*Credit* LAMOST)

A black hole, as you may know, is the final resting place for a star. A stellar black hole is a type of black hole created from the collapse of a star. The initial mass of the star, as well as its metallicity and mass-loss mechanism, determine the maximum size of the black hole to which it can collapse. The term "metallicity" refers to the abundance of all elements in a star other than hydrogen and helium. When a massive star has a low metallicity, its star wind loss rate is reduced, which results in a larger mass for the black hole that is generated when the core collapses. As a result, the statement that "the higher the initial mass of a star, the larger the black hole that can be produced" is demonstrably false (Foglizzo 2017) (Fig. 9.7).

The textbook theory of stellar evolution predicts that even if a star's mass reaches 100 times that of the Sun at the metallicity of the Sun, the stellar black hole it forms will not exceed 25 times the mass of the Sun due to a succession of material losses (Özel et al. 2010). This means is that a stellar black hole can't theoretically grow larger than 25 times the mass of the Sun. Several prior observations have confirmed this theory. Around 20 stellar black holes have been discovered, and their masses are all within a factor of 20 times the Sun's mass, which agrees perfectly with theoretical expectations.

This conventional textbook explanation of stellar evolution, however, is in crisis due to a discovery made by Chinese astronomers.

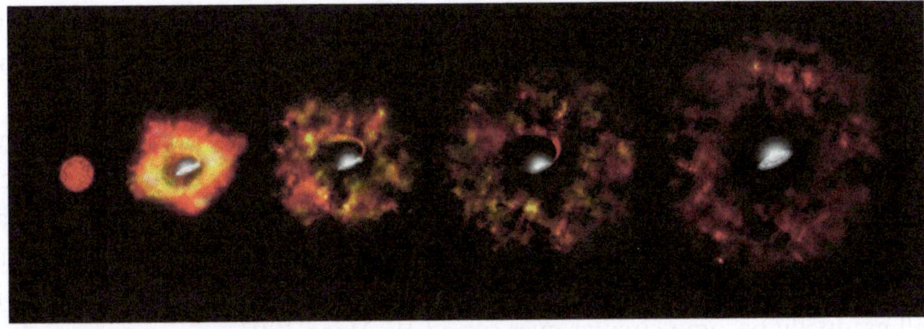

Fig. 9.7 How a star becomes a black hole (*Credit* NASA, ESA, and P. Jeffries (STScI))

LAMOST discovered a unique star in the sky close to the constellation Gemini. Based on the spectral properties of this star, it may be deduced that it and another invisible black hole combine to form a binary system and they rotate around a common center of mass. A new black hole in the Milky Way is, as we've mentioned, a huge scientific breakthrough in and of itself. Naturally, this made astronomers quite delighted. The black hole was the first of its kind to be detected by LAMOST, hence the name LB-1.

Using the mass of stars and the period of their orbits, astronomers can estimate the mass of black holes. The astronomers calculated the mass of LB-1 to be around 70 times that of the Sun (Liu et al. 2019).

As the mass was determined, however, scientists were left scratching their heads, wondering how such a massive black hole could exist. Its extreme mass significantly surpasses the limit for stellar black holes, which is shocking.

When astronomers discover a binary system consisting of a star and a black hole, the most common explanation is that the black hole was formerly part of the primordial binary system. To rephrase, initially, there were two stars in orbit around each other, but one of them eventually became a black hole. The two stars in a binary star system often share a similar metallicity since they both formed from the same cloud of cosmic dust (Tauris and van den Heuvel 2006).

The LB-1 companion star shares the Sun's metallicity. It makes sense that the star that merged to form the LB-1 black hole would have had a metallicity not too dissimilar from the Sun's; the current understanding of stellar evolution predicts that LB-1's mass should be no more than 25 times that of the Sun.

The hypothesis that LB-1 was born from a primordial binary system has to be abandoned as a result. The remaining possibility is that the companion star has been captured by the black hole. That is to say, the LB-1 black hole developed to its present mass after wandering across space, devouring several stars. Then it ran into the present-day star and snatched it up (Fig. 9.8).

However, more findings of LAMOST have strongly contradicted this hypothesis. There is substantial evidence that the LB-1 black hole and its companion star follow a roughly circular orbit around each other. Let me give you my reasons.

Imagine a black hole, alone and lost in space. The black hole's gravity acts like an invisible hand clutching at an approaching star. Since stars and black holes are subject to the same universal gravitational rule, we can deduce that their initial orbits must be eccentric circles rather than perfect circles. Orbital circularization, as predicted by celestial dynamics models, shows that the orbit will gradually become circular over time (Ali Alpar 2001).

Fig. 9.8 When a black hole meets a bright star (*Credit* Depositphotos/eevl)

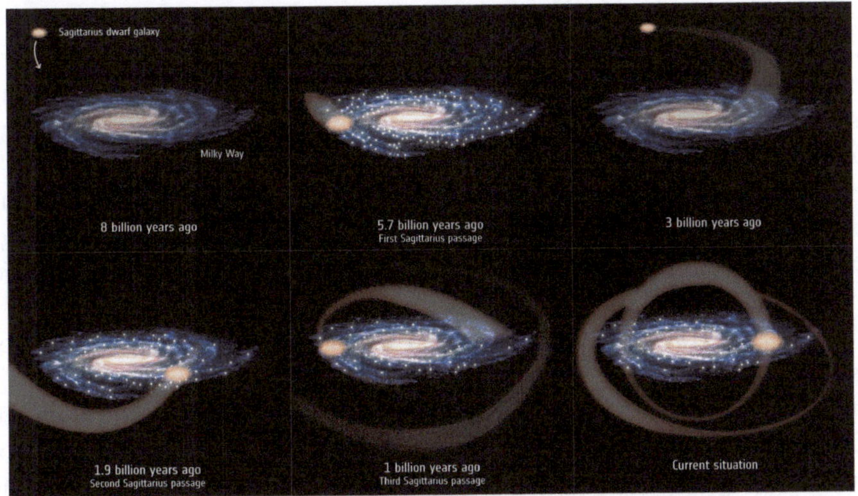

Fig. 9.9 Merger of the Milky Way and a dwarf galaxy (*Credit* ESA)

The strength of the tidal force produced by the black hole on the star determines how quickly its orbit will become circularized. The rate of rounding increases with increasing tidal force and decreases with decreasing tidal force. Since Newton's time, humans have perfected this science of motion, or astrophysics. Everything has been thoroughly tested, and there is no way to get it wrong. We can now run precise computer simulations of the phenomenon.

Yet, Chinese scientists have shown through computer simulations that the period required to circularize the orbits of LB-1 and its companion star is greater than the age of the universe. To phrase it another way, the capture theory cannot possibly be correct.

This made LB-1 an anomaly that couldn't be accommodated by any accepted astronomical theory. Due to the significance of the finding, the LAMOST team spent three years meticulously checking the data and enlisted the assistance of two other telescopes—the 10.4-m Gran Telescopio Canarias in Spain and the 10-m Keck Telescope in the United States—before they were able to definitively establish their finding (Science Communication and Outreach Unit 2019). According to researcher Liu Jifeng, the reviewers for the magazine *Nature* were very inquisitive when they submitted the article. The paper underwent extensive editing and proofreading before being released.

One of the major contributions made by Chinese astronomers to modern human astronomy is this difficult but substantial accomplishment. For this simple reason, any finding that threatens the accepted textbook theory is bound to be a huge boost for the advancement of human knowledge.

Astronomers are currently facing the LB-1 conundrum. It is still a mystery of the universe that has not been resolved to this day. This is an ongoing astronomical discovery, and we may wait with bated breath until the day when LB-1's mysterious origin is revealed.

In addition to its groundbreaking work in the study of black holes, the LAMOST Telescope has also contributed to the advancement of knowledge in the studies of galactic structure and evolution, stellar physics research, and the hunt for rare celestial bodies.

In April 2019, LAMOST identified a star that piqued the interest of astronomers due to its abundance of heavy elements. Together with their Japanese counterparts at the National Astronomical Observatory of Japan, Chinese scientists compared the star's chemical composition with those of other dwarf galaxy stars and the stars in the Milky Way's Galactic halo. The star's chemical composition came out to be similar to those of stars in dwarf galaxies, but distinct from those of stars in the Milky Way's halo (Xing et al. 2019). The fact that this star originated from a dwarf galaxy rather than the Milky Way is convincing evidence that the Milky Way and the dwarf galaxy had merged. Such stars with a high abundance of heavy elements are a great resource for learning more about the merger history of the Milky Way and how galaxies develop and evolve.

Using information from LAMOST and the Gaia Space Telescope, the National Astronomical Observatory of China identified 591 high-speed stars in December 2020 (Li et al. 2020). These stars were traveling at incredibly high speeds compared to other stars. Forty-three of them had the potential to free themselves from the Milky Way's gravitational pull and eventually leave the Galaxy altogether. This finding had more than doubled the number of high-speed stars discovered, and the total number directly exceeded 1,000. This had substantially expanded the sample of high-speed stars, providing more information for astronomers to study them, which is of tremendous significance in their research history. Previously, scientists only found around 550 high-speed stars through telescopes (Fig. 9.10).

In the study of galaxies and stars, LAMOST ranks among the most powerful and fruitful telescopes. These accomplishments, however, are simply the beginning of its tale (Deng et al. 2021). As part of the second phase of the Guoshoujing Telescope project, China will relocate the LAMOST Telescope to the more favorable geographic conditions of Lenghu Town in Qinghai Province (Lu 2021). This phase of the project calls for a number of upgrades, including a larger aperture for LAMOST (from 4.9 to 8.4 m), more optical fibers (from 4,000 to 12,000), a higher limiting magnitude (from 17.8 to 21), and a larger survey sample (from 10 million spectra to 100 million spectra). When the second phase of the project is finished, the Super Telescope LAMOST will not only be a "watcher" of the Milky Way but also a "photographer" of the entire universe. This will be possible due to an increase in its spectral number as well as its limiting magnitude. It should come as no surprise that data from LAMOST's second phase of observations will renew humanity's perspective on the cosmos (Fig. 9.11).

Fig. 9.10 High-speed stars "escaping" the Milky Way (*Credit* NASA, ESA, and G. Bacon (STScI))

Fig. 9.11 Lenghu Town in Qinghai Province, China (*Credit* Depositphotos/wirestock)

China's historical astronomical prowess was in large part due to the country once boasting the world's finest stargazing stations and the most diligent and exceptional astronomical recorders. After entering the modern era, China's astronomy, on the other hand, definitely fell behind the developed countries of the Western world. But in the last ten years or so, China's astronomy industry has made significant strides, and one after another celestial artifacts have sprouted up, LAMOST being only one of them. Major Chinese observatories will start to show up in the subsequent chapters. The completion of one astronomical artifact after another on Chinese soil is gradually shifting the center of the world's astronomical community from the West to the East. The new era of Chinese astronomy is approaching as the curtain of time's transition opens.

References

Ali Alpar, M.: Superfluid dynamics and energy dissipation in neutron stars. In: The Neutron Star—Black Hole Connection, pp. 57–70. Springer Netherlands (2001). https://doi.org/10.1007/978-94-010-0548-7_4

Carroll, B.W., Ostlie, D.A.: An Introduction to Modern Astrophysics. Addison-Wesley (2007)

CAS. (n.d.): Guoshoujing Telescope (LAMOST). Retrieved April 12, 2023, from Chinese Academy of Sciences: https://lssf.cas.cn/en/facilities-view.jsp?id=ff8080814ff56599014ff5a807d70057

Deng, L., Yang, F., Chen, X., He, F., Liu, Q., Zhang, B., Pan, J.: Lenghu on the Tibetan Plateau as an astronomical observing site. Nature **596**(7872), 353–356 (2021). https://doi.org/10.1038/s41586-021-03711-z

Foglizzo, T.: Explosion physics of core-collapse supernovae. In: Handbook of Supernovae, pp. 1053–1073. Springer International Publishing (2017). https://doi.org/10.1007/978-3-319-21846-5_52

Guo, Y., Wang, F., Bai, Z., Li, J.: Design of LAMOST data processing and production database. SPIE Proceedings **10**(1117/12), 925959 (2012)

Hu, H., Zhai, C., Li, W., Chen, H., Chu, J., Xing, X.: Implementary scheme of parallel controllable optical fiber positioning system for LAMOST. SPIE Proc. (2006). https://doi.org/10.1117/12.670786

IAU. (n.d.).: ShouGuan WANG. Retrieved April 12, 2023, from International Astronomical Union: https://www.iau.org/administration/membership/individual/2951/

Janesick, J.R.: Scientific Charge-Coupled Devices. SPIE Press (2001)

Li, Y.-B., Luo, A.-L., Lu, Y.-J., Zhang, X.-S., Li, J., Wang, R., Zhang, S.: 591 high-velocity stars in the galactic halo selected from LAMOST DR7 and Gaia DR2. Astrophys. J. Suppl. Ser. **252**(1), 3 (2020). https://doi.org/10.3847/1538-4365/abc16e

Liu, J., Zhang, H., Howard, A.W., Bai, Z., Lu, Y., Soria, R., Isaacson, H.: A wide star–black-hole binary system from radial-velocity measurements. Nature **575**, 618–621 (2019). https://doi.org/10.1038/s41586-019-1766-2

Lu, H.: China Focus: Chinese Astronomers Discover Unexpected Huge Stellar Black Hole (2019). Retrieved from Xinhuanet: http://www.xinhuanet.com/english/2019-11/28/c_138589985.htm

Lu, C.: First World-Class Astronomical Observatory Site in the Eastern Hemisphere Found in Lenghu, Qinghai Province—Guo Shou Jing Telescope to Relocate There (2021). Retrieved from Shanghai Observer—Science and Technology Daily: https://export.shobserver.com/toutiao/html/435116.html

NAOC. (n.d.): Large Sky Area Multi-Object Fiber Spectroscopic Telescope (LAMOST). Retrieved April 12, 2023, from National Astronomical Observatories, Chinese Academy of Sciences: http://english.nao.cas.cn/Research2015/Facilities2015/Telescopes2015/201701/t20170120_173586.html

NIAOT. (n.d.): CAS and CAE Memebers—Su Dingqiang. Retrieved April 12, 2023, from Nanjing Institute of Astronomical Optics & Technology, National Astronomical Observatories, CAS: http://sourcedb.niaot.cas.cn/yw/pe/casmem/201806/t20180608_5024004.html

Özel, F., Psaltis, D., Narayan, R., McClintock, J.E.: The black hole mass distribution in the galaxy. Astrophys. J. **725**(2), 1918–1927 (2010). https://doi.org/10.1088/0004-637x/725/2/1918

Science Communication and Outreach Unit: Astronomers Discover a Giant Black Hole that Challenges Current Models of Stellar Evolution (2019). Retrieved from Instituto de Astrofísica de Canarias: https://www.iac.es/en/outreach/news/astronomers-discover-giant-black-hole-challenges-current-models-stellar-evolution

Tatum, E.T.: Introductory Chapter: Black Holes, The Singularity Problem, and The Universe. In New Ideas Concerning Black Holes and the Universe, IntechOpen (2020). https://doi.org/10.5772/intechopen.90071

Tauris, T.M., van den Heuvel, E.P.: Formation and evolution of compact stellar X-ray sources. In: Compact Stellar X-ray Sources, pp. 623–666. Cambridge University Press (2006). https://doi.org/10.1017/cbo9780511536281.017

Xing, Q.F., Zhao, G., Aoki, W., Honda, S., Li, H.N., Ishigaki, M.N., Matsuno, T.: Evidence for the accretion origin of halo stars with an extreme r-process enhancement. Nat. Astron. **3**(7), 631–635 (2019). https://doi.org/10.1038/s41550-019-0764-5

Xinhua: Chinese Telescope Collects Largest Data Set on Stars (2019). Retrieved from China Daily: https://global.chinadaily.com.cn/a/201903/29/WS5c9dd02ba3104842260b3586.html

Zhao, G.: LAMOST spectral survey—an overview. Res. Astron. Astrophys. **12**, 723 (2012). https://doi.org/10.1088/1674-4527/12/7/002

10

The ALMA Telescope in Chile: Witness to the Birth and Death in the Universe

The ALMA Telescope in Chile

A major news conference took place at 9 a.m. EDT on April 10, 2019, in seven different cities around the world. In the main venue in Washington, USA, astrophysicist Sheperd Doeleman, director of the Event Horizon Telescope at the Center for Astrophysics (CfA), Harvard & Smithsonian, could not contain his excitement. "We have achieved something presumed to be impossible just a generation ago," he exclaimed (Dunham 2019). After pausing, he spoke louder: "We have seen what we thought was unseeable. We have seen and taken a picture of a black hole. Here it is" (Borenstein 2019). Suddenly, the crowd erupted in loud yells and applause (Fig. 10.1).

After that, an image was shown to the entire planet. A doughnut-shaped celestial body could be seen against the pitch-black background. The picture might not be fantastic to look at because it was blurry, out of focus, and somewhat empty, but it is a priceless record of human history. This is the first time that a photograph of a black hole has been obtained by humans. Its emergence marks a new era in humanity's quest to learn more about the universe. The protagonist of our story in this chapter, the Atacama Large Millimeter/submillimeter Array (ALMA) (ESO, What is the Atacama Large Millimeter/submillimeter Array (ALMA)? n.d.), located in the Atacama Desert in northern Chile, has made a significant contribution to this picture. ALMA is labeled as "the witness to the life and death in the universe."

If you've read this far, then you probably know a lot about telescopes. Astronomical telescopes are like extra eyes that allow us to look further into space. Millimeter-band radio telescopes have long been of great interest to

J. Wang, *Eye Beyond the Sky*, Astronomers' Universe,
https://doi.org/10.1007/978-981-99-9818-0_10

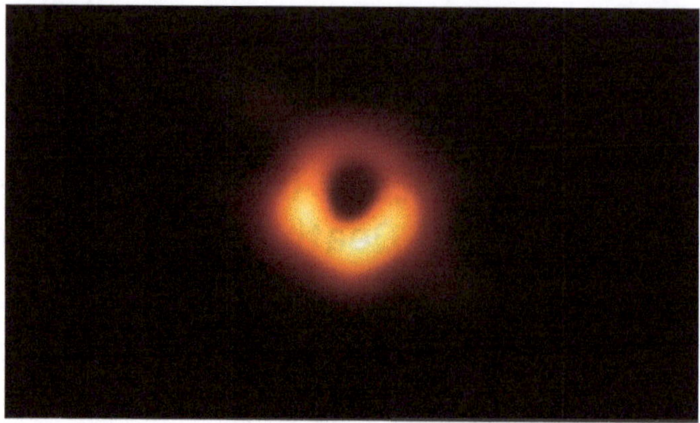

Fig. 10.1 The first picture of a black hole in human history (*Credit* EHT collaboration)

astronomers from all around the world. Molecular matter is what radio telescopes detect best in this frequency range. Those powerful radio telescopes can detect the presence of water and organic materials in addition to tracking the orbits of the stars. Construction of a millimeter-wave radio telescope, while possible, is challenging (Wall et al. 2012).

Millimeter-wave radio telescopes necessitate more antenna and emitting surface accuracy than their centimeter- and meter-wave counterparts. It is common to compare the receiving antenna of a telescope to the "cauldron." If the antenna's receiving surface is not perfectly smooth or there is even a tiny variation in the design angle, the telescope will be out of focus and poor observation results will ensue. The mirrors of meter-class radio telescopes can be made from metal mesh, while millimeter-wave radio telescopes require extremely smooth and precise metal plates or coatings. In addition, the size of the antenna has an effect on the amount of information that can be observed with the radio telescope. In theory, observational precision improves in proportion to telescope aperture size (Marr et al. 2015). This means that researchers and engineers will have to strike a compromise between the precision of observations and the complexity of antenna production (Fig. 10.2).

It can be observed with greater precision by increasing the aperture, but it can also be achieved by arranging a large number of small antennas in an array according to a specified configuration rule (Thompson et al. 2001). A floor mirror, like a mirror in a fitting room, can provide a precise reflection of the model's body, and a mirror constructed from multiple smaller mirrors can do the same. As one might expect, building a small antenna is much simpler than creating a massive one (Fig. 10.3).

Fig. 10.2 64-m Parkes telescope (*Source* CSIRO ATNF) © Copyright CSIRO Australia, (September 1, 2016))

Fig. 10.3 The Submillimeter Array (SMA) (*Credit* Depositphotos/fthuerig)

The number of small antennas is inherently required for achieving the capability of a large-aperture telescope, which subtly increases the cost of construction despite the fact that the difficulty of building has been lessened. Spending billions of dollars at the time, the United States constructed the Submillimeter Array (SMA) consisting of eight 6-m diameter radio telescopes in Hawaii (Blundell 2007) at the turn of the century. Even the following annual maintenance costs were as high as $7 million. This is a substantial amount of money, even for a superpower like the United States. America

Fig. 10.4 The Submillimeter Array (SMA) on Mount Mauna Kea (*Credit* Afshin Darian)

had planned to construct a larger millimeter-wave radio telescope once the SMA project was completed, but the new project was halted due to the huge expense involved (Fig. 10.4).

Similarly, the European Southern Observatory had been unable to move on with its own plans for a millimeter-wave radio telescope due to a lack of finance. Given the challenges of independent engineering, the two observatories (NRAO and ESO) started talking about how they might collaborate to construct a unique array of millimeter-wave radio telescopes. There was a great deal of mutual eagerness to work together. Still, it wasn't until the turn of the millennium that the planning of a massive radio telescope array was settled, and the National Astronomical Observatory of Japan (NAOJ) was brought into the fold (ALMA 2021).

The ALMA project began construction in 2003 following years of planning. The project was estimated to cost a total of $1.4 billion (Courtland 2013), and it would involve the construction of 66 small antennas, 25 of which would be assembled by the American part, 25 deployed by the ESO, and the remaining 16 manufactured by the NAOJ. Yet, the location of the most advanced radio telescope in the world is neither in the United States nor in Europe; rather, it is in the Atacama Desert in Chile, South America, also the location where the array is named after.

The Atacama Desert is extremely arid (Peppas 2012) because of its high elevation (over 4,000 m on average (Cunningham 2021)) and its lack of exposure to the monsoon's moistening clouds. The extreme conditions mean that there is no human settlement for thousands of miles around; therefore, there will be no interference from man-made electromagnetic radiation in any observations. It's the most Martian environment I've ever encountered on the Earth if I were to sum it up in a single sentence. When ALMA's construction began, it was more akin to establishing a human settlement on Mars than erecting a telescope (Fig. 10.5).

After more than 5 years of infrastructure development, the first antenna was installed on this red soil in 2008. The installation of all 66 of the small antennas was expected to take another 5 years. According to astronomers, that's a really significant amount of time. While just 12 ALMA antennas were erected in 2011, scientists nonetheless gave it a significant mission in the hopes that Atacama would yield unexpected results.

The first challenge for ALMA was to work in tandem with the "Eye in the Sky," the Hubble Space Telescope, which was no easy feat considering that only a fifth of the construction had been finished. An exciting event was taking place 70 million light-years away in the constellation of Corvus, where two galaxies were passing one another and an unprecedented number of stars were being generated. Galaxies like these are known as starburst galaxies. Gravity was ripping apart and destroying countless old stars, and at the same time, numerous new stars were being born in the collision.

Hubble soon revealed a stunning image: two galaxies were seen drawing closer together yet maintaining some distance from each other in the area of visible light. It was beautiful, but it didn't surprise scientists, who were keener on seeing what the Atacama could see in the millimeter range.

Fig. 10.5 Scientists ultimately chose to build ALMA on the Chilean Chajnantor plateau. (*Credit* ESO/NAOJ/NRAO)

Not only had ALMA not let down astronomers but also had revealed an even more shocking truth: although the two galaxies did not appear to have collided, their dense clouds of cold gas had already mixed together. Countless stars were poised to be born anew in this turmoil, whose combined mass was billions of times that of the Sun (ESO, ALMA Opens Its Eye 2011) (Figs. 10.6 and 10.7).

The early strong performance of the ALMA had provided scientists reasons to have high expectations for its continued success. "*ALMA Opens Its Eyes*" as the ESO official report titled (ESO, ALMA Opens Its Eyes 2011). The year 2012 marked the beginning of a more ambitious agenda for scientists all across the world. The "Event Horizon Telescope" Project had the goal of capturing images of black holes using a virtual Earth-sized telescope (EHT n.d.). For the time being, however, scientists would have to wait for ALMA to finish construction and completely open its "eyes" before they can implement this ambitious plan.

After years of expectation, on March 14, 2013, ALMA, one of the most sensitive radio telescopes in human history, finally began its mission of discovery with the installation of its 66th small antenna. The ideal approach to mark the completion of this magnificent endeavor was to release a publication using ALMA observation data simultaneously in *Nature* (Vieira et al. 2013) and *Astrophysical Journal* (ESO, ALMA Rewrites History of Universe's

Fig. 10.6 A starburst galaxy in the constellation Corvus as imaged by Hubble (*Credit* ESO)

Fig. 10.7 A starburst galaxy in the constellation Corvus as imaged by ALMA (*Credit* ALMA (ESO/NAOJ/NRAO). Visible light image: the NASA/ESA Hubble Space Telescope)

Stellar Baby Boom 2013). Through the research presented in these publications, humans have a deeper comprehension of the cosmos again (Vieira et al. 2013) (Fig. 10.8).

ALMA's observations of 26 cosmological areas have led to a startling discovery: Star creation occurred considerably earlier in the universe than scientists had previously believed. Assuming the previous theory is correct, stars first appeared in vast quantities around 3 billion years after the Big Bang. On the other hand, Atacama showed us through observational evidence that stars began to develop in great quantities only 2 billion years after the Big Bang. Even by astronomical standards, a separation of a billion years is an

Fig. 10.8 ALMA went into full operation. (*Credit* ESO/B. Tafreshi (twanight.org))

Fig. 10.9 ALMA under the Milky Way (*Credit* ESO/B. Tafreshi (twanight.org))

extremely lengthy duration. It stands to reason that given the early emergence of stars, the appearance of planets would also be much ahead of schedule.

It was in a galaxy around 13 billion light-years from the Earth that ALMA detected the most distant water yet identified by humans, along with evidence of the existence of carbon monoxide (Tobin et al. 2023). As the fundamental ingredients for the formation of life have been found to be present earlier and more frequently than was predicted in the past, and since the seemingly ordinary Earth can create intelligent life, the question arises again as to whether other planets may also yield living beings. Thanks to its own discoveries, ALMA has at least made this fantasy a little more realistic (Fig. 10.9).

The time has come to actually photograph black holes, thanks to the sensitivity of ALMA, which has already allowed us to see stars being born. Although Einstein's general theory of relativity theoretically revealed the mystical existence of black holes (Einstein 2005), the debate regarding their existence has persisted ever since. Is it possible for a dying star to transform into something so bizarre that not even light can leave? Black holes have recently become not only the hottest subject for research in astrophysics but also a popular topic of conversation on the streets and alleys.

Having humans observe a black hole firsthand would be the strongest evidence of its existence. Nevertheless, this situation creates a paradox: if black holes are so massive that not even light can escape, how can their existence be observed by humans?

Nonetheless, scientists had not been confounded by this paradox. The presence of black holes is not immediately observable, but it can be inferred from other observations. Though we cannot see or feel the wind, we may confirm its existence through the movement of the flag. There's always a black hole in the core of a supermassive galaxy. While black holes may swallow everything, the process of devouring creates luminous accretion disks around the edges. In this way, the black hole can be viewed indirectly by checking if there is an accretion disk, provided the radio telescope is pointed in the direction of the galactic center (Chakrabarti 2017).

The plan to observe black holes has finally begun to be implemented with the help of ALMA. ALMA incorporates the efforts of seven additional radio telescopes located in a variety of areas across the planet. Similar to how ALMA is made up of 66 individual antennas, a global network of smaller radio telescopes can be used to create a single instrument with a much bigger aperture. Scientists have successfully converted the entire planet into one massive telescope this time. The observation period saw eight telescopes pointed in the same direction, marking the first time that humans had met the distance and precision requirements for observing black holes. A supergiant elliptical galaxy 55 million light-years away, designated M87, soon became the focus of all telescopes (Fletcher 2018).

On April 5–14, 2017, eight of the best radio telescopes on the Earth, from Greenland in the Arctic to the highest point of the Antarctic ice sheet, were pointed at M87, 55 million light-years away. After ten days of observations, scientists had collected a massive amount of information—so much so that the data produced by a single telescope was as high as 7 PB (Mearian 2015), or enough to store the information contained in 100,000 Blu-ray copies of Interstellar. Data was copied to hard drives and continuously transported to two data processing centers in the United States and Germany because the volume of data was so big that it exceeded the limit of international broadband transmission (Fig. 10.10).

Following more than two years of data analysis, scientists were able to produce the first-ever photograph of a black hole. The entire panorama spans a distance of 100 billion kilometers. Since it is located 55 million light-years away, the black hole at the image's center seems quite small in the sky, taking up only 42 micro-arcseconds (Temming 2019). The accuracy of this observation is comparable to being able to read the information on a credit card located on the surface of the Moon from the Earth. ALMA's participation allows us to accomplish this almost unachievable goal (Fig. 10.11).

Even though ALMA had only been around for 5 years by the time May of 2018, the 1,000th peer-reviewed paper based on its observation data was officially released (ESO, ALMA reaches 1000 published papers 2018). Humans benefit from its perceptive "vision" since it allows us to see further into the cosmos. ALMA, the deepest "eye" of humanity, will keep watching the birth and death in the universe with unruffled vision for a long time to come.

When viewed against the vastness of space, humanity pales in comparison. Perhaps human civilization will vanish from the face of the universe one moment. Our existence is but a blink in the grand scheme of things. Despite this, I am certain that there will be evidence of the existence of human civilization somewhere in our galaxy or somewhere in the universe. This

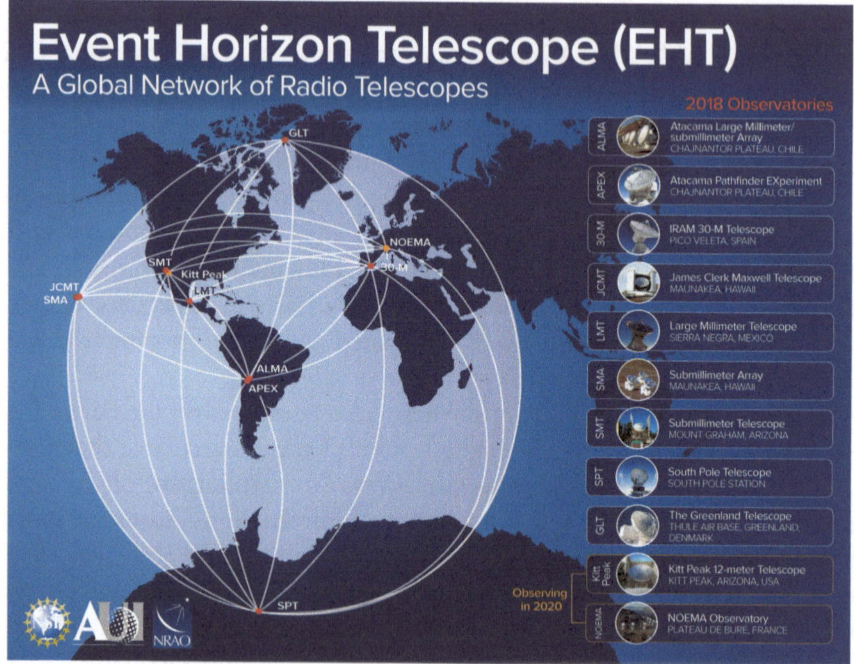

Fig. 10.10 Distribution map of the Event Horizon Telescope as of 2022, where 30-M, Kitt Peak, and NOEMA did not participate in the 2017 black hole imaging. (*Credit* NRAO)

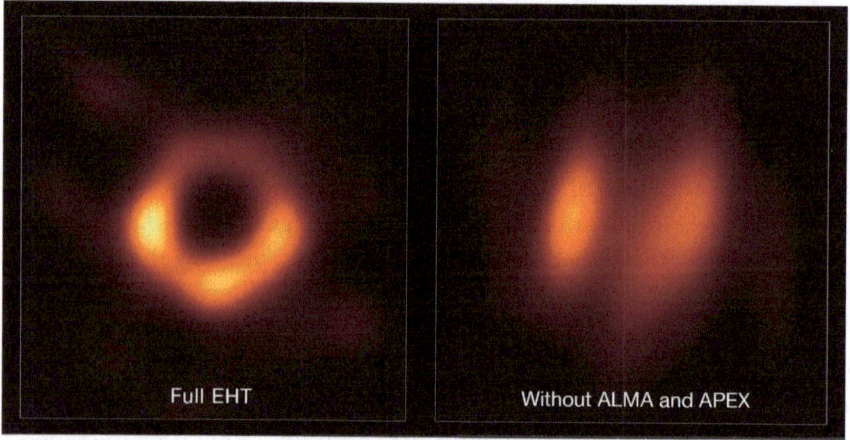

Fig. 10.11 Comparison of the full EHT black hole image (Left) with the one that would have been obtained without ALMA and APEX (*Credit* EHT collaboration)

Fig. 10.12 ALMA under night sky (*Credit* ESO/B. Tafreshi (twanight.org))

conviction fills me with undiluted pride for being a member of the comparatively puny human race. "I could be bounded in a nutshell and count myself a king of infinite space," Shakespeare once famously stated (Fig. 10.12).

References

ALMA: Origins (2021). Retrieved April 13, 2023, from ALMA: https://almaobservatory.org/en/about-alma/origins/

Blundell, R.: The submillimeter array. In: 2007 IEEE/MTT-S International Microwave Symposium (2007). https://doi.org/10.1109/mwsym.2007.380132

Borenstein, S.: Seeing the "unseeable": Scientists Reveal First-Ever Image of a Black Hole (2019). Retrieved from The Times of Israel: https://www.timesofisrael.com/seeing-the-unseeable-scientists-reveal-first-ever-image-of-a-black-hole/

Chakrabarti, S.: Aspects of Gravitational Collapse and the Formation of Spacetime Singularities (2017). Retrieved from arxiv.org: https://arxiv.org/pdf/1709.01512.pdf

Courtland, R.: Powerful ALMA Telescope Makes High-Profile Debut (2013). Retrieved from IEEE Spectrum: https://spectrum.ieee.org/alma-makes-a-highprofile-debut

Cunningham, S.: 5 of Earth's Highest Deserts (2021). Retrieved from HowStuffWorks: https://science.howstuffworks.com/environmental/earth/geology/high-deserts.htm

Dunham, W.: "Seeing the unseeable": Scientists Reveal First Photo of Black Hole (2019). Retrieved from Reuters: https://www.reuters.com/article/us-space-exploration-blackhole-idUSKCN1RM1OP

EHT. (n.d.): Event horizon telescope. Retrieved April 13, 2023, from Event Horizon Telescope Collaboration: https://eventhorizontelescope.org/

Einstein, A.: Die Feldgleichungen der Gravitation. Albert Einstein: Akademie-Vorträge 88–92 (2005). https://doi.org/10.1002/3527608958.ch5

ESO: ALMA Opens Its Eye (2011). Retrieved from ESO: https://www.eso.org/public/news/eso1137/

ESO: ALMA Rewrites History of Universe's Stellar Baby Boom (2013). Retrieved from ESO: https://www.eso.org/public/news/eso1313/

ESO: ALMA Reaches 1000 Published Papers (2018). Retrieved from ESO: https://www.eso.org/public/announcements/ann18029/

ESO. (n.d.): What is the Atacama Large Millimeter/submillimeter Array (ALMA)? Retrieved April 13, 2023, from ESO: https://www.eso.org/public/teles-instr/alma/

Fletcher, S.: Einstein's Shadow: The Inside Story of Astronomers' Decades-Long Quest to Take the First Picture of a Black Hole. HarperCollins (2018)

Marr, J.M., Snell, R.L., Kurtz, S.E.: Fundamentals of Radio Astronomy: Observational Methods. CRC Press (2015)

Mearian, L.: Massive Telescope Array Aims for Black Hole, Gets Gusher of Data (2015). Retrieved from Computerworld: https://www.computerworld.com/article/2972251/massive-telescope-array-aims-for-black-hole-gets-gusher-of-data.html

Peppas, L.: The Atacama Desert (Deserts Around the World). Crabtree Classics (2012)

Temming, M.: How Scientists Took the First Picture of a Black Hole (2019). Retrieved from Science News Magazine: https://www.sciencenews.org/article/event-horizon-telescope-black-hole-picture

Thompson, A.R., Moran, J.M., Swenson Jr, G.W.: Radio interference. In: Interferometry and Synthesis in Radio Astronomy, Second Edition, pp. 613–626. WILEY-VCH Verlag GmbH & Co. KgaA (2001). https://doi.org/10.1002/9783527617845.ch15

Tobin, J.J., Hoff, M.L., Leemker, M., Dishoeck, E.F., Paneque-Carreño, T., Furuya, K., Cieza, L.: Deuterium-enriched water ties planet-forming disks to comets and protostars. Nature **615**(7951), 227–230 (2023). https://doi.org/10.1038/s41586-022-05676-z

Vieira, J.D., Marrone, D.P., Chapman, S.C., Breuck, C.D., Hezaveh, Y.D., Weiβ, A., Brodwin, M.: Dusty starburst galaxies in the early Universe as revealed by gravitational lensing. Nature **495**, 344–347 (2013). https://doi.org/10.1038/nature12001

Wall, W. F., Carramiñana, A., Carrasco, L., Goldsmith, P.F.: Millimeter-Wave Astronomy: Molecular Chemistry & Physics in Space: Proceedings of the 1996 INAOE Summer School of Millimeter-Wave Astronomy held at INAOE, Tonantzintla, Puebla, Mexico, 15–31 July 1996. Springer Science & Business Media (2012)

11

China's "Eye of Heaven" (FAST): The New King Taking Its Place Atop Radio Telescope

China's "Eye of Heaven" (FAST)

A feeling of uneasiness filled an office at Beijing's National Astronomical Observatory that afternoon in 2002, and a group of eminent professors huddled over a conference table to discuss an issue. Pictures of the Arecibo Telescope, the world's largest radio telescope at the time, littered the table.

In the far corner of the meeting table, Professor Ren Gexue (Ren n.d.) studied a magnified picture of the feed cabin for several minutes. In a radio telescope, the feed cabin is the main piece of hardware, and it is often installed directly above the telescope to receive cosmic signals picked up by the telescope's reflection panel (Condon and Ransom 2016) (Fig. 11.1).

Ren handed the picture to Professor Nan Rendong (1945–2017) (NAOC, Deep Grief for the Death of Prof. Nan Rendong, the FAST Chief Scientist and Chief Engineer 2017), then chief engineer of the FAST Project and said: "Arecibo's reflection panel is a totally stationary sphere. It exploits the principle of spherical and paraboloid similarity to roughly observe distant objects and gather signals using the Gregory three-mirror-reflection feed cabin. Such a complicated feed cabin, coupled with a stainless steel casing, weighs up to 900 tons" (Fig. 11.2).

Ren paused for a moment, seemingly overcome by emotion, and then continued, "If FAST adopts the same design, the feed cabin will be close to 10,000 tons, and according to the conventional design, this weight may not be put directly on a beam with a span of more than 500 meters."

Professor Nan reflected for a moment and then stated, "We may have to break with this history and create FAST with a new telescope design concept

J. Wang, *Eye Beyond the Sky*, Astronomers' Universe,
https://doi.org/10.1007/978-981-99-9818-0_11

Fig. 11.1 The feed cabin of Arecibo telescope (*Credit* Depositphotos/alexusha2008)

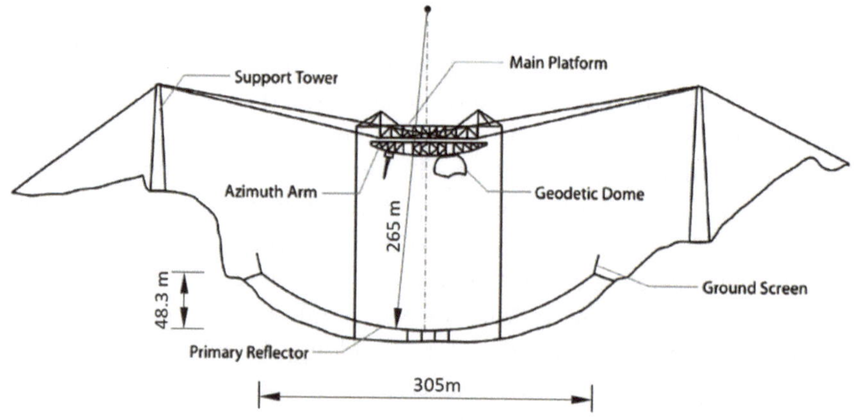

Fig. 11.2 Design of the Arecibo telescope (*Credit* Andrzej Kulowski)

that has never been adopted before." He then proceeded to point to the picture in his hand and said, "Arecibo's feed cabin serves like a fixed TV antenna that can only be adjusted. This design is too traditional to be suited for large-aperture telescopes like FAST (Nan et al. 2003)."

A researcher named Zhu Wenbai (Zhu n.d.), who was standing nearby, opened his mouth wide and remarked, "Gramps, it is already exceedingly tough to make a cauldron with a circumference of 500 meters, and now we have to turn the feed cabin into eyeballs. In order to concentrate the signals on the feed cabin, is it even necessary to maintain the cauldron in its distorted state? This conception appears to be straight out of science fiction! "

Professor Nan took a sip, inhaled deeply, and stated, "We need not only deform FAST but also control the precision to the millimeter level; that is, the deformation error cannot exceed the thickness of a nail cover."

Everyone at the time felt the design was extraordinary but unachievable, and the likelihood of successfully building it in accordance with such design specifications was exceedingly low. However, it is precisely because of this wonderful design that China's Five-hundred-meter Aperture Spherical radio Telescope (FAST), now the largest of its kind in the world, has earned the title "Eye of Heaven." So, how did the team of Chinese astronomical scientists break the engineering limit of traditional telescopes' 100-meter aperture and complete the construction of this astronomical telescope? It's necessary to rewind the narrative to the beginning of the process in order to grasp it.

At the International Union of Radio Science General Assembly held in Kyoto in 1993 (Kellermann 1993), radio astronomers from ten different countries, including China, jointly launched an initiative to construct a new generation of large radio telescopes. The common objective had been to finish the related construction around the world before the Earth's electromagnetic wave environment was destroyed. During the conference, attendees came up with two opposing points of view: one advocated for the construction of a huge telescope with a single, large aperture as proposed by China, while the other supported the construction of a large array of telescopes each with a smaller aperture (Ekers 2012). Except for China, other nations declined to implement the former strategy because they considered a huge radio telescope with a single aperture to be fraught with danger and the locations for construction to be out of sight.

But astronomers like Nan Rendong disagreed. Across China's huge expanse and varied topography—from plateau to basin, desert to mountain—plenty of locations fit the criteria for installing these massive radio telescopes. It is entirely feasible that we construct a number of large-aperture telescopes in the future and then combine them into more potent telescope arrays.

However, this could only be a brilliant concept at the time. Back in 1993, the radio telescopes located in Shanghai Sheshan Mountain (SAO 2014) and Urumqi Nanshan (XAO 2013) were the largest in China, both measuring 25 meters in diameter, which is less than one-tenth the size of the Arecibo.

Unexpectedly, more than 20 domestic universities and research organizations joined the National Astronomical Observatory 2 years later to form China's Propulsion Committee and began developing a pilot program for China's telescope arrays. An ambitious plan needs a manageable "small target" to get started on the road to success. This led the committee to propose erecting a 500-meter-diameter telescope as the first step.

However, this initial step's goal had far surpassed the then-world's leading Arecibo Telescope. The "small target" went far beyond common expectations, even the phrase "boldly daring" might underestimate its ambition (Fig. 11.3).

The project's first challenge was finding an appropriate location for construction. Large astronomical telescopes cannot be built in an unplanned location. The optimum construction topography should be a geomorphic area that satisfies the following conditions. First, it has to be remote and sparsely populated so that it can avoid the interference of human radio. Second, it can be modified according to the ground, preferably a depression raised on four sides, so as to minimize the difficulties of project construction and effectively screen out external electromagnetic interference. Third, it should not be prone to stagnant water (Schock 2004).

Since its initial assessment in 1994 and subsequent site selection in 2002, Da Wodang in Pingtang County, Guizhou Province, has been designated as FAST's permanent home. Concurrent with the process of finding a suitable location, the project team ran into a new obstacle: how to lighten the load of the feed cabin by shifting the reflector. This design, after all, is unprecedented in the history of astronomy (Li et al. 2012) (Fig. 11.4).

Fig. 11.3 The 25-m Radio Telescope of Urumqi Astronomical Station (*Credit* the Xinjiang Astronomical Observatory (XAO))

Fig. 11.4 Pingtang County of Guizhou Province, China (*Credit* LkJidm)

Large telescopes like Arecibo have permanent, non-adjustable reflector cauldrons. However, the night sky is perpetually changing. Arecibo makes use of a spherical cauldron, which precisely takes advantage of the feature that all angles on a sphere have the same optical reflection. This allows Arecibo to examine various targets. The combination of the similar spherical and paraboloidal shapes always allows it to track the movement of celestial bodies as they move across the sky (Phillips and Wolszczan 2005) (Fig. 11.5).

A closer look reveals flaws in this otherwise flawless plan. In contrast to the paraboloid, the spherical microwave convergence is not a point but a line, making collecting signals more challenging, while the paraboloid can converge the microwave to a focus, making it more convenient for the receivers to function. By employing Gregory light path reflection in the feed cabin with secondary and third paraboloid mirrors, Arecibo is able to focus and converge the microwave signals so that it can obtain all signals at once (Murdin 2001).

Because such a complex design would definitely result in heavy cabins, a technique must be developed to not only make excellent use of the advantages of the sphere but also change the sphere into a paraboloid when necessary, thus tackling this challenge in the most effective way possible. Getting FAST on the move is a feasible solution for observation.

The reflector updated its conventional design to 4,450 aluminum reflector panels with side lengths from 10.4 m to 12.4 m to facilitate its smooth motion. These devices are joined to 6,670 cables. An actuator located below the cable shapes it into a paraboloid. Therefore, the feed cabin can be brought down to a manageable 30 tons in weight. Wherever the feed cabin moves, the reflector unit below it changes shape to a paraboloid with a diameter of 300 meters, while other areas simultaneously revert to a spherical shape (Gu et al. 2014).

To further demonstrate the scheme's viability, astronomers have conducted an extensive preliminary study and verified experimental prototype telescopes of several diameters, including 20 and 50 meters.

Fig. 11.5 The feed cabin of Arecibo telescope (*Credit* Courtesy of the NAIC—Arecibo Observatory, a facility of the NSF)

The National Development and Reform Commission ultimately gave the go-ahead for the FAST initiative in July 2007, after years of effort, with the intention of digging out a sizable "eye socket," accompanying civil project at the site commenced in March 2011.

The project seemed to be moving along smoothly in the direction that all people involved had envisioned. A renewed sense of enthusiasm pervaded.

At this moment, however, an incident occurred that nearly derailed the entire project. The research team led by Nan Rendong ran into extraordinary difficulties in the design of the active reflector.

The active reflector's cable network has been the source of the issue. Depending on the different positions of the feed cabin, the most difficult section of the cable needs to have a stretch of 47 centimeters. In order to match the specifications of FAST, team member and researcher Jiang Peng tested more than a dozen different high-strength steel cables from home and abroad. However, all these cables snapped after only a few uses, and none of them measured up to requirement.

The project crew was taken aback by the unexpected twist of events. This issue was completely disregarded because the paraboloid diameter of the prototype telescope is not up to 300 meters, the tension distance of the cable is not that long, and the tension force is not that significant. It was at this point that excavation work had already begun on the site. If a solution hadn't been found, the entire project would have been scrapped.

At this point, it was of the utmost importance to determine the level of cable tension and the fatigue performance that was required by FAST (Fig. 11.6).

Jiang first simulated and assessed FAST's trajectories for the next 30–50 years in light of these concerns. The trajectories of one month were essentially represented by a line, but 1 year's trajectories turned into a black mass. With these trajectory data, it was possible to carry out a large-scale mechanical simulation, and an exact analysis of the range and number of tensions that each cable would experience over the course of the next 30 years could be performed.

The analysis's conclusion, however, only increased Jiang's apprehension. About 30 percent of the cables would be pulled by the tension of more than 300 MPa, which is similar to having the weight of two automobiles rest on your fingernail. In extreme cases, steel cables might have to endure 445 MPa, which had been far beyond the capacity of standard steel cables.

These cables function similarly to FAST's eye "muscles." To ensure that these "muscles" continue to function effectively for 30 or even 50 years, the team established an extremely stringent safety criterion. That is, each cable could endure 500 MPa of tension and 2 million stretches, which had

Fig. 11.6 Scale model of this "Cauldron" (*Credit* N509FZ)

surpassed twice the fatigue strength of conventional cables and for which there was no suitable prefabricated option.

The circumstance pushed the National Astronomical Observatory to collaborate with a wide range of businesses and academic institutions to conduct research and development, ultimately leading to the largest and most systematic cable fatigue test ever conducted, including everything from anchoring damage to monofilament wear. All experiments that could cause damage have been conducted.

In addition, millimeter forming accuracy was necessary for these cables because they nearly never had identical specifications in terms of length or thickness. There would be a difference of more than 60 mm to the edge of the cable, even if its machining accuracy deviated less than 1 mm. Therefore, if there existed a deviation in the machining precision of any cable, the entire project would run the danger of failing.

The project team even constructed a constant temperature room for these cables to minimize the effect of environmental temperature changes during processing, hence controlling the accuracy error during production. As the most rigorous quality assurance process ever implemented, it required real-time filming of every cable to help pinpoint and fix future issues.

After more than two years and an innumerable number of setbacks, the first steel cable that was eventually acceptable for use in the construction of the "Eye of Heaven" was manufactured (Kong et al. 2013).

With the unrelenting effort of the engineers, this unique steel cable's material was created, which served as the bedrock upon which FAST had been built. In addition to its use in the FAST project, this type of high-fatigue-resistance steel cable continued to contribute significantly to the development of the motherland through its application in subsequent mega-projects, such as the Hong Kong-Zhuhai-Macao Bridge, Beijing-Shanghai high-speed railway, South-to-North Water Diversion Project, and so on.

By August 2, 2015, the first "retinal" reflection panel had been assembled, all thanks to this breakthrough in the cable network that finally allowed FAST's "eye socket" to support the load-bearing skeleton in 2014 (Lu and Ren 2007) (Fig. 11.7).

FAST was opened to the public for the first time on September 25, 2016, after more than 5 years of construction. People clustered on the viewing platform and stared at the feed cabin in this cauldron's exact center. At this moment, a ray of sunshine that had been resting on the side of the cauldron began to gently slide down until it reached the bottom. Simultaneously, the feed cabin began to slowly rise, and the cauldron gave off the appearance of

Fig. 11.7 FAST is based at Ping Tang County, Guizhou. (*Credit* Depositphotos/bill8713)

being infused with life before beginning to dance in a manner that resembled a large eye peering into the depths of the universe. Thunderous applause erupted accompanying tears of joy rolling down.

Now that this "cauldron" is finally finished, "Eye of Heaven" can officially begin its mission and set out on its trip.

Since August 2017, FAST has officially launched the Commensal Radio Astronomy FAST Survey (CRAFTS). When examining the data obtained on August 22, 2017, researchers discovered that a 52.4-second drift scan contained a highly rated candidate pulsar PSR J1859-01 with a pulse period of 1.832 seconds, 16,000 light-years from the Earth. The research team requested a second confirmation observation from a third-party astronomical telescope immediately.

The first pulsar detected by FAST was confirmed on September 10, 2017, when the Australian Parkes Telescope observed the same region of the sky and recorded microwave signals with the same period and dispersion (Sun 2021). The semaphores gathered by the two telescopes suggest that Parkes would need to "gaze" for 2,100 seconds to observe the same degree of detail as is captured by FAST (Fig. 11.8).

Once again in 2019, FAST and Parkes collaborated to observe PSR J1926-0652 (Zhang et al. 2019). This pulsar's uneven pulse signal has aroused the curiosity of scientists since it may challenge the standard merry-go-round model.

These findings, which have been highly recognized by international scholars, represent only a small portion of FAST's accomplishments from its official operation in 2018 through 2021. During this period, the number of pulsars discovered by FAST surpassed 500, which was more than four times the total number of other telescopes' discoveries in the same period, and even

Fig. 11.8 Completion of FAST's construction (*Credit* Depositphotos/Tuchong-Microstock9)

exceeded the observational results of America's Arecibo Observatory in a time span of 15 years (Wu 2022).

Peking University's Professors Li Kejia and Dong Subo, together with co-authors Dr. Xu Heng and Dr. Chen Ping, published an article in the world-famous academic magazine *Nature* on September 21, 2022, discussing the changes that occur in the core magnetic field during the burst of the Fast Radio Burst (FRB) (Xu et al. 2022). As a sporadic astronomical occurrence, FRB owns the potential to quickly unleash tens of billions of times more energy than the yearly electrical generation of the entire Earth. However, its genesis and mechanism continue to be a source of consternation for astronomers (Fig. 11.9).

In the spring of 2021, the team led by Professor Li Kejia utilized FAST's large-aperture advantage to examine the source of FRB 20201124A and discovered that the magnetic field intensity and density surrounding it would continue to vary dramatically. On the basis of this, they hypothesized that this FRB originated in a binary system no further than one astronomical unit (the distance between the Earth and the Sun) away. The binary system comprises of a magnetar (a neutron star with an exceptionally powerful magnetic field) and a Be star encircled by a magnetic field. When the magnetar is in motion, its surrounding Be star will experience dramatic changes to its magnetic field. Magnetic polarization will be created by radio radiation when a magnetar lies directly between a Be star and an observer.

Fig. 11.9 The "Heavenly Eye" in the Mountains (*Credit* Pierre Markuse from Hamm, Germany)

Since the first FRB was discovered in 2007, this has been the first explanation for their origin. At the very least, some of them can be traced back to the similar binary systems with which we are familiar. This significant discovery has garnered much attention. Two days after the related paper (Wang et al. 2022) was published, CNN also reported on the matter (Rogers 2022).

On April 1, 2021, FAST formally stated that it was available to the international scientific community (Qi 2021). Heavenly Eye should not only belong to China but also to all of humanity as well. As Professor Nan Rendong put it: "Human beings stand out because there is a spirit of exploration for the unknown. (CCTV 2021)" And this exploration spirit should transcend national boundaries.

Jiang Peng took over as chief engineer of the FAST project after Professor Nan left (Pang and Qin 2023). On September 15, 2022, Jiang received Tencent Foundation's Scientific Exploration Award (NAOC, Jiang Peng, a researcher at the National Astronomical Observatory, has won the 2022 "Scientific Exploration Award" 2022) for his exceptional work on the FAST project. He is one of fewer than fifty people in the world to get this award each year, and this award is only presented in recognition of the awardees' achievements in ten subfields of basic scientific research and cutting-edge technology.

Despite its prominence, FAST is actually just the pilot telescope in China's ambitious SKA initiative (SKAO n.d.). The Square Kilometer Array (SKA) is often referred to as the "International Radio Telescope of the 21st Century." The initiative has been to establish a radio telescope array with a reception area of one square kilometer around the Earth. Even though China wasn't chosen to host the hub of this global super radio telescope, we declined to let our ambitions fade.

The KARST project, which is China's version of the SKA, is essentially divided into four areas. Ten telescopes will be constructed in the core area, where FAST is located. A center area will be developed within a radius of 2.5 kilometers outside the core area, with 15 telescopes to be constructed. And outside the central area is the outer area, with a radius of around 150 kilometers, where there will erect 25 telescopes. The outermost is a remote area with a radius of 3,000 kilometers, with a total of 11 telescopes, including those from Shanghai Sheshan, Beijing Miyun, and Urumqi (Liu et al. 2005) (Fig. 11.10).

Looking back at this period in history, Chinese astronomers designed a magnificent and one-of-a-kind telescope despite having no prior expertise from which to draw inspiration. If ever there were miracles in the annals of architecture, they would definitely include this mega-structure because it was beyond human imagination at first and came into being by human wisdom and spirit at last.

The successful construction of FAST is the product of the labor of multiple generations of Chinese astronomers, enabling the entire human race to

Fig. 11.10 The Shanghai Sheshan Tianma Telescope that will participate in the China SKA project (*Credit* Voice of Science Copyright © Voice of Science)

observe an older universe. As the largest single-aperture telescope in the world, its significance is self-evident.

Many junior astronomers have no regrets about contributing to the cause, and this was made possible by the experience of many veteran astronomers from earlier generations. We need more ambitious young people to join in if we are determined to advance China's astronomical cause and keep exploring the vast, uncharted cosmos for the benefit of China as a nation, and human race as a whole.

References

CCTV: Chinese Stories | These two letters to Nan Rendong are lost forever in the story... (2021). Retrieved from CCTV: https://m.news.cctv.com/2021/02/01/ARTIrwufRe15K7w3BIY6pQNP210201.shtml

Condon, J.J., Ransom, S.M.: Essential Radio Astronomy. Princeton University Press (2016)

Ekers, R.: The history of the square kilometre array (SKA)—born global. Instrumentation and Methods for Astrophysics (2012). https://doi.org/10.48550/arXiv.1212.3497

Gu, X., Zhao, B., Wang, Q.: Design and Manufacturing Project task Book of FAST Reflector Unit[R]. National Astronomical Observatories of China, Bejing (2014)

Kellermann, K.: URSI 24TH GENERAL ASSEMBLY, KYOTO, JAPAN, 25 AUGUST—2 SEPTEMBER 1993 (1993). Retrieved April 14, 2023, from NRAO/AUI Archives: https://www.nrao.edu/archives/items/show/13961

Kong, X., Jiang, P., Wang, Q.: Investigation on fatigue life of FAST cable-net under high stress amplitude produced by long-time active[J]. Eng. Mech. **30**, 169–174 (2013). https://doi.org/10.6052/j.issn.1000-4750.2012.04.S038

Li, D., Nan, R., Pan, Z.: The Five-hundred-meter Aperture Spherical radio telescope project and its early science opportunities. Proc. Int. Astron. Union **8**(S291), 325–330 (2012). https://doi.org/10.1017/s1743921312024015

Liu, H., Song, J., Wang, W., Peng, B., Nan, R.: A configuration of ska siting in Guizhou, China. Geology-geochemistry, 69–74 (2005).https://doi.org/10.1360/gs050303

Lu, Y., Ren, G.: Simulation of the cable mesh reflector for the large radio telescope FAST[J]. Eng. Mech. **24**(10), 165–169 (2007)

Murdin, P.: Arecibo observatory. In: Encyclopedia of Astronomy & Astrophysics. CRC Press, Boca Raton (2001).https://doi.org/10.1201/9781003220435

Nan, R., Ren, G., Zhu, W., Lu, Y.: Adaptive cable-mesh reflector for the FAST. Acta Astronom. Sinica **44**, 13–18 (2003)

NAOC: Deep Grief for the Death of Prof. Nan Rendong, the FAST Chief Scientist and Chief Engineer. Retrieved from National Astronomical Observatories,

Chinese Academy of Sciences (2017). http://english.nao.cas.cn/focus2015/201709/t20170917_183187.html

NAOC: Jiang Peng, a researcher at the National Astronomical Observatory, has won the 2022 "Scientific Exploration Award" (2022). Retrieved from NAOC: https://www.bao.ac.cn/news/zh/202209/t20220916_6513408.html

Pang, X., Qin, H.: Jiang Peng feels the spirit of "Chinese Heavenly Eye": Every step we have taken is down-to-earth (2023). Retrieved from People's Daily: http://lianghui.people.com.cn/2023/n1/2023/0307/c452599-32638975.html

Phillips, J.A., Wolszczan, A.: Pulsar astronomy at meter and decameter wavelengths: Results from Arecibo. In: Low Frequency Astrophysics from Space, pp. 175–186. Springer, Berlin, Heidelberg (2005). https://doi.org/10.1007/3-540-52891-1_122

Qi, J.: "The Chinese Heavenly Eye" will officially open to the global scientific community on April 1st (2021). Retrieved from Xinhua News Agency: http://www.gov.cn/xinwen/2021-01/04/content_5576862.htm

Ren, G.: Gexue Ren (n.d.). Retrieved April 14, 2023, from IEEE Xplore: https://ieeexplore.ieee.org/author/37087190246

Rogers, K.: A bizarre fast radio burst came from a binary system, astronomers say (2022). Retrieved from CNN: https://www.cnn.com/2022/09/23/world/fast-radio-burst-20201124a-origin-study-scn/index.html

SAO: The 25m Radio Telescope Observing Station (2014). Retrieved from Shanghai Astronomical Observatory, Chinese Academy of Sciences: http://english.shao.cas.cn/fs/201410/t20141008_128932.html

Schock, M.: Site selection and characterization for giant telescopes. SPIE Proc. (2004). https://doi.org/10.1117/12.551847

SKAO: A brief history of China's involvement with the SKAO (n.d.). Retrieved April 14, 2023, from SKAO: https://www.skao.int/en/partners/skao-members/372/china

Sun, C.: Two new pulsars have been discovered using FAST (2021). Retrieved from FAST: https://fast.bao.ac.cn/cms/article/1/

Wang, F.Y., Zhang, G.Q., Dai, Z.G., Cheng, K.S.: Repeating fast radio burst 20201124A originates from a magnetar/Be star binary. Nat. Commun. **13**(1) (2022). https://doi.org/10.1038/s41467-022-31923-y

Wu, X.: About 500 pulsars have been found by China's Heavenly Eye (2022). Retrieved from People's Daily: http://www.xhby.net/index/202201/t20220106_7376327.shtml

XAO: History (2013). Retrieved April 14, 2023, from Xinjiang Astronomy Observatory, Chinese Academy of Sciences: http://english.xao.ac.cn/au/hi/

Xu, H., Niu, J.R., Chen, P., Lee, K.J., Zhu, W.W., Dong, S., … Yang, Y.P.: A fast radio burst source at a complex magnetized site in a barred galaxy. Nature **609**, 685–688 (2022).https://doi.org/10.1038/s41586-022-05071-8

Zhang, L., Li, D., Hobbs, G., Agar, C.H., Manchester, R.N., Weltevrede, P., … Zhi, Q.: A pulsar with interesting emission properties discovered at fast. Astrophys. J. **877**, 55 (2019).https://doi.org/10.3847/1538-4357/ab1849

Zhu, W.: Scientist Wenbai Zhu (n.d.). Retrieved April 14, 2023, from CAS Members: http://sourcedb.naoc.cas.cn/en/enaoexpert/200907/t20090706_2000366.html

12

Large High Altitude Air Shower Observatory (LHAASO): Contributing to the Resolution of a Century-Long Mystery

Large High Altitude Air Shower Observatory (LHAASO)

The horizon was a pale yellow on July 4, 1054 (the first year of Zhihe of the Song Dynasty) at 4:00 a.m. A Chinese official in charge of monitoring the astronomical events yawned subconsciously as dawn approached and a cool wind blew. A strange light appeared near the star of Tianguan (Zeta Tauri), and a very bright point of light came out, like a thunderbolt piercing the silent night sky, completely dissipating the sleepiness of the officials in the Department of Heavenly Supervision, just as they were about to finish a night of celestial observations and pack up their things to get off work.

"An unexpected appearance of a rare guest star! (Li 656)" Immediately, the leading official realized that this was probably a guest star of the type that only arrives in the sky once per century or so and then vanishes. The guest star was as bright as Venus for the subsequent 23 days, making it visible even during the daytime, and it was visible in the night sky for the whole 643 days.

Formerly known as Tianguan Guest Star, the Song Dynasty discovery is today known as a supernova, after the type of star that explodes violently at the end of its life (Stephenson and Green 2002). Song Dynasty astronomers and historians left accurate, detailed, complete, and exquisite astronomical records, which had served as a crucial foundation for modern astronomy's study of supernovae (Martzloff 2016), whereas European astronomers left behind few important documents due to religious control during the Middle Ages. The ancient Chinese records of celestial observation had long been the most accurate and complete of their kind (Fig. 12.1).

J. Wang, *Eye Beyond the Sky*, Astronomers' Universe,
https://doi.org/10.1007/978-981-99-9818-0_12

Fig. 12.1 Crab Nebula, the remnant of Supernova 1054 (*Credit* ESO)

By the year 2020, this guest star had evolved into the Crab Nebula, a brilliant source of high-energy radiation.

Located in Daocheng County, Sichuan Province, at an elevation of 4,410 m above sea level, on the barren and remote Haizi Mountain, the Large High Altitude Air Shower Observatory (LHAASO) (Cyranoski 2017) is an important piece of China's scientific and technological infrastructure, and it takes up a disk-shaped area on the mountaintop of 1.36 km^2.

Early in the month of April 2020, LHAASO was pointed at the same part of the sky studied by astronomers from the Song Dynasty. Then, as was her custom, Wang Lingyu, an associate researcher at the Institute of High Energy Physics of the Chinese Academy of Sciences, reviewed the data by LHAASO in great detail, and she quickly made a surprising discovery (Ni 2021). Abnormal signals indicated toward the Crab Nebula, where a "super wanderer" had crashed to Earth after traveling at the speed of light for 6,500 years, repeatedly clashing with atmosphere molecules. She double-checked everything, caught her breath, and turned to her coworkers with excitement, saying, "LHAASO seems to have uncovered ultra-high-energy gamma photons, which may be a world-shaking discovery!" (Fig. 12.2).

Fig. 12.2 Aerial view of LHAASO (*Credit* Institute of High Energy Physics, Chinese Academy of Sciences)

So, what exactly is this groundbreaking finding that spans millennia? Cosmic rays are the place to begin with.

Will you still gently massage your drowsy eyes and hunt for your pants and socks in a stupor if, one day, you open your eyes and find that the room is being machine-gunned and bullets are flying? Absolutely not. There's probably not much of a difference between going out clothed or unclothed when your options are life or death. Don't be shocked if I suggest that incidents like these occur frequently in the world around us. The Earth, like this room, is constantly being pelted by high-energy particles traveling at close to the speed of light from all directions in deep space (Bothmer and Daglis 2007).

Cosmic rays, the name given to high-energy particles originating in space, have two defining features. To begin, the energy is exceptionally high; the energy of certain particles can reach 10^{20} electron volts (Gaisser 1990). This is a tremendously high value. When air is at normal temperature, each molecule has an energy of only 0.04 electron volts. Even at its maximum strength, Europe's Large Hadron Collider (LHC) can only accelerate particles to 10^{13} electron volts (Jones 2008). The state of Utah in the United States was the location where a particle with such a high energy was discovered in 1991. It could be a proton, after all. It traveled at a speed of 0.999 999 999 999 999 999 999 995 1 times the speed of light. Its kinetic energy had been measured to reach 3×10^{20} electron volts. A single particle had as much energy as a baseball moving at 100 km per hour (Bird et al. 1995). Therefore, the "Oh-My-God" particle became the moniker for this elusive particle (Fig. 12.3).

As for the second feature, most cosmic rays have a charge. The percentage of nuclei that are either hydrogen or helium or those of other heavy elements exceeds ninety-nine percent. Each one carries a positive charge. The

Fig. 12.3 Details of Europe's Large Hadron Collider (LHC) (*Credit* Vieamusante)

remaining one percent is made up mainly of electrons, with a negative charge. Only the extremely rare gamma rays and neutrinos are completely neutral (Sachs 2010).

Cosmic rays have a lot of energy and are still charged. Isn't it a huge risk for us to be here on Earth? The area we currently reside in is, in reality, rather secure. First, the likelihood of being exposed to cosmic ray particles with energies greater than 10^{20} electron volts is extremely low; most people won't even encounter one throughout their entire lifetime. High-energy cosmic ray particles are extremely uncommon compared to their lower-energy counterparts.

The protective role of the Earth's atmosphere and magnetic field is, on the other hand, well-established. The atmosphere acts like a "bulletproof vest," deflecting the impact of cosmic rays by converting their high-energy primary particles into secondary particles with much lower energy, much like how large raindrops break apart into smaller ones (Dorman 2013). In the same way that most of the energetic particles of cosmic rays are charged, they will also change direction and be deflected to the poles of the Earth due to the action of the Earth's magnetic field. After that, the particle is obviously not as effective as a direct hit to the ground or a direct drive into the target (Balogh et al. 2007) (Fig. 12.4).

More than a century has passed since the discovery of cosmic rays in the twentieth century, during which time numerous advancements have been

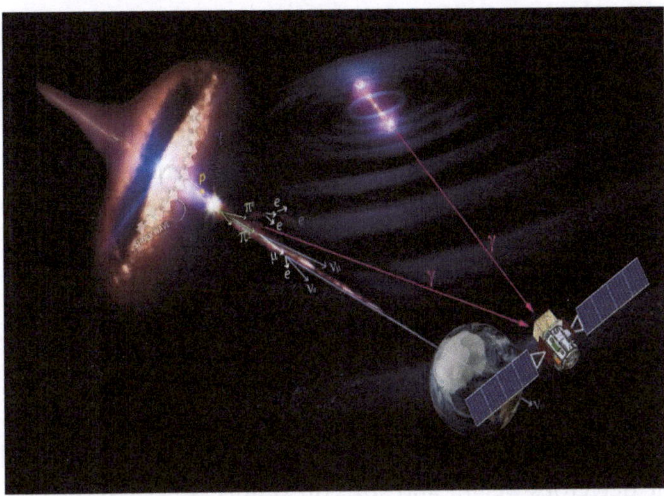

Fig. 12.4 Illustration of a dazzling star sending a beam of cosmic rays toward the Earth (*Credit* NASA)

achieved in the field, and five Nobel Prizes have been awarded as a result. Even so, the concept of cosmic rays is still unfamiliar to the average person. The origin of cosmic "super wanderers" has been labeled the "mystery of the century" by astronomers, and no one knows for sure where they originate.

The fact that the vast majority of cosmic rays are made up of charged particles like hydrogen nuclei, helium nuclei, heavy element nuclei, and electrons is a major factor. Since the orientation of charged particles is altered by the ubiquitous magnetic fields in the universe, which act like a blender, these particles swarm around like "headless flies," rather than moving in a straight line (Choi 2013). The majority of cosmic ray particles have already strayed significantly from their initial route by the time they reach the Earth, making it impossible for humans to determine the direction from which they originated. It's like getting that cold feeling in your back when you know someone is pointing a gun at you but you have no idea who it is, where it is, or whether he will shoot in the next second.

Fortunately, less than one percent of cosmic rays are gamma rays, which are neutral particles and are not deflected by the Earth's magnetic field. Therefore, many researchers employ gamma rays as a first step in investigating cosmic rays (Berezinskiĭ et al. 1990). Like the visible light, gamma rays are electromagnetic waves, but their energy is significantly greater. A single gamma photon can have more than a billion times the energy of an ordinary light photon (Rosswog and Brüggen 2011).

Gamma rays can be utilized in the role of a probe in a variety of charged particle accelerators. Some interactions between high-energy particles and ordinary matter generate gamma rays with energies that are detectable by our instruments; these gamma rays traveling in a straight line may "expose" the stealthy charged particles. As a result, the observation of gamma rays with high energies is an essential part of the process of researching the source of cosmic rays, and the higher the energy, the more successful the research will be. With the investigation of the radiation mechanism and the direction of gamma rays, we are able to ascertain the corresponding celestial bodies and investigate the source of cosmic rays (Harwit 2019) (Fig. 12.5).

So, how can we identify these high-energy gamma rays? Two primary approaches exist. One is through direct measurement, where a spacecraft or a high altitude balloon launches a detector into space to collect data. As a result of the detector's weight and size constraints, as well as the fact that only a small fraction of the particles being studied have very high energies, this

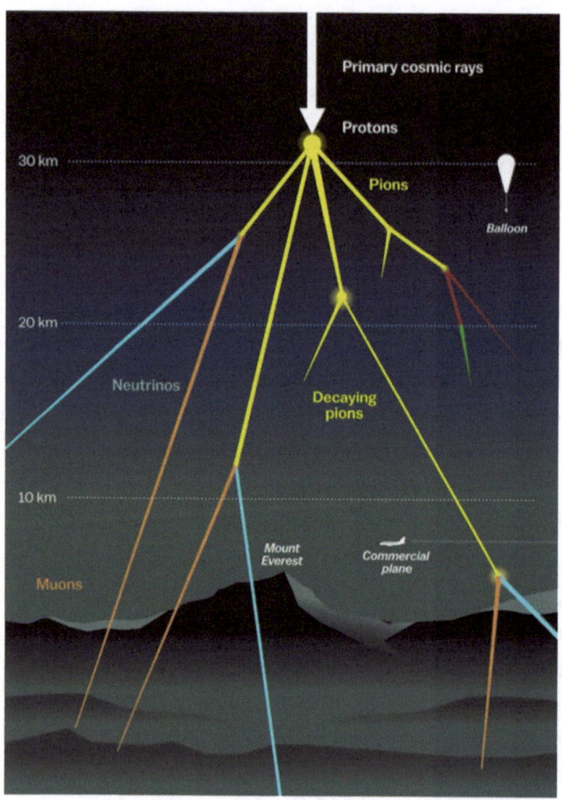

Fig. 12.5 Gamma ray air shower (*Source* NASA)

technique is unable to collect an adequate amount of data and can only be used in the very low-energy zone. Secondly, we have indirect measurement as an alternative. On the ground, a massive array of detectors is installed. Atmospheric showers, triggered by the entry of high-energy gamma rays, cause air molecules to react and release a plethora of secondary particles. Tertiary particles will be created as a result of further interactions between these secondary particles and molecules in the air. A "particle shower" composed of electrons, positrons, and low-energy photons will be rained down upon the Earth in this fashion. The features of high-energy particles entering the atmosphere, such as their originating direction, temporal distribution, energy distribution, and so on, can be uncovered by receiving these particles through vast arrays and then analyzing the characteristics of these particle showers through computerization (Longair 2011).

The Yangbajing International Cosmic Ray Observatory was established in 1989 in Tibet by the Chinese government. In 2019, the Crab Nebula was identified as the source of the most powerful gamma ray ever seen on the Earth (Xinhua 2019), with an energy of 0.45 PeV (4.5 X 10^{14} electron volts), more than five times greater than the previous maximum value. Although being the most sensitive gamma ray observatory in the world at the earlier time, the Yangbajing Observatory was rendered helpless in the face of the higher-energy gamma photons due to its limits in scale, accuracy, and detecting technology (Fig. 12.6).

Fig. 12.6 Yangbajing International Cosmic Ray Observatory (*Credit* Institute of High Energy Physics, Chinese Academy of Sciences)

Chinese scientists first put forward the proposal of constructing a new generation of cosmic ray observatory in 2008; the project was finally authorized in 2015 and given the name LHAASO (Large High Altitude Air Shower Observatory). The southeastern edge of the Qinghai-Tibet Plateau, sometimes known as the "roof of the world," was identified as the ideal location for LHAASO. This location is ideal for detecting secondary particles because of its high altitude (4,410 m), thin air (less absorption by the atmosphere), flat terrain, and ample water supplies. Being 10 km from Daocheng Aden Airport and within easy driving distances of Chengdu, Kunming, and other major cities makes it ideal for transporting a wide variety of materials. The local government places a premium on this, which has been a huge help in every way possible (Cao et al. 2019) (Fig. 12.7).

Launched in June 2017, LHAASO's main project had been finished in 2021. It consisted primarily of three components: a ground shower particle array with 5,195 electromagnetic particle detectors and 1,188 muon detectors; a 78,000 m^2 water Cherenkov detector array; and an array of 18 wide-angle Cherenkov telescopes. Using different technologies, cosmic rays can be detected in every direction and in numerous dimensions (Fig. 12.8).

Detectors per square kilometer per year can only capture two ultra-high-energy gamma photons from the brightest spot, and these two photons are typically buried in hundreds of thousands of cosmic ray signals, making it

Fig. 12.7 Mount Haizi (*Credit* Zhangzhugang)

Fig. 12.8 Illustration of Cherenkov Telescope Array (*Credit* Gabriel Pérez Diaz (IAC)/ Marc-André Besel (CTAO)/ESO/ N. Risinger (skysurvey.org))

like looking for a needle in a haystack. However, LHAASO has two secret weapons. The first is a large detection area (1.3 km^2), which is 20 times larger than the Yangbajing Observatory and even 60 times larger than HAWC, an observation station of the same type in the United States. The second is the ability to distinguish with a golden eye. Atmospheric showers can be created by the entry of gamma rays and other high-velocity charged particles into the atmosphere. As a rule of thumb, the number of fast-moving charged particles can be over ten thousand times more than the number of high-energy gamma photons. The "golden eye" of the 1,188 muon detectors, each covering 36 m^2 of underground space, has a detection probability of one in a million for gamma photons (LHAASO Collaboration, A future project at Tibet: the large high altitude air shower observatory (LHAASO) 2010). LAHHSO's sensitivity in the ultra-high-energy range is significantly stronger than that of any other equipment, even including CTA (Acharya 2013), the next generation ground-based observatory for gamma ray astronomy at very high energies scheduled for construction in 2027.

In addition, LHAASO was constructed and put into operation simultaneously, and a groundbreaking finding was made even before the project was fully implemented. Photons with an energy of 1.1 PeV were discovered by LHAASO in the Crab Nebula (LHAASO Collaboration et al. 2021), which is about 6,500 light-years away and was the location where officials in charge of heavenly supervision in the Song Dynasty identified the guest star (Supernova 1054). Chinese astronomers of the Song Dynasty a thousand years ago located this supernova, leaving behind the world's most precise, detailed, and comprehensive astronomical record. The highest-energy gamma photon was discovered by Chinese scientists a millennium later, ushering in the era of ultra-high-energy gamma photon astronomy for the study of cosmic rays. Over the course of a millennium, stellar objects have dispersed into nebulae. The fact that these discoveries are at the forefront of astronomical fields worldwide and that they were made by Chinese researchers remains unchanged.

This finding provides further evidence that supernova remnants like the Crab Nebula are a source of cosmic rays. The electrons around the highly-rotating pulsar (a type of neutron star) at the remnant's center have a high kinetic energy because of the pulsar's powerful magnetic field. Ultra-high-energy gamma photons may be formed when electrons with strong kinetic energy transfer a portion of the energy to them, turning these low-energy photons into high-energy ones. It took 6,500 years for these ultra-high-energy photons to travel to the Earth, but when they arrived, LHAASO was ready to catch them (Fig. 12.9).

Moreover, LHAASO found 11 additional sources of extremely high intensity gamma rays within the Milky Way. The most powerful photon among them has a maximum energy of 1.4 PeV (Cao et al. 2021) and originates from a young cluster of massive stars in the constellation Cygnus, which boasts an abundance of massive stars with short lifetimes. Stars are born and die within star clusters in a matter of millions of years, and the high levels of stellar activity present there may be ideal for the acceleration of cosmic rays.

A century has passed since the discovery of cosmic rays. Theoretically, these events should be linked to high-energy astrophysical occurrences like supernova explosions, huge celestial object collisions, etc. (Ginzburg and Syrovatskii 1964). However, for a considerable amount of time, neither

Fig. 12.9 Data from the Chandra X-ray Observatory revealed the partial remnants of Supernova 1987a (*Credit* NASA/CXC)

experiments nor observations had been able to corroborate these theories. Assuming that young massive star clusters, pulsar clouds, and supernova remnants are the sources of ultra-high-energy cosmic rays in the Milky Way, LAHHSO's stunning performance and impressive discoveries have at least confirmed the hypotheses above. To scientists, the origin of cosmic rays has been a mystery for over a century, but recent discoveries have shed new light on the subject.

LHAASO received acceptance of performance technology on October 24, 2021, indicating that the project has been finished and will continue to wait for the "messenger" from space on the Qinghai-Tibetan Plateau. When the fully armed one finally makes its debut in the C spot, we can rest assured that it will continue to impress us with its upcoming findings. For a very long time, it will be the cutting edge of studying cosmic rays and learning about their mysteries. Both the potential for scientific discoveries and the availability of cosmic rays will never be depleted. As LHAASO incorporates new telescope arrays to improve spatial resolution, we will get closer to the full picture of cosmic rays and their origins.

References

Dorman, L.: Cosmic Rays in the Earth's Atmosphere and Underground. Springer Science & Business Media (2013)

Acharya, B.S.: Introducing the CTA concept. Astropart. Phys. **43**, 3–18 (2013). https://doi.org/10.1016/j.astropartphys.2013.01.007

Balogh, A., Lanzerotti, L.J., Suess, S.T.: The Heliosphere Through the Solar Activity Cycle. Springer Science & Business Media (2007)

Berezinskiĭ, V.S., Bulanov, S.V., Dogiel, V.A.: Astrophysics of Cosmic Rays. Elsevier Science & Technology (1990)

Bird, D.J., Corbato, S.C., Dai, H.Y., Elbert, J.W., Green, K.D., Huang, M.A., ... Thomas, S.B.: Detection of a cosmic ray with measured energy well beyond the expected spectral cutoff due to cosmic microwave radiation. Astrophys. J. **441**, 144 (1995). https://doi.org/10.1086/175344

Bothmer, V., Daglis, I.A.: Space Weather: Physics and Effects. Springer Science & Business Media (2007)

Cao, Z., Aharonian, F.A., An, Q., Axikegu, Bai, L.X., Bai, Y.X., ... Chen, L.: Ultrahigh-energy photons up to 1.4 petaelectronvolts from 12 γ-ray Galactic sources. Nature **594**, 33–36 (2021). https://doi.org/10.1038/s41586-021-03498-z

Cao, Z., Chen, M.-J., Chen, S.-Z., Hu, H.-B., Liu, C., Liu, Y., … Zhang, S.-S.: Introduction to large high altitude air shower observatory (LHAASO). Introduction to Large High Altitude Air Shower Observatory (LHAASO) **43**(4), 457–478 (2019). https://doi.org/10.1016/j.chinastron.2019.11.001

Choi, C.Q.: Cosmic rays zap a planet's chances for life (2013). Retrieved from Phys.Org: https://phys.org/news/2013-11-cosmic-rays-zap-planet-chances.html

Cyranoski, D.: Chinese mountain observatory to probe cosmic-ray origins. Nature **543**(7645), 300–301 (2017). https://doi.org/10.1038/543300a

Gaisser, T.K.: Cosmic Rays and Particle Physics. Cambridge University Press (1990)

Ginzburg, V.L., Syrovatskii, S.I.: The origin of cosmic rays. In: The Origin of Cosmic Rays, pp. 186–282. Elsevier (1964). https://doi.org/10.1016/b978-0-08-013526-7.50011-6

Harwit, M.: Cosmic Discovery: The Search, Scope, and Heritage of Astronomy. Cambridge University Press (2019)

Jones, D.G.: Large Hadron Collider (LHC) (2008). Retrieved from Encyclopedia Britannica: https://www.britannica.com/technology/Large-Hadron-Collider

Collaboration, L.H.A.A.S.O.: A future project at tibet: the large high altitude air shower observatory (LHAASO). Chin. Phys. C **34**(2), 249–252 (2010). https://doi.org/10.1088/1674-1137/34/2/018

LHAASO Collaboration, Cao, Z., Aharonian, F., An, Q., Axikegu, Bai, L.X., … Chen, L.: Peta–electron volt gamma-ray emission from the Crab Nebula. Science **373**(6558), 425–430 (2021). https://doi.org/10.1126/science.abg5137

Li, C.: Yi Si Astrology (656). Retrieved from https://zh.wikisource.org/wiki/%E4%B9%99%E5%B7%B3%E5%8D%A0

Longair, M.S.: High Energy Astrophysics. Cambridge University Press (2011)

Martzloff, J.C.: Astronomy and Calendars—the Other Chinese Mathematics: 104 BC–AD 1644. Springer (2016)

Ni, S.: LHAASO Breakthrough: The Milky Way's particle acceleration ability exceeds expectations (2021). Retrieved from LHAASO, CAS: http://www.ihep.cas.cn/lhaaso/cmsm/202105/t20210518_6020392.html

Rosswog, S., Brüggen, M.: Introduction to High-Energy Astrophysics. Cambridge University Press (2011)

Sachs, M.: Physics of the Universe. World Scientific (2010)

Stephenson, F.R., Green, D.A.: Historical Supernovae and their Remnants. Oxford University Press (2002). https://doi.org/10.1093/acprof:oso/9780198507666.001.0001

Xinhua: Endurance in the Snowy Regions—Capturing Mysterious Cosmic Rays (2019). Retrieved from Xinhuanet: http://www.xinhuanet.com/politics/2019-10/21/c_1125133408.htm

13

Square Kilometer Array (SKA): The Upcoming Storm in Astronomy

Square Kilometer Array (SKA)

One night in September 2019, Astronomer Anna D. Kapinska (IAU 2018) sat down in front of her computer and swiftly scrolled through the array radio telescope's sky survey images. Anna's work often involved spotting anomalies in images captured by radio telescopes. She must select the most instructive pictures and present them for discussion at the next scheduled meeting.

Anna found that gazing at these space images at night gave her the same feeling as looking up at the stars. Therefore, that's when she preferred to conduct this work.

Anna's hand on the mouse fell suddenly. She focused intently on the computer screen, and her pupils widened. A big, eerie, blue-green face appeared in the center of the screen. There was a blur around the weird face (Norris et al. 2020), but Anna could make out two eyes in the black area just on it, peering through the screen at her (Figs. 13.1 and 13.2).

"Oh, for God's sake, what is this? The remnant of a supernova?" Anna's mind worked slowly.

Anna always made a remark in the photo caption on what she thought about strange celestial bodies. However, Anna hesitated this time. The blue-green weird "face" in front of her coincided with a huge number of stars in the Milky Way, which had been odd because supernova remnants were typically far away from most stars. There shouldn't be a supernova (Mandow 2021).

"Could it be that this peculiar object is not located deep within our galaxy?" Instead of continuing to operate the mouse, Anna put it down,

J. Wang, *Eye Beyond the Sky*, Astronomers' Universe,
https://doi.org/10.1007/978-981-99-9818-0_13

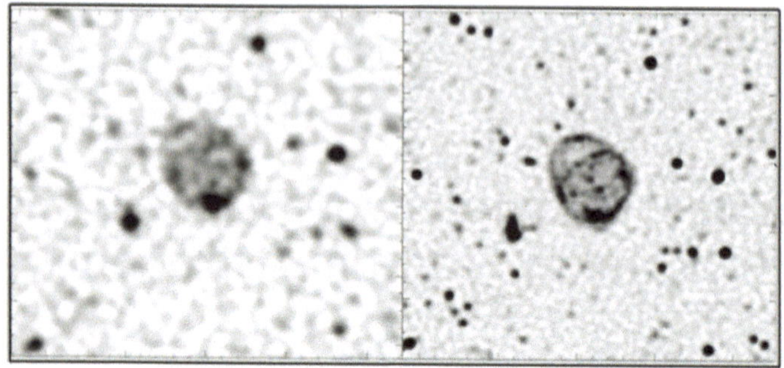

Fig. 13.1 The original discovery in ASKAP data (*Source* CSIRO, © The EMU team, using ASKAP and MeerKAT radio continuum data)

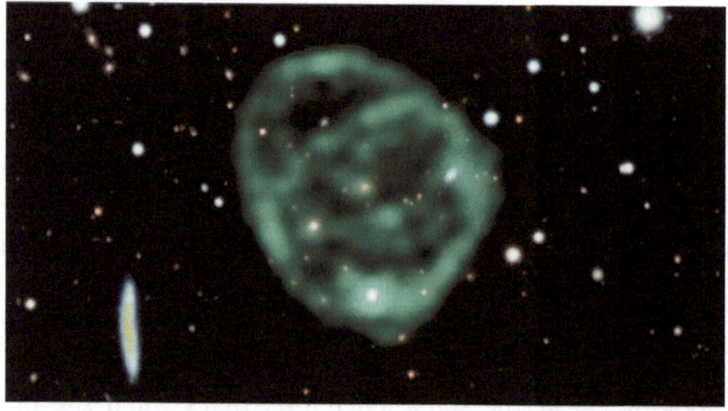

Fig. 13.2 MeerKAT radio telescope data (green) of the "WTF?" (*Credit* Jayanne English MeerKAT))

cupped her cheek with the other hand, and peered into the strange face's large, empty eyes as if it could answer for her.

"This weird object will likely take center stage at tomorrow's team meeting." Anna whispered to herself. She wasn't used to having a blank caption, so she just dubbed it "WTF?" (What The fuck?).

The next day, this picture with the caption "WTF?" was the topic of intense debate. Many bold hypotheses were proposed by the astronomers at the meeting regarding the "weird face," but they were rapidly disproven by the analyses that followed.

After the meeting, scientists uncovered deep space pictures from the same region that had been obtained using visible light telescopes in order to decipher the significance of the blue-green "mask". Conversely, the "weird face" was not found in the corresponding visible light picture, as if it had never existed.

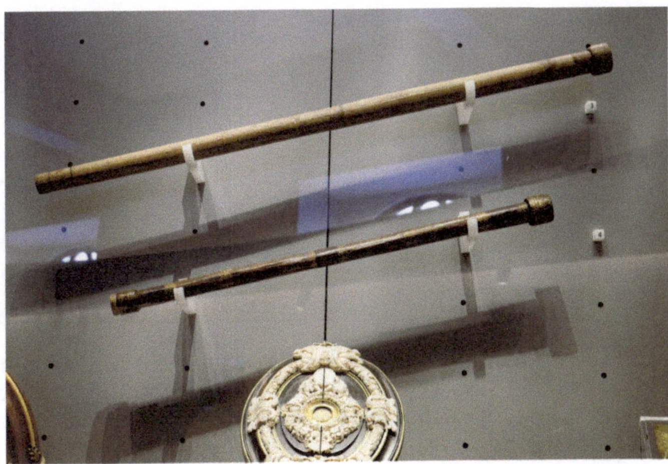

Fig. 13.3 Galilei telescopes (*Credit* Zde)

This blue-green "weird face" was perplexing. Exactly what information might Anna and her coworkers glean from the mysterious appearance? We can learn this by first looking at the blueprints for a massive international science project called the Square Kilometer Array (SKA) (Fig. 13.3).

There has been a long-running rivalry between optical telescopes and radio telescopes in the field of astronomical study. With Galileo's 1609 invention of the first astronomical telescope, optical telescopes became the primary instruments of choice for studying the cosmos. It wasn't until 1931 that the concept of a radio telescope was conceived, 322 years after the optical telescope was invented (Verschuur 1987).

Radio astronomy, on the other hand, emerged on the scene with a vengeance and quickly caught up. The discoveries of pulsars (Lyne and Graham-Smith 2012), quasars (Kellermann 2014), cosmic microwave background radiation (Wilkinson and Peebles 2001), and interstellar organic molecules (Guélin and Cernicharo 2022) in the 1960s were hailed as the four key discoveries of radio astronomy in the 20th century (Fig. 13.4).

When their equipment deteriorated, many early-era radio telescopes were shut down in the 1990s, causing a temporary lull in radio astronomy study. In contrast, optical telescopes experienced a golden decade beginning with the launch of the Hubble Space Telescope in 1990 and continuing with the construction of the massive Keck Telescope, Subaru Telescope, and ESO's Very Large Telescope (VLT).

This, however, was not the finish line for radio astronomy studies. Scientists had noticed that radio antenna array techniques, which had relied heavily on computing and then data processing, were becoming increasingly possible

Fig. 13.4 From Left to right: Replicas of the first equipment to detect radio waves from space made by Karl G. Jansky in 1932 (*Credit* Astrochemist) and the first dish radio telescope built by Grote Reber in 1937 (*Credit* Jarek Tuszyński)

due to the rapid growth of computer technology. A multinational partnership led by countries from Europe and the Commonwealth eagerly seized this ideal opportunity to leapfrog and take the lead in the field.

With its high signal-to-noise ratio and powerful signal-gathering capabilities, a single-aperture radio telescope is an attractive option. More of the incredibly weak signals from outer space can be gathered by a single-aperture radio telescope of greater size. One clear drawback of single-aperture telescopes is that as long as they grow even slightly in size, radio telescopes will encounter enormous engineering challenges (Stanimirovic 2002). A radio telescope with a single aperture that may rotate freely has a maximum aperture of 100 m. It would be necessary to construct the telescope in accordance with the topography of the mountain if we were to succeed in breaking through this barrier. Such constructions are the Arecibo Radio Telescope and the Chinese "Heavenly Eye" FAST Radio Telescope.

The advantages of array radio telescopes are particularly striking when compared to those of single-aperture telescopes. Through interferometry (Steel 1985), they may tailor the telescope's spatial resolution without increasing the size of any individual antenna. When it comes to the array radio telescope, more is better, and larger is better if we ignore the complexity of data transmission, computation, and subsequent data processing (Lobanov et al. 2015). To create a virtual telescope with an aperture of two astronomical units, we can even put a radio telescope into orbit around the Earth.

Even though the technical challenges of constructing a telescope in Earth orbit are insurmountable now, it is still possible to construct a super telescope array on the surface of the Planet.

At the 1993 conference of the International Union of Radio Science in Kyoto, Japan, astronomers from ten countries, including the United

Kingdom, China, Australia, and Italy, proposed constructing a world-class giant radio telescope array. When finished, this radio telescope would have a reception area that was one square kilometer in size. Hence, the Square Kilometer Array (SKA) became the official name for this endeavor (Hall 2006). Moreover, this had been the most advanced astronomical instrument ever built.

When the SKA was finally built, its sensitivity would be raised by fifty times, its survey speed would be increased by approximately ten thousand times, and it would be able to accomplish wide band coverage ranging from 0.1 to 30 GHz. These improvements were made in comparison to the then-world's largest radio telescope. It would produce 10 petabytes (PBs) of data every second due to the millisecond sampling accuracy with which it would operate. What does "10 PBs" mean? Since the typical capacity of a mobile hard disk was 1 TB, you would need more than 10,000 mobile hard disks to fill the space. Keep in mind that this would be only the information it generated in a single second (Scaife 2020) (Fig. 13.5).

Information about the SKA project can be found in the China SKA Scientific Report (Wang 2020):

> By revealing the first generation of celestial bodies born in the universe, SKA will reproduce the historical process of the universe from darkness to light; by using hydrogen, the most abundant element in the world, as a messenger, SKA will draw the greatest three-dimensional structure of the universe; by examining the fundamental properties of dark matter and dark energy, SKA will help dispel the two dark clouds hanging over the natural sciences in the 21st century; by discovering practically all pulsars in the Milky Way and discovering the first black hole-pulsar pair, SKA will accurately verify the gravitational theory; by accurately measuring hundreds of millisecond pulsars, SKA will identify gravitational waves from supermassive black holes; by reconstructing

Fig. 13.5 From Left to Right: The SKA-MPI dish prototype in South Africa (*Credit* SKAO) and SKA-Low prototype antennas Australia (*Credit* ICRAR)

> the structure of the cosmic magnetic field, SKA will detect its source; by revealing the cradle of primitive life, SKA will hunt for a bosom companion. The SKA will also conduct research into the uncharted parts of the universe and bring back some ground-breaking discoveries. If any of these challenges can be addressed, it will spark a scientific revolution.

The SKA is a ground-breaking engineering project that has never been attempted before. As a precautionary measure, scientists designed a series of pathfinding initiatives to be carried out at the target sites as a confirmatory pre-project for SKA.

The 21CMA project, led by China, is the first SKA pathfinding effort. Affectionately dubbed "the first light of the universe," the project's objective was to use digital techniques to capture low-frequency photographs of the cosmos.

The first large-scale phased telescope array dedicated to the detection of low-frequency signals in the universe was established in August 2003 at the Wulasitai base between the northern and southern Tianshan Mountains in Xinjiang, China.

When finished, 21CMA had been an antenna array having a width of 4 km from north to south and a length of 6 km from east to west, for a total of 10,287 antennas. The digitization achieved by 21CMA set it apart as the SKA project's pathfinder from more conventional array antennas. The radio wave signals collected by each antenna unit can be processed by computers, allowing them to keep an eye on several targets at once without the need for rotation (Huang et al. 2016) (Fig. 13.6).

It is anticipated that the 21CMA will contribute its experience and success to the SKA project, considering that it functioned on the same premise as the low-frequency array antenna that the SKA would construct.

In anticipation, the second SKA Pathfinder project was initiated in December 2009. The Australian Square Kilometer Array Pathfinder (ASKAP) was being built in the Shire of Murchison, Western Australia, a remote desert region with a very low population density (Johnston 2008). Around $300 million had been spent on infrastructure in Australia, and a dedicated Communications and Media Authority was established to safeguard the "radio silence zone" in Murchison, so that optimal ionospheric conditions could well be achieved for the project.

The ASKAP radio telescope array featured thirty-six 12-m dishes spaced no more than 6 km apart, for a total signal receiving area of around 4,000 m^2 (Fig. 13.7).

Fig. 13.6 Planning graph for China's 21CMA project (*Credit* National Astronomical Observatories of Chinese Academy of Sciences)

Fig. 13.7 ASKAP in Australia (*Credit* CSIRO © Copyright CSIRO Australia)

The data revealed that ASKAP's antenna count and size were not particularly remarkable. However, its superior design earned it the designation of the SKA project's pathfinder.

Scientists have developed multi-beam technology, in which numerous receivers (also called feeds) are placed on the focal plane of an existing reflector telescope to enhance its field of vision. Multi-beam technology is superior to single-beam methods because it can take pictures of multiple objects simultaneously, much like the compound eye of insects. Beam-forming, the core in "phased array feed" technology, has been regularly updated to the point where it can fully sample the focal plane and give a continuous vast field of vision (DeBoer et al. 2009).

Following Australia's initial proposal in 2006, the ASKAP site saw the debut of the prototype of the phased array feed in 2012. Each feed included 188 independent receiving components, allowing them to effectively achieve an instantaneous field of view of around 30 square degrees, nearly 30 times greater than the field of view of the typical receiver. This resulted in a significant improvement in the effectiveness of sky surveying and opened a new era for the application of phased array feed receivers in the area of radio astronomy (Fig. 13.8).

The issue with the field of view had been resolved. Data processing was a second challenge that needed to be addressed. An increase in ingest capacity by a factor of 30 would result in a data volume that was 30 times greater. The computing power required to handle the data would reach 5,000 trillion floating-point operations per second, nearly three times the processing capability of China's "Sunway TaihuLight" supercomputer.

The SKA Pathfinder project's computational capability was put to the test for the first time in its history. To put things in perspective, ASKAP's computational requirement was barely a tenth as large as the SKA. This had been both a difficulty and a possibility. As one Chinese scientist and SKA supercomputer expert, An Tao, put it: "The way SKA processes data will undermine the existing model and precipitate a dramatic change in astronomical data processing. (Liu 2019)"

The Murchison Widefield Array, a radio telescope array (Tingay 2012), was finished in the desert not far from ASKAP in 2012. With its 30-degree ultra-wide field of vision, this array could scan a vast swath of the sky in a single snapshot. It could be used to map the halos and relics of extragalactic systems and examine the ionosphere of the Earth and the heliosphere of the Sun (Fig. 13.9).

The HERA radio interferometric array, located in the Karoo Desert of South Africa, was completed in 2015. Hydrogen Epoch of Reionization Array

Fig. 13.8 ASKAP in Australia, single antenna (*Credit* CSIRO © Copyright CSIRO Australia)

is what HERA stands for. Neutral hydrogen atoms eventually turned into plasmas, illuminating the universe with light, from the Big Bang until the birth of the first generation of stars. HERA's mission was to investigate the ionized gas between the stars and solve the universe's first enigma. This telescope had a reception area of 54,000 m^2, which was comparable to that of the Arecibo, and was comprised of 350 hexagonal fixed antennas with a diameter of 14 m each (Fagnoni and De Lera Acedo 2016) (Fig. 13.10).

In 2016, the Karoo Array Telescope, another SKA pathfinder project, was finished in the desert not far from the HERA. Karoo's sixty-two 13.5-m antennas were connected to a digital receiver from the very start of the process. Following synchronization with an atomic clock, data from the

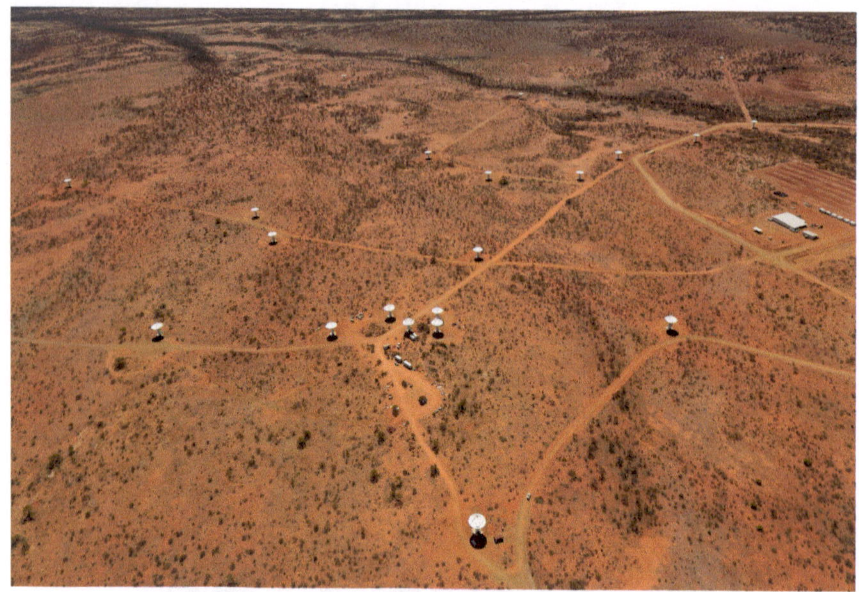

Fig. 13.9 An aerial view of ASKAP in Australia (*Credit* CSIRO © Copyright CSIRO Australia)

Fig. 13.10 The HERA radio interferometric array in South Africa (*Credit* HERA Partnership)

receiver could be transmitted by optical fiber to an underground supercomputer facility (Woudt et al. 2013). To that end, this was the flagship initiative that came closest to realizing the SKA's ultimate vision (Fig. 13.11).

Tianlai, or "Sound of Heaven," Project, China's second SKA pathfinding effort, was approved by the research team and put into operation at the

Fig. 13.11 The Karro Array Telescope in South Africa (*Credit* South African Radio Astronomy Observatory (SARAO))

end of 2016 (Department of Sci. & Tech. of Hunan Province 2017). When fully implemented, Tianlai's outstanding design would enable it to utilize two distinct types of antennas: the more conventional dish array and a more novel cylindrical antenna. For the purpose of observing the cosmos' large-scale structure, Tianlai intended to exploit Earth's rotation to quickly survey the sky over the northern hemisphere (Fig. 13.12).

Fig. 13.12 Tianlai, or "Sound of Heaven" Project (*Credit* National Astronomical Observatories of Chinese Academy of Sciences)

After the primary technical issues were resolved one at a time, all the precursor projects reached the scientific verification phase.

In spite of the fact that researchers had held out longstanding optimism for ASKAP and other equipment that represented the next generation of radio telescopes, they had been confronted with a variety of unexpected findings.

The Fast Radio Bursts discussed earlier in this book are a good illustration. Though incredibly powerful, most of the FRBs reported are isolated events. This made pinpointing transient FRBs extremely challenging (Stentiford 2021).

On September 24, 2018, using the "real-time playback" feature of a killer APP, ASKAP, which was in the verification phase, recorded a single FRB designated FRB180924. Later, it pinpointed the location of its host galaxy with the help of the Keck, Southern Gemini, and the ESO's VLT. This new finding, published in the June 2019 issue of *Science* (Bannister et al. 2019), would aid in the resolution of the more vexing "cosmic baryon missing conundrum" by showing that the electron column density during the explosion of fast radio bursts was consistent with the intergalactic medium model.

To put ASKAP to the test, 739 astronomers from 208 institutions around the world created the "Evolutionary Map of the Universe (EMU) (SKAO 2021)" program, an expansive scientific survey. Regular tasks in this study included the selection of valuable photographs from numerous sky surveys, as depicted in this chapter's opening story, whereby an astronomer named Anna D. Kapinska performed these duties.

That blue-green "weird face" in space was one of the remarkable sights that ASKAP caught that had previously eluded other telescopes.

A few days later, Emil Lenk, another astronomer who worked with Anna to compile the photographs, looked through the data and found a similar thing, which was labeled as "Odd Radio Circle (Woods 2022)". ORC was quite congruent with the depiction of these heavenly things in pictures, and it was interesting to note that orc signifies orcish or demon in English.

Most intriguingly, these odd radio circles can be seen by nothing except radio telescopes. There will be no traces of these eerie circles in X-ray, infrared, or optical telescopes. Unfortunately, astronomers have not yet been able to determine the cause of the five separate ORCs that have been found.

Astronomer Ray Norris from Western Sydney University suspects these mysterious circles are shock waves caused by the collision of supermassive black holes in faraway galaxies. Telescopes have picked up on faint radio wave signals emanating from these enigmatic rings, which behave like an expanding bubble and excite electrons in space. Three of the five circles discovered so far feature black holes at their centers, which is Norris's main piece of evidence. Perhaps these mysterious ORCs are the cosmic fingerprints of gravitational waves (Willett 2016).

Clearer images of these enigmatic circles were captured once the Karoo Array Telescope was finished. Unfortunately, there is still no consensus among scientists about what generates these peculiar circles.

Observations with the SKA continue to usher in new discoveries thanks to its great sensitivity, extremely wide field of view, and efficient sky survey capability.

Over the course of just ten days in December 2020, ASKAP mapped 3 million galaxies throughout the southern sky, including 1 million that had never been observed before. In the years that followed, ASKAP collaborated with the German-Russian eROSITA Space Telescope and observed a massive filamentary structure up to 50 million light-years long, providing crucial support for the "Cosmic Web" theory (Hale et al. 2021) (Fig. 13.13).

Another phenomenon that had never been seen before by astronomers was the sight of two "dancing ghosts (Western Sydney University 2021)" holding hands in outer space, which was detected by ASKAP in August 2021 and shed spotlight on the mysterious "interstellar wind" problem (Fig. 13.14).

In addition to these pioneering efforts, the Square Kilometer Array Project has benefited from the confirmatory research of 17 other prominent radio

Fig. 13.13 ASKAP mapped 3 million galaxies across the entire southern sky in a record-breaking ten days at the end of 2020. (*Credit* CSIRO © Copyright CSIRO Australia)

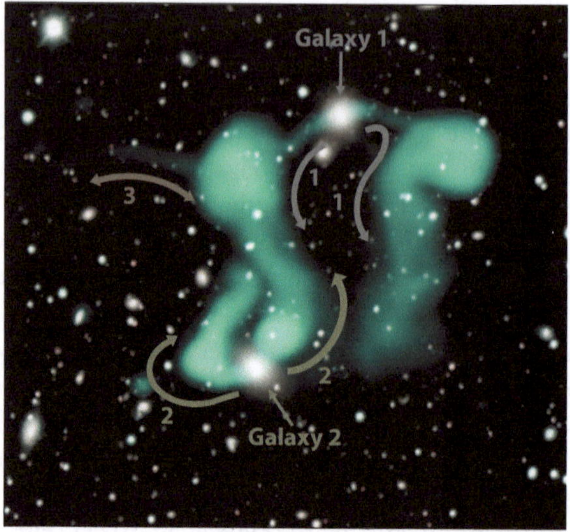

Fig. 13.14 "Dancing ghosts" observed by ASKAP in 2021 (*Credit* Jayanne English and Ray Norris using data from EMU and the Dark Energy Survey) © Copyright CSIRO Australia

telescopes, including Arecibo, VLA, and FAST. When a new generation of telescopes finally emerges, all the partners will rejoice.

We used to believe that more powerful telescopes would only provide us with a more expansive universe and a greater number of stars. Once SKA opened its eyes, the once-familiar sky had been taken over by a weird realm of flying ghosts and orcs.

Where will all of these recent discoveries take us? The answer still needs to be clarified. This journey from the unknown to the known, followed by the discovery of even more of the unknown, must be an amazing experience. As the creator of the SKA project and the backbone of its scientific study, the Chinese SKA team is ushering in a new epoch of astronomy right in our presence.

References

Bannister, K.W., Deller, A.T., Phillips, C., Macquart, J.-P., Prochaska, J.X., Tejos, N., … Burchett, J.: A single fast radio burst localized to a massive galaxy at cosmological distance. Science **365**, 565–570 (2019).https://doi.org/10.1126/science.aaw5903

DeBoer, D.R., Gough, R.G., Bunton, J.D., Cornwell, T.J., Beresford, R.J., Johnston, S., … Chippendale, A.: Australian SKA Pathfinder: a high-dynamic range wide-field of view survey telescope. Proc. IEEE **97**(8), 1507–1521 (2009).https://doi.org/10.1109/jproc.2009.2016516

Department of Sci. & Tech. of Hunan Province: Tianlai Array Completion Lays Foundation for Domestic Dark Energy Detection. Retrieved from Department of Science and Technology of Hunan Province (2017). http://kjt.hunan.gov.cn/kjt/xxgk/gzdt/kjzx/201701/t20170122_3965637.html

Fagnoni, N., De Lera Acedo, E.: The "Hydrogen Epoch of Reionization Array" (HERA)—Improvement of the antenna response with a matching network and scientific impacts. 2016 International Conference on Electromagnetics in Advanced Applications (ICEAA) (2016). https://doi.org/10.1109/iceaa.2016.7731474

Guélin, M., & Cernicharo, J.: Organic molecules in interstellar space: latest advances. Front. Astron. Space Sci., 9 (2022).https://doi.org/10.3389/fspas.2022.787567

Hale, C.L., McConnell, D., Thomson, A.J., Lenc, E., Heald, G.H., Hotan, A.W., ... Whiting, M.T.: The rapid ASKAP continuum survey paper II: First stokes I source catalogue data release. Publications of the Astronomical Society of Australia, 38 (2021). https://doi.org/10.1017/pasa.2021.47

Hall, P.J.: The Square Kilometre Array: An Engineering Perspective. Springer Science & Business Media (2006)

Huang, Y., Wu, X., Zheng, Q., Gu, J., Xu, H.: The radio environment of the 21 Centimeter Array: RFI detection and mitigation. Res. Astron. Astrophys. **16**(2), 016 (2016). https://doi.org/10.1088/1674-4527/16/2/036

IAU: Anna D. Kapinska. Retrieved from International Astronomical Union (2018). https://www.iau.org/administration/membership/individual/16472/

Johnston, S.: Science with ASKAP. The Australian square-kilometre-array pathfinder. Exp. Astron. **22**(3), 151–273 (2008). https://doi.org/10.1007/s10686-008-9124-7

Kellermann, K.I.: The discovery of quasars and its aftermath. J. Astron. History Heritage **17**(03), 267–282 (2014). https://doi.org/10.3724/sp.j.1440-2807.2014.03.03

Liu, P.: Preliminary Construction of China's SKA Data Center Prototype Completed. Retrieved from China News (2019). https://www.chinanews.com.cn/gn/2019/09-08/8950782.shtml

Lobanov, A.P., Gómez, J.L., Bruni, G., Kovalev, Y.Y., Anderson, J., Bach, U., ... Voytsik, P.A.: RadioAstron space VLBI imaging of polarized radio emission in the high-redshift quasar 0642+449 at 1.6 GHz. Astron. Astrophys., 583 (2015). https://doi.org/10.1051/0004-6361/201526335

Lyne, A., Graham-Smith, F.: 1 The discovery of pulsars. In Pulsar Astronomy, pp. 1–15. Cambridge University Press (2012). https://doi.org/10.1017/CBO9780511844584.002

Mandow, R.: "What is that?"—Astronomers Find a New Odd Radio Circle IN SPACE (2021). Retrieved from Space Australia: https://spaceaustralia.com/feature/what-astronomers-find-new-odd-radio-circle-space

Norris, R., Intema, H., Kapińska, A., Koribalski, B., Lenc, E., Rudnick, L., . . . Whiting, M.: Unexpected circular radio objects at high galactic latitude. Publ. Astron. Soc. Australia, 38 (2020).https://doi.org/10.1017/pasa.2020.52

Scaife, A.M.: Big telescope, big data: towards exascale with the Square Kilometre Array. Philosophical Transactions of the Royal Society A: Mathematical, Physical and Engineering Sciences, 378(2166) (2020). https://doi.org/10.1098/rsta.2019.0060

SKAO: One stage to go before Australia's ASKAP begins full survey science (2021). Retrieved from issuu: https://issuu.com/ska_telescope/docs/contact_-_issue_09/s/14022788

Stanimirovic, S.: Single-Dish Radio Astronomy: Techniques and Applications : Proceedings of the Naic-Nrao Summer School Held at National Astronomy and Ionosphere Center, Arecibo Observatory, arecib. Astronomical Society of the pacific (2002)

Steel, W.H.: Interferometry. Cambridge University Press (1985)

Stentiford, H.: A brief history: What we know so far about fast radio bursts across the universe (2021). Retrieved from CSIRO scope: https://blog.csiro.au/what-we-know-so-far-about-fast-radio-bursts-across-the-universe/

Tingay, S.: Realisation of a low frequency SKA precursor: the murchison widefield array. Proc. Sci. 163 (2012). https://doi.org/10.22323/1.163.0036

Verschuur, G.: The Invisible Universe: The Story of Radio Astronomy. Springer Nature (1987)

Wang, F.: China SKA Scientific Report (Selected) (2020). Retrieved from Science Press: https://blog.sciencenet.cn/blog-528739-1214808.html

Western Sydney University: Dancing ghosts point to new discoveries in the cosmos (2021). Retrieved from phys.org: https://phys.org/news/2021-08-ghosts-discoveries-cosmos.html

Wilkinson, D.T., Peebles, P.J.: Discovery of the cosmic microwave background. Part. Phys. Univ., 136–141 (2001).https://doi.org/10.1142/9789812810434_0016

Willett, K.: Radio galaxy zoo: Host galaxies and radio morphologies for large surveys from visual inspection. Proceedings of The Many Facets of Extragalactic Radio Surveys: Towards New Scientific Challenges — PoS(EXTRA-RADSUR2015) (2016). https://doi.org/10.22323/1.267.0008

Woods, P.: Odd radio circles ring galaxies. Nat. Astron. **6**(4), 419–419 (2022). https://doi.org/10.1038/s41550-022-01663-6

Woudt, P.A., Fender, R.P., Armstrong, R.P., Carignan, C.: Early science with the Karoo Array Telescope test array KAT-7. S. Afr. J. Sci. **109**(7/8), 2 (2013). https://doi.org/10.1590/sajs.2013/a0030

14

Viking Program: Exploration of the Red Planet

Viking Program

The NASA JPL's Theodore von Kármán Auditorium was packed from the get-go on July 20, 1976. Closed-circuit television was set up in the control room, and 400 journalists from around the world were allowed to watch the event with 1,800 guests (Taback and Goodlette 1992). Mission planner Albert Hibbs (1924–2003) (NASA 2023) was currently in the midst of an explanation. Together, they were anxious about the possibility that the Viking 1 spacecraft, which was currently above Mars, would be able to safely conduct a gentle landing. The designers were fully aware of the difficulties presented by the scheme, given that the Soviets had attempted it four times previously without success.

This extremely challenging task came with an extremely high expense. The price tag for the pair of Viking spacecraft was $1 billion (NASA, Viking 1 Orbiter 2022), or $4.5 billion in today's dollars, 12 times that of Mariner 4. A significant amount of cash was spent by the Americans on the Viking (see Fig. 14.1).

The much-anticipated and long-awaited Viking 1 had been making an attempt to touch down on Mars, but it would take another 19 min for the signal to reach the Earth. Everyone was on edge and impatient during the whole suffocating 19 min. If the landing were successful, Viking 1 would go down in history as the first spacecraft in human history to perform a soft landing on the surface of another planet. Nevertheless, if the landing went unsuccessful, all of the genuine gold and silver would be "wasted."

J. Wang, *Eye Beyond the Sky*, Astronomers' Universe,
https://doi.org/10.1007/978-981-99-9818-0_14

Fig. 14.1 Viking model (*Credit* NASA/JPL-Caltech/University of Arizona)

Let's take a quick look at the exploration of Mars prior to the Viking program before moving further with the story.

Italian astronomer and observatory director Giovanni Schiaparelli (1835–1910) spotted stripes on Mars's surface stretching hundreds of kilometers in 1877. He claimed that "channels," later misidentified as "canals," were discovered on Mars (Schiaparelli 1905).

Percival Lowell (1855–1916) spent 15 years observing Mars in a barren desert and painstakingly constructing a series of maps. He made his first trip there in 1894. From the looks of these images, Mars was home to not only hundreds of "canals," but also to enormous oasis-like regions (Lowell 1906). There had been near-universal agreement, from the layperson to the expert, that Martians resided on the red planet. In those years, interest in Mars had skyrocketed.

Yet, the photographs of Mars taken by Mariner 4 in 1965 let down many observers. A total of twenty-one photos, all in grainy black and white, demonstrated that Mars is a desolate, crater-filled wasteland (Hamilton 1998). It broke the hearts of numerous people who had held onto romantic notions about our Martian neighbors. Imagined to be bustling with life, Mars instead appeared as lifeless as the Moon (see Fig. 14.2).

Scientists, though, were not as disheartened. They were curious as to whether life existed beneath the surface of Mars. To find the answer, NASA invested a significant amount of resources into the development of Viking, the most advanced and complicated planetary probes ever created at the time.

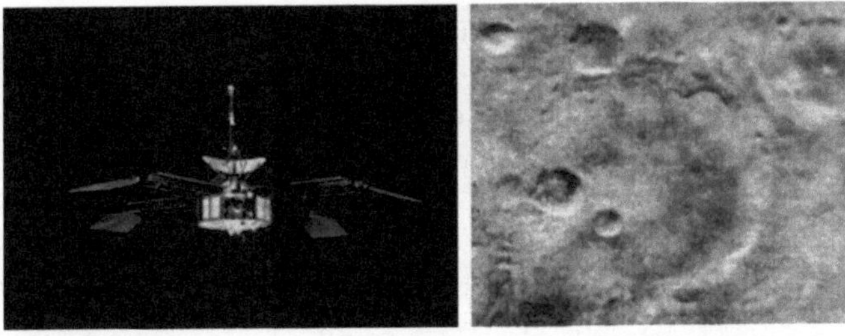

Fig. 14.2 Mariner 4 and one of its images of Mars (*Credit* NASA/JPL)

Viking 1 and 2 serve as the primary focus of this chapter. No matter how well or how poorly they did, it would touch the hearts of countless individuals.

Viking 1 was successfully launched from Cape Canaveral on August 20, 1975, while Viking 2, which served as a double insurance, was launched on September 9, 1975, more than a half month after the initial launch. Each of the two probes consisted of a lander in addition to an orbiter. The two Viking probes arrived over Mars in quick succession after 10 months of traveling in space, prepared to make a landing (Marietta 1975).

On July 20, 1976, Viking 1 came out of the cloudless apricot-yellow sky of Mars and burst into the Martian thin atmosphere, which was glowing with a hot protective layer. The parachute deployed, the thermal shield was jettisoned, and the three landing gears extended like claws at an altitude of around 6.4 km. At a height of around 1.6 km, the recoil rocket ignited, and a minute later, Viking 1 dropped down to a speed of about 10 km per hour, touched the surface of Mars with a small wobble, and fell straight to the Chryse Planitia, or the Golden Plain (Howell 2012).

Success for the Viking! Since this is the first time humans have ever successfully landed a spacecraft on the surface of another planet, it represents a watershed moment in the development of space travel. This timeless picture is still the subject of conversations decades after its initial release (see Fig. 14.3).

The anxious scientists in the control room, as well as the reporters, guests, and general public waiting outside, were ecstatic to receive the signal of success that had been transmitted back to Earth. Everyone gave one another hearty hugs and cheered the accomplishment. The Viking landing significantly stoked the global dream of setting foot on Mars (Atkinson 2015). NASA's Gentry Lee (Stooke 2012) commented about his feelings at the time: "The Viking team did not grasp the atmosphere of Mars, we knew nothing about the surface and rock state of Mars, and even our soft landing was exceedingly irresponsible. We were terrified and thrilled at the same time, and when we realized the landing had gone off without a hitch, our joy and relief overflowed all at once" (see Fig. 14.4).

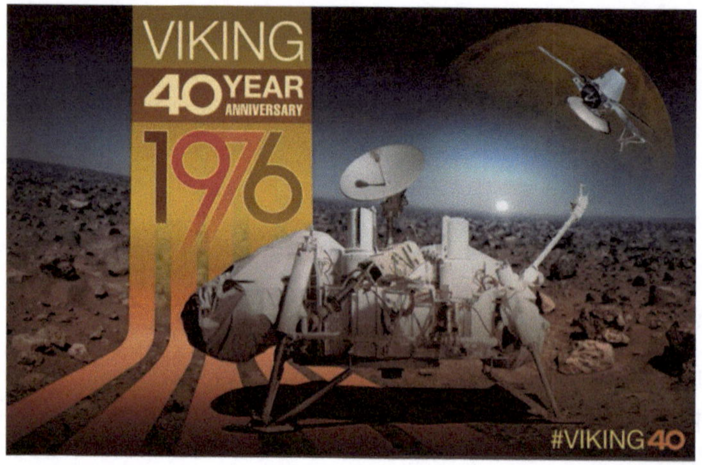

Fig. 14.3 In 2016, the 40th anniversary of the Viking landing was commemorated. (*Credit* NASA)

Fig. 14.4 Photo taken by Viking 2 after its landing (*Credit* NASA/JPL)

Although the Viking hardware system was only intended to endure 90 days, it had been highly durable. Both Viking orbiters continued functioning for extended periods of time; Viking 1 for about two years and Viking 2 for over four. The two landers did even better on Mars' surface; Viking 1 was still going strong after six years until a minor human error in a software update led the antenna to retract and lose contact with the Earth; Viking 2 quit functioning after three and a half years (see Fig. 14.5).

What had the Viking probes been doing all this time on Mars? The advancements made in their work over the past few years could be simply summed up in three aspects. First, take photographs of the Martian surface to reconstruct the planet's original topography. Then, investigate whether there was evidence of water on the surface of Mars. Finally, seek indicators of whether there was life on Mars. The final issue, which concerned the investigation of whether or not Mars ever supported life, stirred up debate as soon as the official finding was made public (Soffen 1976).

Fig. 14.5 Viking 2 working on the Martian surface (*Credit* NASA/JPL-Caltech)

Let's begin with the first aspect, which is photographing the Martian surface. Two orbiters were mostly responsible for this. Both orbiters diligently worked to capture images of Mars from orbit, sending back over 50,000 pictures that collectively covered 97% of the planet's surface in visible and infrared light. These photographs had good resolution as well, allowing us to make out details on Mars's surface at a range between 150 and 300 m. These two sets of "eyes" allow us to observe Mars's many geological characteristics, such as its enormous volcanoes, wrinkly lava plains, steep canyons, and numerous wind erosion structures. This demonstrates that Mars has a geologically complex and varied environment (Kieffer 1992) (see Figs. 14.6, 14.7, 14.8 and 14.9).

The second is proof of water on Mars, discovered by Vikings. Regrettably, Vikings discovered signs indicating that water existed on Mars in the past but not in the present. Mars has a thin atmosphere that is only about 1% as dense as Earth's, and it is also extremely freezing. In just a few seconds, a glass of water left on Mars's surface will evaporate. The Vikings' photographs show enormous river valleys with several elements resembling cobwebbed rivers.

Fig. 14.6 The first human image of the Martian surface was sent back by Viking 1 shortly after the soft landing (*Credit* NASA/JPL)

Fig. 14.7 Sunset on Mars photographed by Viking 1 (*Credit* NASA/JPL)

Fig. 14.8 Sunrise on Mars captured by Viking 2 (*Credit* NASA/JPL/LaRC)

Several grooves, resembling those eroded by water on Earth, can also be seen on the side walls of volcanoes. These facts point to the conclusion that water was once present on Mars (Baker 1982) (see Figs. 14.10 and 14.11).

Finally, Vikings' biological research packages were primarily responsible for searching for signs of life on the Martian surface. The Vikings' robotic arms collected soil samples from the Martian surface and then subjected them to a series of four separate experiments. Experiment data are available to the public, but how those data should be interpreted has become a hotly debated topic in the scientific community (see Fig. 14.12).

The first experiment involved heating samples of Martian soil to varying temperatures and then using the gas chromatography-mass spectrometer included in the research package to determine the molecular weight of different components in the released gas, down to concentrations of a few

Fig. 14.9 A map of Mars based on images sent back by Viking 1 (*Credit* NASA/JPL-Caltech)

Fig. 14.10 Bahram Vallis, an ancient river valley observed by Viking (*Credit* NASA)

parts per billion. No carbon-based or organic compounds were detected in the analysis. There is no convincing evidence that life ever existed on Mars.

The second was a gas exchange experiment, in which nutrient solution and water were administered to the soil and the concentrations of oxygen, methane, and other gases in the sealed container were measured and analyzed using a gas chromatography-mass spectrometer. These gas concentrations might shift if Martian life were present, but there was no sign of any such shift.

A pyrolysis release experiment was the fourth one. In the experiment, Martian soil was placed in a container with water and a mixture of carbon monoxide and carbon dioxide containing radioactive carbon, and then exposed to light. The radioactive carbon was thought to be absorbed by any

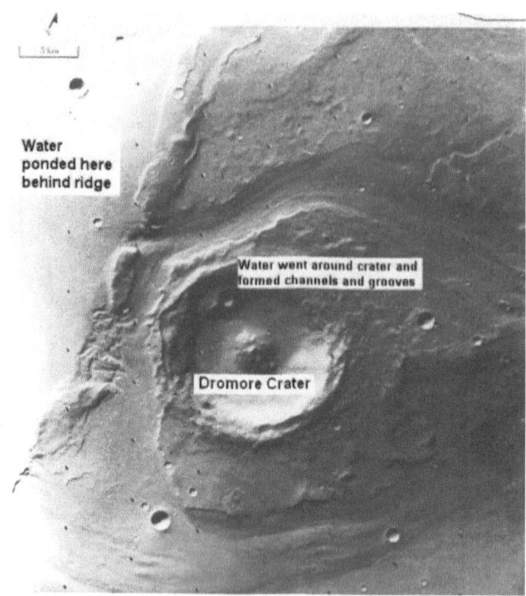

Fig. 14.11 Images captured by Viking show that it takes a lot of water to form similar erosion (*Credit* NASA)

Fig. 14.12 A rock sample from Viking 2 (*Credit* NASA/JPL)

photosynthesizing life in the Martian soil. The soil samples were baked at 650°C after being incubated for several days with the gas removed from the container. In the event that photosynthetic life did exist in the Martian soil, radioactive carbon-containing gases would be produced during the baking process. The end outcome, though, is awful (Klein 1978).

You might have noticed that I skipped the third experiment entirely. The first two experiments, as well as the fourth, all came to a definitively negative result about the possibility of life on Mars, but the conclusion of the third experiment has been open to a wide variety of interpretations and turned into the subject of considerable controversy. This is the release test for the marking.

Adding water-soluble nutrients to radioactive carbon-labeled soil samples was all that had been required for this experiment. The unexpected finding of radioactive carbon dioxide in the air above the samples provided further evidence that these nutrients played a role in microbial metabolism and were thus released. The outcome was the same for both Viking landers. A week later, however, when the experiment was repeated, no radiocarbon was detected. As a result, NASA was unable to conclude anything positive from these findings. According to the official statement, "Every experiment has findings, but these results can be interpreted in a variety of ways, and there is no conclusion that there is life on Mars."

Gilbert Levin (1924–2021) was an ardent skeptic regarding the validity of this conclusion. Levin, who was in charge of the Viking mark release studies that year, was adamant that seven out of nine experiments conducted at the two Martian landing sites showed signs of biological activity. His argument lay in that the presence of microbes in the Martian soil samples had been the sole explanation for the marker release experiment's results, and that no single chemical reaction could adequately replicate them (Levin and Straat 1977).

Rafael Navarro-Gonzalez (1959–2021) held the same viewpoints as him (Navarro-González et al. 2010). Rafael and his colleagues went to some of the world's highest and driest regions, like the Atacama Desert in South America and the dry valleys in Antarctica, to replicate the work Vikings accomplished decades ago. An acidic soil containing organic matter and microorganisms was discovered in northern Chile, which had evaded the detection by Viking sensors. In other words, even though Earth's soil is closest to Mars's, Vikings' detectors are not yet sensitive enough to detect organics or living beings.

To what extent, then, can they persuade you to accept their viewpoint? Furthermore, you might not believe that Earth is the primary source of understanding the Vikings' biological studies. In recent decades, we have been continually astounded by life's capacity for change and the sheer variety of Earth's biota. Certain forms of life are able to thrive in extremely acidic or caustic settings, while others can endure extremely high levels of ultraviolet radiation and continue to multiply. Extremophile is a term that has been coined to describe those who can survive in extremely harsh environments (Wharton 2002). This term was not recognized in 1970. This suggests that life on Mars would have been "unconventional" due to the planet's drastically different chemical and physical environment from the Earth's. Vikings' one-size-fits-all detection method might have overlooked life with a unique molecular foundation.

After reading this, you might conclude that the contributions of Spirit and Curiosity considerably outweigh those of Vikings, whereas the contributions of Vikings are nothing spectacular and there are still many debates. If so, you must have a lower opinion of the Viking. The Gaia hypothesis owes a debt of gratitude to the Viking exploration of Mars; it is this mission that made us reflect on the possibility of life on Earth. Some researchers think that the Gaia hypothesis will go down in history as one of the most significant discoveries made in the twentieth century.

In 1965, James Lovelock, a British scientist, worked at the NASA Jet Propulsion Laboratory. While planning the experiment to detect life on Mars for the lander, he had a sudden epiphany: When we want to know if there are microorganisms on Mars, we need to look at the atmosphere, because if a planet supports life, its atmosphere must be changed by the biosphere. Is Earth's atmosphere, then, a natural outgrowth of Earth's biosphere, or at least a by-product of it?

That realization would later form the backbone of the Gaia hypothesis, and his paper detailing the idea would be published in *Nature* that same year (Lovelock 1965). At a later date, he co-proposed the Gaia hypothesis with his fellow biologist Lynn Margulis (1938–2011) (Lovelock and Margulis, Atmospheric homeostasis by and for the biosphere: The gaia hypothesis 1974). Bacteria and other organic substances, according to the Gaia hypothesis, have a much larger impact on Earth than was previously thought. Not only do they consume and replenish oxygen, methane, and other components in the atmosphere, but they also play an important role in the temperature and livability of the biosphere through rock weathering and the cycling of carbon, oxygen, sulfur, and other chemicals. In other words, the Earth's atmosphere is a complex system interacting with the lithosphere, hydrosphere, and biosphere, and organisms have the potential to alter the planet's atmospheric and geological features.

In the early 1970s, it was widely accepted that the atmosphere was mostly unaffected by life because it simply "borrowed" some gases from the atmosphere and then returned them unchanged. In contrast, the Gaia hypothesis holds that the atmosphere is a living extension of the biosphere, which

Fig. 14.13 The Viking data was stored on microfilm, and NASA's archive team is continuing its efforts to digitize it (*Credit* David Williams, NASA)

went against conventional wisdom but is being widely recognized by scientists today. Yet, the Gaia hypothesis is sometimes misunderstood. A popular misconception is that the Earth is a living creature of which we are all a part (Lovelock, The Ages of Gaia: A Biography of Our Living Earth 1995) (see Fig. 14.13).

In any event, the hypothesis suggests that living things have an impact on the surrounding environment. Our carbon footprint has been threatening the world, and Lovelock was terrified that global warming might spell the end of human civilization. After all, Venus is right in front of us as an example of what could happen when the greenhouse effect went out of control and raised temperatures to an extreme where no life could live. A similar scenario could play out on Earth if we did not reduce our carbon emissions (Kasting 1988).

During the 2006 celebration of the Viking program's 30th anniversary, Mars Ares Mission chief scientist Joel Levine (NASA, Joel S. Levine—Research scientist 2009) observed that many scholars were still poring over the mission's 50,000 photographs. According to Levine, they continued to be "one of the most intriguing and productive data sources ever developed. (NASA, Images taken by the Viking Mission n.d.)" Nevertheless, mounting evidence suggests Viking's findings need further investigation, and its legacy will undoubtedly lead to discoveries that take us by surprise.

References

Atkinson, J.: "Old vikings" recall NASA's first successful Mars landing, 3 Aug 2015. NASA: https://www.nasa.gov/langley/old-vikings-recall-nasas-first-successful-mars-landing

Baker, V.R.: The Channels of Mars. Univ of Texas Pr. (1982)

Hamilton, J.: The Mariner Missions to Mars. ABDO & Daughters (1998)

Howell, E.: Viking 1: First U.S. Lander on Mars, 26 Oct 2012. Space: https://www.space.com/18234-viking-1.html

Kasting, J.F.: Runaway and moist greenhouse atmospheres and the evolution of Earth and Venus. Icarus **74**(3), 472–494 (1988). https://doi.org/10.1016/0019-1035(88)90116-9

Kieffer, H.H.: Mars: Maps. University of Arizona Press (1992)

Klein, H.P.: The Viking biological experiments on Mars. Icarus **34**(3), 666–674 (1978). https://doi.org/10.1016/0019-1035(78)90053-2

Levin, G.V., Straat, P.A.: Recent results from the Viking labeled release experiment on Mars. J. Geophys. Res. **82**(28), 4663–4667 (1977). https://doi.org/10.1029/js082i028p04663

Lovelock, J.E.: A physical basis for life detection experiments. Nature **207**(4997), 568–570 (1965). https://doi.org/10.1038/207568a0

Lovelock, J.E.: The Ages of Gaia: A Biography of Our Living Earth. W. W. Norton & Company (1995)

Lovelock, J.E., Margulis, L.: Atmospheric homeostasis by and for the biosphere: the Gaia hypothesis. Tellus **26**(1–2), 2–10 (1974). https://doi.org/10.1111/j.2153-3490.1974.tb01946.x

Lowell, P.: Mars and Its Canals. The Macmillian Company, London (1906)

Marietta, M.: Viking Mission to Mars (The). Martin Marietta Corp. (1975)

NASA: Joel S. Levine—Research scientist, 9 Nov 2009. NASA: https://www.nasa.gov/audience/foreducators/robotics/careercorner/Joel_Levine.html

NASA: Viking 1 Orbiter (2022). NASA—NSSDCA: https://nssdc.gsfc.nasa.gov/nmc/spacecraft/display.action?id=1975-075A. Accessed 14 Apr 2023

NASA: Al Hibbs (1924–2003) (2023). NASA: https://solarsystem.nasa.gov/people/860/al-hibbs-1924-2003/. Accessed 14 Apr 2023

NASA: Images taken by the Viking Mission (n.d.). NASA: https://photojournal.jpl.nasa.gov/mission/Viking. Accessed 14 Apr 2023

Navarro-González, R., Vargas, E., de la Rosa, J., Raga, A., McKay, C.P.: Reanalysis of the Viking results suggests perchlorate and organics at midlatitudes on Mars. J. Geophys. Res. **115**(E12) (2010). https://doi.org/10.1029/2010je003599

Schiaparelli, G.V.: Astronomy in the Old Testament. Clarendon Press, Oxford (1905)

Soffen, G.A.: Scientific results of the Viking missions. Science **194**(4271), 1274–1276 (1976). https://doi.org/10.1126/science.194.4271.1274

Stooke, P.J.: The International Atlas of Mars Exploration: Volume 1, 1953 to 2003: The First Five Decades. Cambridge University Press (2012)

Taback, I., Goodlette, J.: The Viking lander—then and now. In: Mars: Past, Present, and Future, pp. 37–51. American Institute of Aeronautics and Astronautics (1992). https://doi.org/10.2514/5.9781600866173.0037.0051

Wharton, D.A.: Introduction: extreme life. In: Life at the Limits, pp. 1–26. Cambridge University Press (2002). https://doi.org/10.1017/cbo9780511541568.002

15

Mars Exploration Rovers (MER): Spirit and Opportunity Outweighed Their Capabilities

Mars Exploration Rovers (MER)

There was a room at Cornell's Clark Hall known as the Mars Room where rolls of photographic paper were stacked haphazardly in 1977. Graduate student Steve Squyres casually opened one of the rolls in order to rush through it and finish the assignment. But the unusual scenes depicted in the photographic paper immediately piqued his interest. For the next four hours, he studied the pictures of the Martian landscape transmitted back by Vikings like a man possessed. In retrospect, he saw a beautiful and frightening barren environment in these photographs, which he says changed the path of his life. "I left that room knowing exactly what I had to accomplish with my life. (Squyres 2005)" (Fig. 15.1)

After working in the field for over twenty years, Squyres has risen to the position of Scientific Principal Investigator for the Mars Exploration Rovers (MER) Project (Squyres 2023). In fact, it was his Mars expedition that paved the way for the development of the Spirit and Opportunity rovers.

It's hard to believe, but by the end of the 20th century, Earthlings had mapped out Mars more precisely than the Moon. "Now, the deserts of Mars and the rings of Saturn are as recognizable to us as the odd landscapes of our own planet," said Serge Brunier, the famous science writer and astrophotographer (Brunier 2009). We owe a great deal to the spacecrafts Mariner, Viking, the Mars Global Surveyor, and others for providing us with such high-quality images of Mars.

What were the following steps that humans needed to take on Mars? Obviously, we were on the lookout for life. Even if there is no water on the

J. Wang, *Eye Beyond the Sky*, Astronomers' Universe,
https://doi.org/10.1007/978-981-99-9818-0_15

Fig. 15.1 Image of the Martian landscape from Viking 2 (*Credit* NASA, The Viking Project, M. Dale-Bannister (Washington University))

surface of Mars presently, there may still be water in the past, especially liquid water that can support life, so "tracking water" is a crucial strategy in NASA's search for Martian life (Petranek 2015). Spirit and Opportunity, the story's protagonists, undertook this mission (Fig. 15.2).

Similar to their Viking forebears, Spirit and Opportunity were only expected to last for 90 days, yet they had far outlived their projections. They explored Mars for over ten years like two robotic geologists. Their "accounts" of daring deeds have greatly exceeded the creators' wildest dreams. Their exploration itself, together with the events and hardships they encountered, have entered popular consciousness and folklore.

After traveling 480 million kilometers, Spirit slammed into the Martian atmosphere on January 4, 2004, at 19,200 km/h, or about 16 times the speed of sound. Then the "six minutes of terror" occurred as the spaceship made contact with the hot Martian atmosphere. The thrust reverser kicked in and the parachute deployed about 8 km above the Martian surface to halt the lander's descent. And then a picture emerged that looked quite cute. Spirit's six enormous airbags, popping up on each of its four sides, tightly wrapped it like a bunch of big grapes, slammed into the Martian ground and bounced back up. Spirit successfully landed in Gusev Crater after falling and bouncing more than a dozen times (The Azza 2014)! Around two hundred scientists,

Fig. 15.2 Spirit/Opportunity (*Credit* NASA/JPL/Cornell University, Maas Digital LLC)

engineers, and NASA executives erupted in joy in the Pasadena control room (Travis 2010) (Fig. 15.3).

Opportunity arrived at its destination, Meridiani Planum, on the opposite side of Mars three weeks later (Greicius 2017). With the aid of airbags, it gleefully made a "hole-in-one" into Eagle Crater, a crater roughly 22 meters in diameter. It's thrilling to imagine how these two "robot geologists" were about to set off on their first adventure. Let's first take a look at the gear they were carrying before probing into their journey on Mars (Fig. 15.4).

The eyes are the most important sense organ for a geologist working in the field. For "Mars geologists," this is especially true because of the extreme lack of oxygen on the red planet, the absence of taste and smell, and the major role played by the eyes in forming impressions of the environment. The rovers' "eyes," which resemble the eyes of a crab, are mounted on the top of the mast, stand roughly 1.5 meters above the ground, and can rotate in any direction. An immersive stereoscopic picture can be created by combining the outputs of the two cameras (Bell III et al. 2003). Each weighs only 250 grams and can be held in the palm of a man's hand, yet they each have 16 megapixels (Figs. 15.5 and 15.6).

Geologists also need a strong and adaptable set of arms. Both Spirit and Opportunity are "one-armed heroes," but this "arm" is exceptionally

Fig. 15.3 Spirit's "six minutes of terror" (*Credit* NASA)

powerful, having one elbow, one wrist, and four movement modes. Moreover, each "fist" on this "arm" bears a miniature camera, similar to how a geologist might hold a magnifying glass (Herkenhoff n.d.). It goes without saying that having a hammer in one's hand is also essential, and each of their arms has a device on it called a RAT (Rock Abrasion Tool) (Gorevan n.d.). A hole 5 cm broad and 0.5 cm deep can be ground using RAT's diamond-encrusted rock grinder in just two hours, despite the target could be the hardest volcanic rock. After saying "arms", say "legs". Both Spirit and Opportunity have a total of six wheels, each of which is independently driven. This offers the rovers greater adaptability over Mars' varied terrains (Lindemann and Voorhees 2005). The worst thing that can happen to a pricey rover billions of miles from "home" is for it to fall on all fours, at which point

Fig. 15.4 Victoria Crater on Meridiani Planum, where Opportunity luckily bounced away on landing (*Credit* NASA/JPL/UA)

Fig. 15.5 Schematic diagram of Spirit/Opportunity (*Credit* NASA/JPL-Caltech)

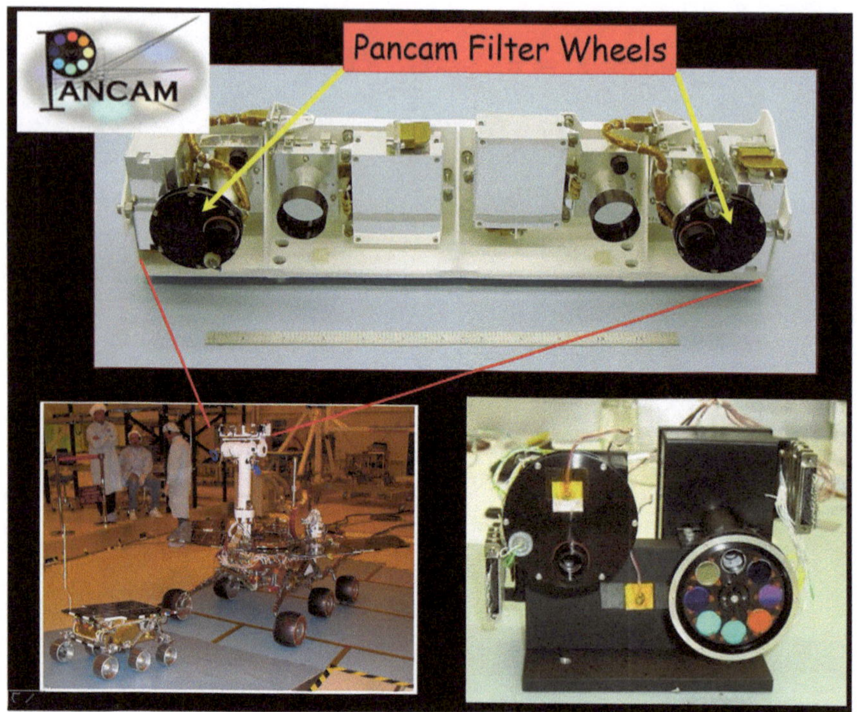

Fig. 15.6 "Eyes" of Spirit and Opportunity (*Credit* NASA)

it will likely become a heap of scrap metal that cannot be relocated in an instant (Figs. 15.7 and 15.8).

One possible mental image of a NASA engineer is one in which his hat goes awry and he frantically smashes the joystick from left to right as the rover begins to roll down a Martian dune. A vivid imagination isn't a good indicator of how serious a situation actually is. On level terrain, the rover can only move as fast as 5 centimeters per second, making its maximum speed at 3 meters per minute (Squyres et al. 2003). The rover's already slow speed is compounded by the fact that it must stop every few seconds to update the hedge software. Here on Earth, a snail might be able to outclimb it in a race to the top.

Spirit and Opportunity had finally begun moving at a snail's pace. As the trip progressed, though, they learned firsthand the weight of decision-making. Although Spirit and Opportunity share the same "gene" in hardware, they have met very different fates due to their locations. One was offered candy, while the other was offered a punch to the face.

Fig. 15.7 Spirit/Opportunity's "Arm" being debugged before expedition (*Credit* NASA/JPL)

Fig. 15.8 "Legs" of Spirit and Opportunity (*Credit* NASA)

In time, Opportunity and Spirit were dubbed "Lucky" Opportunity and "Intrepid" Spirit. Even in their monikers, we can perceive the stark disparity in their destinies (Fig. 15.9).

Opportunity's great life began after the "hole in one." It had only landed when it discovered signs of water on the rocks nearby, and it had been ordered to inspect its thermal shield when it came across a basketball-sized iron-nickel meteorite. The honor of being the first to find meteorites on another planet was readily won (NASA, Iron meteorite on Mars – NASA Mars Exploration 2004) (Fig. 15.10).

Fig. 15.9 Rear view of Opportunity in motion (*Credit* NASA/JPL-Caltech)

Fig. 15.10 On January 6, 2005, Opportunity used its panoramic camera to take a composite image of a near-true color iron-nickel meteorite. (*Credit* NASA/JPL-Caltech/Cornell)

Its good fortune didn't end there. Due to the dust on its solar panels, Opportunity's energy output was severely reduced in 2007 when a massive dust storm on Mars blocked nearly all of the Sun's light. With the help of the wind, Opportunity's solar panels were cleaned and brought back to life, giving the rover a fresh beginning (Spector 2014).

In 2011, Opportunity found a bright dike of hydrated calcium sulfate and a zinc-rich region on an outcrop near the rim of Endeavor Crater (NASA, NASA Mars Rover Finds Mineral Vein Deposited by Water 2011). "The clearest evidence we have uncovered of the existence of liquid water in eight years' exploration," as Squyres put it, best describes the discovery. After 15 years of operation on Mars, the Lucky Opportunity rover finally bid farewell to us in 2019 (Fig. 15.11).

Compared to Opportunity, the path that Spirit had taken would be far more turbulent. Tears abound, as the saying goes.

When Spirit ceased communicating with the mission control system and entered a "restart loop" less than three weeks after landing, it was the first of many mishaps. In fact, something went wrong during the restart, and it had to be done multiple times. The "culprit" was caught, thank goodness. It was a flaw in the flash memory that caused the issue. With time and effort put into treatment, the issue was resolved (WebManager 2004).

It wasn't just the software on Spirit that failed; the hardware did as well. The task must be completed despite the frequent failure of a single wheel. Its only option was to take its tiny hammer, RAT, and ascend the Husband Hill that was higher than the Statue of Liberty. Even the diamond drill head wore

Fig. 15.11 Bright dikes of hydrous calcium sulfate discovered by Opportunity in November 2011 (*Credit* NASA/JPL-Caltech/Cornell/ASU)

Fig. 15.12 Husband Hill captured by Spirit's panoramic camera (*Credit* NASA)

out in this drilling mission. Spirit's front wheels stopped working entirely in 2006, just two years after the rover's arrival on Mars. After that, all it could do was dragging the ineffective wheels and slog through the Martian soil (Petit 2006) (Fig. 15.12).

The worst accident in Spirit's journey happened three years later. Spirit dove into a stretch of white, soft sand on its left flank as it plowed through a dark layer of soil near the base of the Columbia Hills. It was a low-cohesion iron sulfate that was not meant to trap the wheel, but Spirit got stuck in it nevertheless. On November 21, 2009, the operator gave it several orders to advance 5 meters, but it advanced a total of 0.25 cm, essentially the same as not moving at all. As the system continued to age, Spirit started to suffer from memory loss. After sending 1,300 instructions with no response, the rescue attempt was called off on May 25, 2011 (NASA, Mars rover spirit - Mars missions - NASA Jet Propulsion Laboratory n.d.). Finally, our disturbed and ailing Spirit could rest in peace.

The rover's operator can also take a break and relax. They are the true "Martian geologists" working behind the scenes. Spirit and Opportunity have 14 operators, including not only youthful people but also a sizable number of women, dispelling the common misconception that the scientists who drive the rovers are scruffy and unwashed (Jankvist and Toldbod 2007).

Whether they set out to measure a stone of interest or conduct a patrol around an obstacle, they will first assess the outcomes of yesterday's efforts and then formulate a strategy for today's endeavors. The next step, after planning, is to convert those tactics into commands that can be "understood" by the rovers.

Creating commands is more difficult than it seems since it requires simulation, which is reviewed by a team of scientists to eliminate situations that could lead to errors. The list of instructions will be validated twice before it is transmitted to the rover. The stakes are extremely high, and failure can have devastating consequences. Use extreme caution.

And what exactly does it mean when I tell you that a Martian day is approximately forty minutes longer than a day on Earth? The operator will experience a unique "Mars jet lag" in which his or her daily work schedule will begin 40 minutes later than the day before, eventually shifting to midnight, 2 a.m., and 4 a.m in a gradual manner. (Mirmalek 2002) "It's unhealthy for your body, awful for your intellect, and horrible for social interactions," Scott Maxwell, the operator, joked.

After reading this, I bet you're anxious to learn what Spirit and Opportunity have accomplished. They're heading for the "water," presumably. What major findings have been made? There are, I'm happy to report, not just some discoveries, but a great number.

Opportunity has already proven, just a few weeks after landing on Mars, that the Meridian Plateau was originally a wet plain. A rock hurled from the landing site, El Capitan, which possesses a wealth of maritime history, serves as proof. Opportunity also discovered a plethora of tiny, hard rock balls, nicknamed "blueberries," that are dispersed across the Meridian Plateau's surface and can clump together to form larger rocks. Hematite, the iron-rich blueberry's main component, can only occur in the presence of water on Earth (Malik 2004). In time, Opportunity found jarosite, a mineral found exclusively in acidic water on Earth (Elwood Madden et al. 2004) (Figs. 15.13 and 15.14).

Fig. 15.13 Opportunity demonstrated that the Meridian Plateau was originally a wet plain

Fig. 15.14 Blueberry, discovered by Opportunity in 2004. Image width is 3 cm. (*Credit* NASA/JPL-Caltech/Cornell/USGS)

During its history, Spirit has made a number of similar discoveries. The mineral goethite, which can only be produced on Earth in the presence of water, was discovered in a stone block name Clovis (NASA, Deep hole in "Clovis" – NASA Mars Exploration 2004). A high proportion of salt was also detected by Spirit in a soil sample, which provides further circumstantial evidence for the existence of water on Mars (Sullivan et al., 2008) (Fig. 15.15).

Evidence both vast and little lends weight to the theory that water once flowed freely throughout ancient Mars' surface.

While the notion that water once flowed on Mars may sound exciting, the reality is that it has become a source of headache for scientists due to the additional complexity and difficulty this notion has introduced.

This is because climate models and data from Martian meteorites disprove the theory that the planet's surface was "warm and moist" a billion years ago. Previous findings not only fail to provide supporting evidence, but also contradict one another. In order to melt the water on the surface of Mars, the greenhouse effect needs to raise the Martian temperature to at least 65 °C (or higher; remember that the early Sun was much less bright than it is now). While carbon dioxide might have been in Mars' early atmosphere, it clearly wouldn't have been enough on its own; the greenhouse effect would only be noticeable if the Martian atmosphere were significantly denser. Given the

Fig. 15.15 In August 2004, Spirit ground a 9-mm hole in a rock called Clovis, the deepest hole ever ground in Martian rock. (*Credit* NASA/JPL-Caltech/Cornell)

current state of knowledge, this seems highly improbable. Hence, the reason for the climate shift on Mars remains unknown (Haberle, Clancy, Forget, Smith, and Zurek 2017).

However, Mars remains the only planet in the solar system where humans are expected to leave permanent footprints in the not-too-distant future. There are still many mysteries on Mars, and this has ignited the interest of humans in investigating the Red Planet. Not only do scientists hope for this, but the general public eagerly anticipates it (Fig. 15.16).

As Squyres put it, “Spirit and Opportunity are our replacements, our robotic forerunners, and one day astronauts will tread the boot marks on the tire ruts in Eagle Crater.”

China's Tianwen-1 Mars rover was launched on July 23, 2020. Our Zhurong rover has been operational for longer than its 90-day planned life would have allowed (The Planetary Society n.d.), and has already accomplished a great deal, including mapping new terrain, identifying previously unknown types of rocks and soil, and even monitoring the Martian atmosphere. Zhurong's efforts are much appreciated, and we look forward to the day when the red flag with five stars is erected on Mars.

Fig. 15.16 After Spirit and Opportunity, Curiosity has been exploring Mars since August 2012. (*Credit* NASA/JPL-Caltech)

References

Bell III, J.F., Squyres, S.W., Herkenhoff, K.E., Maki, J.N., Arneson, H.M., Brown, D., … Scherr, L.: Mars exploration rover athena panoramic camera (Pancam) investigation. J. Geophys. Res. Planets **108**(E12) (2003). https://doi.org/10.1029/2003je002070

Brunier, S.: Astrophotographer Serge Brunier (2009). Retrieved from ESO: https://www.eso.org/public/images/eso0932b/

Elwood Madden, M.E., Bodnar, R.J., Rimstidt, J.D.: Jarosite as an indicator of water-limited chemical weathering on Mars. Nature **431**, 821–823 (2004). https://doi.org/10.1038/nature02971

Gorevan, S.: Rock abrasion tool (RAT)—NASA (n.d.). Retrieved April 15, 2023, from NASA Mars Exploration Rovers: https://mars.nasa.gov/mer/mission/instruments/rat/

Greicius, T.: Rover's Landing hardware at Eagle Crater, Mars (2017). Retrieved from NASA: https://www.nasa.gov/image-feature/jpl/pia21494/rovers-landing-hardware-at-eagle-crater-mars

Haberle, R.M., Clancy, R.T., Forget, F., Smith, M.D., Zurek, R.W.: The atmosphere and climate of mars. Cambridge University Press (2017)

Herkenhoff, K.: Microscopic imager (MI)—NASA (n.d.). Retrieved April 15, 2023, from NASA Mars Exploration Rovers: https://mars.nasa.gov/mer/mission/instruments/mi/

Jankvist, U.T., Toldbod, B.: Mars exploration rover: mathematics and people behind the mission. Math. Enthusiast **4**(2), 154–173 (2007). https://doi.org/10.54870/1551-3440.1068

Lindemann, R.A., Voorhees, C.J.: Mars Exploration Rover mobility assembly design, test and performance. 2005 IEEE International Conference on Systems, Man and Cybernetics, 1, 450–455 (2005). https://doi.org/10.1109/icsmc.2005.1571187

Malik, B.T.: Rover reveals Mars was once wet enough for life (2004). Retrieved from NBC News: https://www.nbcnews.com/id/wbna4202901

Mirmalek, Z.: Making Time on Mars. MIT Press (2002)

NASA: Deep hole in "Clovis"—NASA Mars Exploration (2004). Retrieved from NASA Mars Exploration: https://mars.nasa.gov/resources/5820/deep-hole-in-clovis/

NASA: Iron meteorite on Mars—NASA Mars Exploration (2005). Retrieved April 15, 2023, from NASA Mars Exploration: https://mars.nasa.gov/resources/5854/iron-meteorite-on-mars/?site=msl

NASA: NASA Mars Rover Finds Mineral Vein Deposited by Water (2011). Retrieved from NASA: https://www.nasa.gov/mission_pages/mer/news/mer20111207.html

NASA: Mars rover spirit—Mars missions—NASA Jet Propulsion Laboratory (n.d.). Retrieved April 15, 2023, from NASA Jet Propulsion Laboratory (JPL): https://www.jpl.nasa.gov/missions/mars-exploration-rover-spirit-mer-spirit

Petit, C.: AP: Spirit, the Mars Rover with a bad wheel, can't reach preferred winter haven (2006). Retrieved from Knight Science Journalism: https://ksj.mit.edu/tracker-archive/ap-spirit-mars-rover-bad-wheel-cant-reac/

Petranek, S.: How we'll live on Mars. Simon and Schuster (2015)

Spector, D.: The opportunity rover looks nearly unrecognizable after 10 years on Mars (2014). Retrieved from Insider: https://www.businessinsider.com/the-opportunity-rover-solar-panels-dust-covered-2014-1

Squyres, S.: Roving mars: Spirit, Opportunity, and the exploration of the red planet. Hachette Books (2005)

Squyres, S.W.: Steven W. Squyres (2023). Retrieved April 15, 2023, from Department of Astronomy—Cornell University: https://astro.cornell.edu/steven-w-squyres

Squyres, S.W., Arvidson, R.E., Baumgartner, E.T., III, J.F., Christensen, P.R., Gorevan, S., ... Romero, R.A.: Athena Mars rover science investigation. J. Geophys. Res. Planets **108**(E12) (2003). https://doi.org/10.1029/2003je002121

Sullivan, R., Arvidson, R., Bell III, J.F., Gellert, R., Golombek, M., Greeley, R., ... Wray, J.: Wind-driven particle mobility on mars: insights from mars exploration rover observations at "El Dorado" and surroundings at Gusev Crater. J. Geophys. Res. **113**(E6) (2008). https://doi.org/10.1029/2008je003101

The Azza: How we landed on Mars with NASA Spirit [Video] (2014). Retrieved from YouTube: https://www.youtube.com/watch?v=6t3IARmIdOI&t=82s

The Planetary Society: Tianwen-1 and Zhurong, China's Mars orbiter and rover (n.d.). Retrieved April 15, 2023, from The Planetary Society: https://www.planetary.org/space-missions/tianwen-1

Travis, M.: Highlights from landing day of Mars rover Spirit [Video] (2010). Retrieved from YouTube: https://www.youtube.com/watch?v=MWk-umZm86U&t=10s

WebManager: NASA gets new signal from Mars rover (2004). Retrieved from WLTX: https://www.wltx.com/article/news/local/fyi/nasa-gets-new-signal-from-mars-rover/101-384244577

16

Voyager Program: Around the Solar System and Beyond

Voyager Program

In February of 1990, while Jurrie van der Woude (1935–2015) (IMDb n.d.), the Image Coordinator of Jet Propulsion's Laboratory's Public Affairs Office (Rayl 2002), was sitting at his desk, he was faced with an issue. Voyager 1 took 60 photographs in the direction of the Sun just now, when it was around 6 billion kilometers away from the Earth, but it was unable to send those photographs back. At the time, the US Deep Space Network was focused on the Magellan and Galileo spacecrafts, while mostly neglecting Voyager 1, a "veteran" spacecraft that had been in operation for 13 years.

You might be dismissing this as little more than a collection of snapshots taken by a "veteran," especially when contrasted to the findings of the most recent scientific research mission. By the end of March, all but six of the photographs had been transmitted to Earth before a downpour disrupted the poor signal. Unfortunately, the antenna hardware failed in April, delaying their return until May. Only one of the most recent six pictures revealed the Earth.

Unexpectedly, this otherwise unremarkable shot became a viral hit after it was released to the public. It was considered by many to be the most significant image that was captured during the Voyager mission, and even one of the most symbolic photos that have been captured during the history of humans venturing into space. The picture was referred to as "Pale Blue Dot (Sagan and Druyan 2011)". In that case, why did this photograph receive such a high rating? Is it even more iconic than the photograph of the first footstep on the Moon? (Fig. 16.1).

J. Wang, *Eye Beyond the Sky*, Astronomers' Universe,
https://doi.org/10.1007/978-981-99-9818-0_16

Fig. 16.1 "Pale Blue Dot" (*Credit* NASA/JPL-Caltech)

Very little was known about the gas giants Saturn and Jupiter prior to the 1970s. Not until their images were taken during flybys by Pioneers 10 and 11 (NASA 2007), did the public's curiosity about space exploration really take off. Although the photographs were still somewhat blurry, they had been much clearer than those captured by ground-based telescopes. However, they revealed only the tip of the iceberg of the peculiar appearance of gaseous planets. Back then, nobody knew for sure if the moons of gas giants were as inhospitable to life as our own or if the gaseous planets' surfaces could reach temperatures of tens of thousands of degrees Celsius. In 1977, NASA despatched a pair of identical spacecrafts to figure it out on human behalf (Miner 2001). The legendary Voyager probes are our main characters in this chapter (Fig. 16.2).

The identical Voyager 2 and Voyager 1 spacecrafts were launched from Cape Canaveral, Florida, only half a month apart on August 20 and September 5, 1977. Voyager 2 launched earlier than Voyager 1. Their names' rule depends on the fact that Voyager 1 arrived at the destination first since it used a shorter and thus faster route (The Planetary Society 2021).

Each probe, when viewed from the outside, resembles a pot lid with several thin, long "arms" dangling from it. Despite appearances, it weighs close to a car's 800 kg despite being much smaller. The "heart" of each space probe is a complicated decahedron housing numerous electronic components. A 3.7-meter-diameter high-gain antenna is mounted before the "heart" to maintain contact with the Earth. There are ten major scientific instruments mounted

Fig. 16.2 Images of Jupiter from Pioneer-10 in 1973 (*Credit* NASA)

on the two booms that reach out from the "heart," including cameras, spectrometers, and instruments for measuring magnetic fields, charged particles, and cosmic rays (JPL n.d.) (Fig. 16.3).

The two Voyager spacecrafts were on an incredible, extremely lengthy journey, and their first challenge was a lack of energy. They were only able to get as close to Jupiter as their rocket, a Titan III-E Centaur, would allow. Next, they would need to travel at a speed of 17.3 km per second to get to Saturn, and 42.1 km per second to escape the solar system entirely (that's about 150,000 km per hour). What are the secrets to reaching such a breakneck velocity?

It's the gravitational slingshot effect, which some readers may have realized already. Engineers at NASA had been planning for a long time that when Voyager arrived at Jupiter, the "big fat man" would be able to "push" them and hurry them up. By flying consecutively over four gas giants, Voyager 2 was able to "rally" the speed and receive a large gravitational boost at each encounter (Peshin 2018). By performing this intricate "gravitational dance," Voyagers have traveled farther than any other human-made aircraft before (NASA, Voyager 1 n.d.) (Fig. 16.4).

Fig. 16.3 Display of Voyager components (*Credit* NASA)

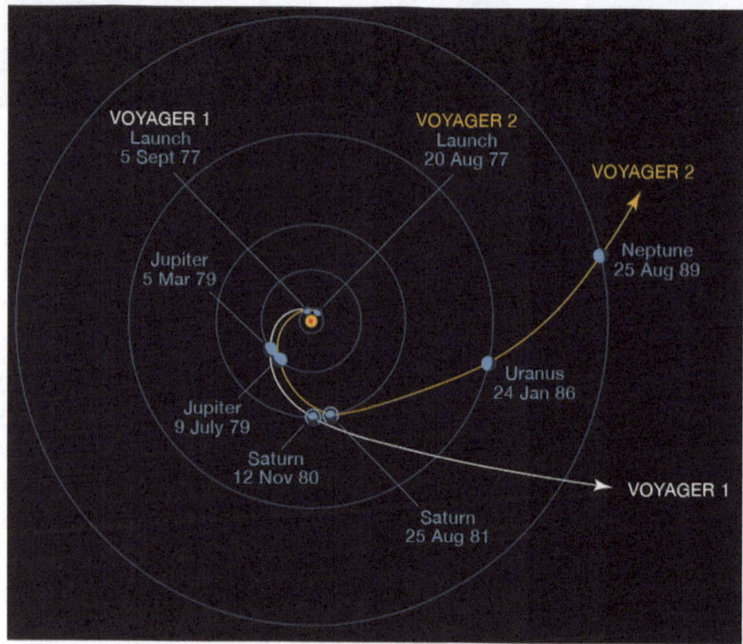

Fig. 16.4 Voyagers' "gravitational dance" (*Credit* NASA/JPL)

Once every 176 years, an opportunity arises to take advantage of the gravitational slingshot effect of four major planets in succession.

Over their lengthy journeys, Voyagers have uncovered numerous scientifically significant findings. Let's begin with Jupiter, which is their initial destination of interest. Many shocks have been delivered to humanity by this gigantic gas planet, which is 320 times (ESA) the mass of Earth.

In 1979, when Voyager 1 and Voyager 2 visited Jupiter, they took pictures of Jupiter's most famous feature: the Great Red Spot. Scientists discovered that the Great Red Spot is a massive anticyclone surrounded by numerous eddies and minor storms and that it is enormous enough to absorb the entire Earth (NASA, Jupiter's Great Red Spot 2008) (Fig. 16.5).

The photographs of Europa they sent back had been the biggest surprise of all. Pixel density is extremely low on the Voyager camera, with only one pixel for every two kilometers. Though the photographs were grainy, they nevertheless shocked scientists: The entire globe was beautifully white and striped, like a cracked egg. The Europa is most likely a big hockey puck made of water ice or dry ice, based on its color (NASA, Europa Clipper Mission n.d.). Further evidence from observations also points to the possibility of an ocean of liquid beneath the ice. People's imaginations cannot help wandering by all the new pieces of information. Due to this "exposure," NASA's attention was suddenly drawn to the previously insignificant Europa (Fig. 16.6).

Fig. 16.5 Jupiter's Great Red Spot recreated from images of Voyager 1 (*Credit* NASA/JPL-Caltech)

Fig. 16.6 Voyager 1 image of Jupiter and its three moons (with the brightest on the far right being Europa) (*Credit* NASA/JPL)

Saturn is the next destination. 500 journalists from across the world covered the "unique event in the history of human space travel" when Voyager 1 made its maiden visit to Saturn in 1980, and an estimated 100 million people watched the live television broadcast. In fact, this strange new globe is the solar system's "star figure," and the mysterious rings are its greatest attraction. Several kinks and pimple-like formations were discovered by Voyager 1 in Saturn's enormous rings, in addition to the expected giant and small circular seams. Even more baffling is the fact that these rings, although being composed of ice and stones, are engaged in a delicate gravitational dance, shifting formation periodically like geese in flight (Briggs 1980). Titan's thick atmosphere was also a surprise to Voyager's scientists as Voyager 1 flew by this moon (NASA, Hazy Titan n.d.). In light of this extraordinary finding, NASA decided to abandon its previous orbit around Uranus and Neptune in favor of the one that passes far closer to Titan (Fig. 16.7).

Voyager 2 was trailing behind as a sort of double insurance policy, which is no doubt a major factor in NASA's final decision. After its stop at Saturn, Voyager 2 went on to Uranus and Neptune.

Fig. 16.7 Voyager 1 image of Saturn taken in 1980 (*Credit* NASA)

Very little was understood about distantly hazy Uranus and Neptune until Voyager 2's groundbreaking flybys (Gray et al. 1982). The only spacecraft to ever pass close to these two faint planets was Voyager 2. What was it able to acquire? According to Voyager 2, the wind speed on Neptune can reach approximately 2,000 km per hour. Equal in size to Jupiter's massive red spot is a similar black spot on Neptune. Voyager 2 also discovered a plethora of intriguing satellites. Miranda, for instance, has a diameter of only 480 km, but it is covered in enormous gorges and tens of kilometers of topography, and it owns some amazing and harmoniously coexistent characteristics from both its ancient and modern eras. For another instance, Triton, the solar system's coldest satellite, has temperatures as low as −235 °C, which is cold enough to solidify any gas (Morris 1986) (Fig. 16.8).

Following Voyager 1's footsteps, Voyager 2 departed Uranus and Neptune and began to "wander" beyond our solar system. This was bound to be a thrilling adventure, since they were going to enter a secret region that had never been reached by aircraft. Although the Pioneer probe was farther from the Sun at the moment, Voyagers would travel further due to their superior

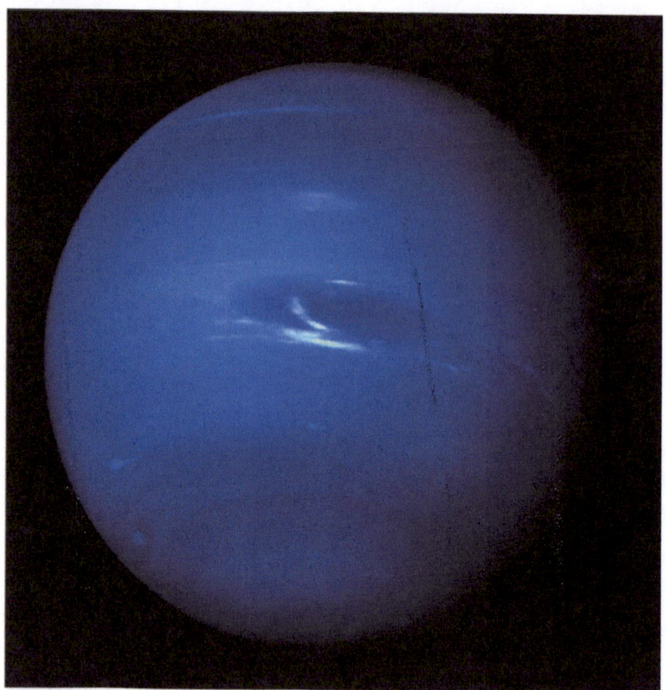

Fig. 16.8 Voyager 2 image of Neptune (*Credit* NASA/JPL-Caltech)

speed. Furthermore, Voyagers marked a turning point in human space exploration because of their continuing contact with the Earth while Pioneer lost contact.

What can be found beyond the solar system? Exactly where does the Solar System end? Opinions on these vary widely. In spite of this, Voyagers did really pass across some sort of mystery boundary. It's no secret that the Sun's upper atmosphere constantly spews out high-energy charged particles in all directions. These solar winds fly through space and over planets at 1.6 million kilometers per hour, yet when they reach a particular spot, they slow down quickly, like hitting a brake. Between the stars and throughout interstellar space is the very thin hydrogen and helium gas. In December of 2004, Voyager 1 made its way through this area; just a few years later, Voyager 2 followed its predecessor into the unknown. Unexpected outcomes came afterward (Matsumoto 2016).

Formerly believed to be a smooth, featureless void, the furthest reaches of our solar system were discovered to be full of chaotic magnetic bubbles measuring around 160 million kilometers across (Stone 2006) (Fig. 16.9).

Six years later, in December 2010, Voyager 1 noticed that the speed of particles in the solar wind surrounding it had reduced to zero. This meant

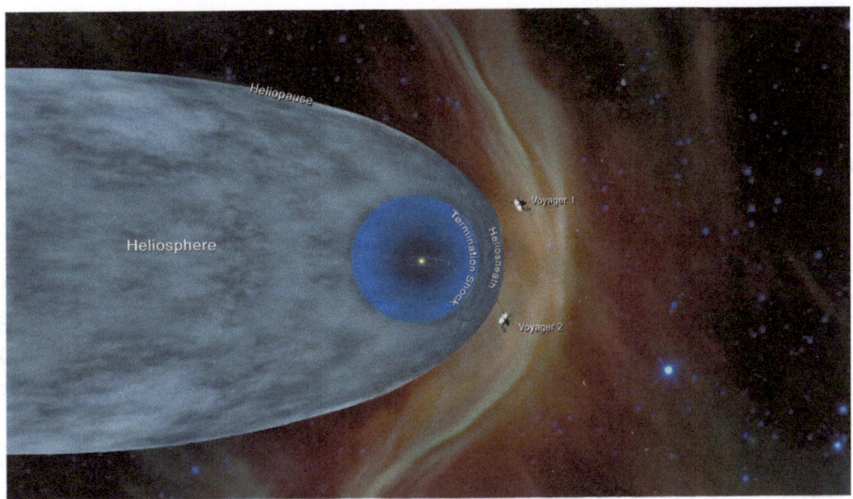

Fig. 16.9 Voyager 1 and Voyager 2 entered interstellar space. (*Credit* NASA/JPL)

that it had reached the edge of a mysterious new world—the vast emptiness between the stars—and solar wind could no longer reach it. Voyager space probes are now well past the age of 40, and they are still galloping. The unexpected turns in this story are not yet over (Hardyman 2016).

I'd like to single out two of Voyagers' many remarkable achievements. The first success is the "Pale Blue Dot" picture that was proposed by Carl Sagan (1934–1996), a member of the Voyager Imaging Science Team and a highly-respected science writer, as shown at the opening of this chapter. Once Voyager left Neptune's orbit and the ecliptic plane, he suggested that it "look back" and snap a photo of the Earth. The phrase "look back" is simple in concept but difficult in practice. Writing commands requires significant mental effort and scarce resources, without guaranteeing a useful scientific outcome. At first, NASA officials were not thrilled. A fortunate turn of events occurred in February 1990, when Voyager 1 completed this seemingly unimportant mission. After traveling around 6 billion kilometers away from the Earth, it turned around and snapped 60 photographs in the direction of the Sun, which took three months to return to Earth. Our home planet Earth can be found in only one of these photos (NASA, Voyager 1's pale blue dot n.d.).

The Earth, like its "Pale Blue Dot" moniker suggests, is obscured by the scattered sunshine against the immense backdrop of the universe. It is unimpressive and relatively faint, occupying such a small space, like a little sesame seed on the blackboard. It is this image that cruelly displays what the Earth truly appears like in the solar system, proclaiming with great visual effect: Hey, human beings, you are too insignificant in the universe (Fig. 16.10).

In light of its potential to alter people's perceptions of the cosmos, this shot has surpassed all others as the most significant event of the Voyager program and maybe the most iconic silhouette in the annals of human space travel (Na 2020).

Similarly well-known is the second accomplishment, the Voyager Golden Record. Humans have given the aliens a genuine, playable, gold-plated copper record as a token of our friendship. Frank Drake was the first to propose using a device like a phonograph record to capture music and photos that could be played via television. If not for physical collisions, the record could survive for a billion years in a vacuum (Fig. 16.11).

Though not the first gift that humanity has ever delivered to aliens, the Golden Record has been the first to include the sounds, music, and images from the Earth. Fascinatingly, it contained greetings in 55 languages or dialects (NASA, Greetings to the Universe in 55 Different Languages n.d.), such as "Hello to you all, and we all miss you." in Mandarin and "Hi, how

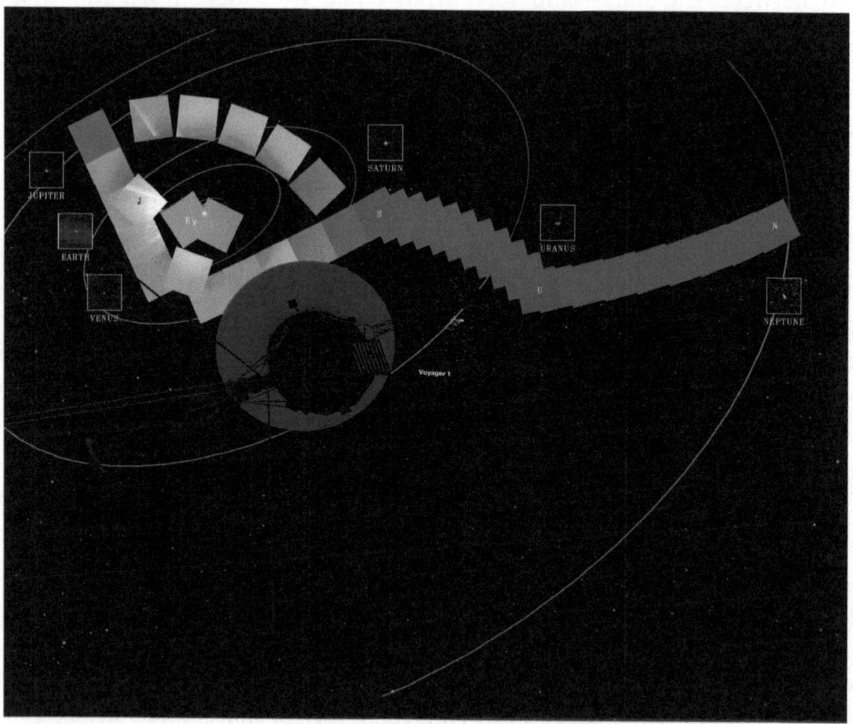

Fig. 16.10 Spots of planets in the solar system corresponding to the pictures taken by Voyager 1 when "looking back" (*Credit* NASA/JPL)

Fig. 16.11 Voyager Golden Record (*Credit* NASA/JPL-Caltech)

are you? I wish you peace, health and happiness." in Cantonese (Wikipedians 2023). Revealing information about the Earth to extraterrestrial civilizations has become increasingly contentious in recent times. But, at the time, nobody saw anything amiss.

The Voyager probes are still traveling through interstellar space, transmitting to us increasingly weak signals. We have high expectations for them, and we know full well that we will probably be let down. The efforts of Voyagers alone have shown the rest of the universe some aspect of humanity: While our bodies are weak and we may perish as a race, our minds are resilient and can maintain optimism even in the face of adversity. It's possible that Homo sapiens' survival depended on an innate lack of fear from the very beginning of our species' existence. Even now, we retain this trait. In this harsh universe, ours is the last vestige of Earthly civilization.

References

Briggs, G.: Voyager 1 encounter with Saturn. EOS, Trans. Am. Geophys. Union **61**(45), 716 (1980). https://doi.org/10.1029/eo061i045p00716-01

ESA. (n.d.): The mass of Jupiter: Calculating the mass of a planet from the motion of its moons. Retrieved April 15, 2023, from CESAR Science Case: https://cesar.esa.int/upload/201811/jupiter_mass_student_guide_gttp.pdf

Gray, D., Cesarone, R., van Allen, R.: Voyager 2 Uranus and Neptune targeting. In: Astrodynamics Conference (1982). https://doi.org/10.2514/6.1982-1476

Hardyman, R.: Voyager probes: Robots on an interstellar mission. The Rosen Publishing Group, Inc. (2016)

IMDb. (n.d.): Jurrie Van Der Woude. Retrieved April 15, 2023, from IMDb: https://www.imdb.com/name/nm1560006/

JPL. (n.d.): Voyager Spacecraft mock-up, full scale. Retrieved April 15, 2023, from National Air and Space Museum: https://airandspace.si.edu/collection-objects/voyager-spacecraft-mock-full-scale/nasm_A19772728000

Matsumoto, S.K.: Voyager interstellar mission: challenges of flying a very old spacecraft on a very long mission. In: SpaceOps 2016 Conference (2016). https://doi.org/10.2514/6.2016-2415

Miner, E.D.: Voyager mission. In: P. Murdin, Encyclopedia of Astronomy and Astrophysics. Boca Raton: CRC Press (2001). https://doi.org/10.1888/0333750888/2207

Morris, R.: Sequencing Voyager II for the Uranus encounter. Astrodyn Conf. (1986). https://doi.org/10.2514/6.1986-2111

Na, M.: Rhetoric of Cosmos—Voyager Golden Record message. Korean J. Rhetor. **38**, 77–114 (2020). https://doi.org/10.31325/kjr.2020.9.38.77

NASA.: Pioneer-10 and pioneer-11 (2007). Retrieved from NASA: https://www.nasa.gov/centers/ames/missions/archive/pioneer10-11.html

NASA.: Jupiter's Great Red Spot (2008). Retrieved from NASA: https://www.nasa.gov/image-article/jupiters-great-red-spot/

NASA. (n.d.): Voyager 1. Retrieved April 15, 2023, from NASA Solar System Exploration: https://solarsystem.nasa.gov/missions/voyager-1/in-depth/

NASA. (n.d.): Voyager 1's Pale Blue Dot. Retrieved April 15, 2023, from NASA Solar System Exploration: https://solarsystem.nasa.gov/resources/536/voyager-1s-pale-blue-dot/

NASA. (n.d.): Hazy Titan. Retrieved April 15, 2023, from NASA: https://www.nasa.gov/mission_pages/voyager/PIA_02238_13172.html

NASA. (n.d.): Europa Clipper Mission. Retrieved April 15, 2023, from NASA: https://europa.nasa.gov/mission/history/#europa-discovered

NASA. (n.d.): Greetings to the Universe in 55 Different Languages. Retrieved April 15, 2023, from NASA: https://voyager.jpl.nasa.gov/golden-record/whats-on-the-record/greetings/

Peshin, A.: Gravitational Slingshot: How Did Gravity Assist Voyager 1 & 2 in Escaping the Solar System? (2018). Retrieved from ScienceABC: https://www.scienceabc.com/innovation/gravitational-slingshot-how-did-gravity-assist-voyager-1-2-in-escaping-the-solar-system.html

Rayl, A.J.: The Stories Behind the Voyager Mission: Jurrie van der Woude (2002). Retrieved from The Planetary Society: https://www.planetary.org/articles/stories_vanderwoude

Sagan, C., Druyan, A.: Pale Blue Dot: A Vision of the Human Future in Space. Ballantine Books (2011)

Stone, E.C.: Voyager explores the final frontier of the solar system. In: 57th International Astronautical Congress (2006). https://doi.org/10.2514/6.iac-06-h.1.1.01

The Planetary Society. The Voyager missions (2021). Retrieved from The Planetary Society: https://www.planetary.org/space-missions/voyager

Wikipedians.: Contents of the Voyager Golden Record (2023). Retrieved from Wikipedia: https://en.wikipedia.org/wiki/Contents_of_the_Voyager_Golden_Record

[illegible]

17

Cassini-Huygens: The World of Rings and Ice

Cassini-Huygens

While it is true that all the outer planets have rings, none of them have captured the imagination of astronomers for as long as Saturn's do. Saturn's rings have a diameter of up to 250,000 km, or 19 times that of Earth's equator, although their thickness ranges from the tens of meters to the thousands (Schütte 2018). This tenuous ring of Saturn dealt Cassini Space Probe an unexpected blow in 2004 and set up a potentially catastrophic scenario (Fig. 17.1).

Cassini successfully navigated its way through Saturn's ring system in July of 2004, when it reached the sky above the planet. It penetrated the F-G ring gap like a needle, traveled parallel to large and small particles at high speed, and could be struck if it was not attentive enough. In space, even a minor scratch may have disastrous implications; imagine if that happened above Saturn, far from home. Cassini had to cut off contact with Earth for a short time, point the high-gain antenna away from Earth, fine-tune the angle, and then fire up the little rocket to slow down with pinpoint accuracy to ensure its safe capture by Saturn (Meltzer 2015). Cassini, after a flurry of exciting activities, successfully navigated the complex rings of Saturn and began its circumnavigation.

Quite a few scientists were shaken by this small incident. The Cassini-Huygens project was, in fact, one of NASA's most difficult and far-reaching space endeavors ever. Both a large and a small probe would be launched together into Saturn's orbit. The primary probe is known as Cassini, while the secondary one is called Huygens. Although Cassini would remain in

J. Wang, *Eye Beyond the Sky*, Astronomers' Universe,
https://doi.org/10.1007/978-981-99-9818-0_17

Fig. 17.1 Artistic illustration of Saturn's rings (*Credit* NASA/JPLCaltech/SSI)

orbit around Saturn, Huygens would attempt a landing on Titan. Starting in 1982, the project incurred staggering costs of $2.6 billion from NASA, $500 million from ESA, and $160 million from the Italian Space Agency. The total cost of the project, split between the three organizations, was around $3.26 billion (Lebreton and Matson 2003). Cassini ultimately demonstrated through its own power that these costs were worthwhile, and it offered humanity surprises beyond anyone's wildest dream. The adventures of Cassini and Huygens, the chapter's heroes, will leave you impressed (Fig. 17.2).

Cassini, a spacecraft with a mass of around 6 tons, was sent on its journey to Saturn in October 1997. Saturn is located approximately one billion kilometers away. It was the French astronomer Gian Domenico Cassini (1625–1712) who uncovered the gaps in Saturn's rings and named four of the planet's moons: Iapetus, Rhea, Tethys, and Dione (The Editors of Encyclopaedia Britannica 1998). In addition, astronomer Cassini collaborated with others in the discovery of the Great Red Spot on Jupiter (Falorni 1987). Hence, the Cassini spacecraft was named after him.

Despite the fact that Saturn and Earth are only ever 1.2 billion kilometers apart at their closest, Cassini traveled 3.5 billion kilometers to reach its destination. What was up with all the deviations that Cassini made? NASA utilized the most powerful rocket available at the time to launch Cassini, but it wasn't capable of propelling the spacecraft past the Sun's gravity and on to Saturn. The Cassini program's planners had to factor in the gravitational slingshot effect while figuring out the most fuel-efficient path for the spacecraft.

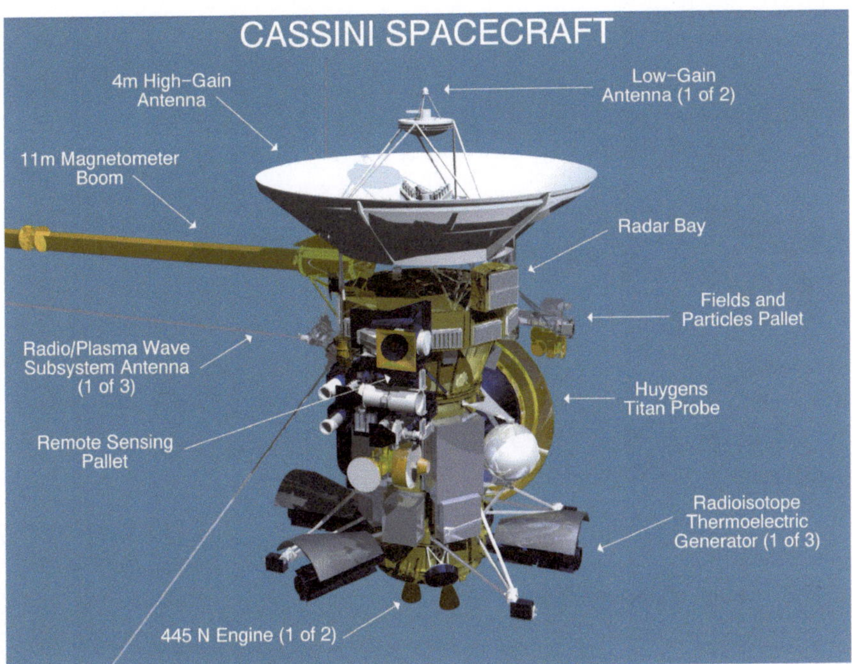

Fig. 17.2 Components of Cassini spacecraft (*Credit* NASA)

The general path went something like this. First, it flew to Venus, where it used the gravitational slingshot effect of Venus to accelerate itself for the first time. Then, after circling the Sun, it encountered Venus again and used Venus to accelerate itself for the second time. Next, with barely enough time to spare, Cassini exploited Earth's gravity as a slingshot to propel itself toward Jupiter. Jupiter was the most important stop along the entire route that Cassini would make. It was the equivalent of a big, burly hammer thrower picking up Cassini and hurling it toward Saturn (Peralta and Flanagan n.d.).

After a lengthy and intricate "throwing game," Cassini had arrived above Saturn. It's not as simple as taking a break and orbiting around Saturn as you might expect. Some throws along the route were only a warm-up for the "gravitational dances" that would follow. Cassini completed 294 orbits (NASA, Cassini-Huygens n.d.) around Saturn during its mission, a feat of mind-boggling intricacy. Cassini altered the size, period, velocity, and tilt angle of its orbit 70 times with the help of Titan's gravity so that it could observe Saturn and its moons from every possible vantage point (Fig. 17.3).

Can you explain why Cassini was looking for assistance from Titan specifically? The rationale for this is obvious: Titan is Saturn's largest moon (NASA, Titan: Facts n.d.), making this "little fat man" the most effective tool to

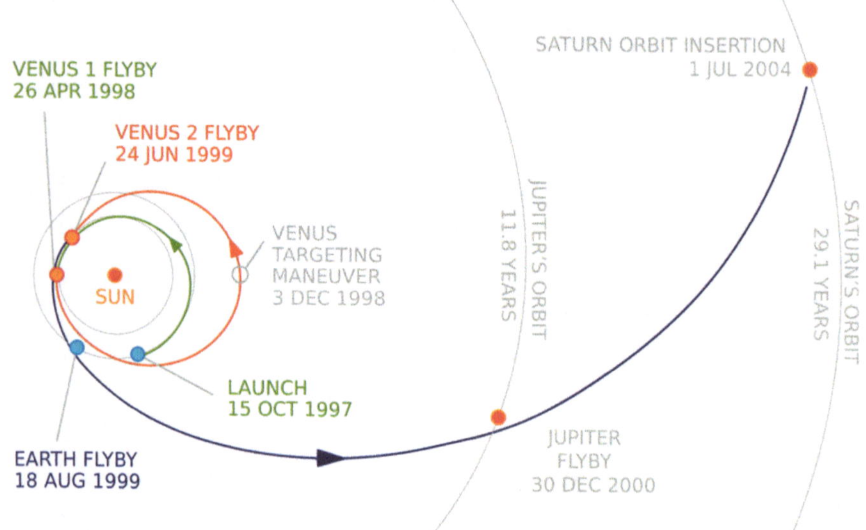

Fig. 17.3 Cassini trajectory (*Credit* NASA)

"direct" Cassini's motion. Overall, NASA's engineers were able to double the mission's duration while only using a fourth of the fuel thanks to a succession of sophisticated and creative orbital designs.

Cassini, despite its limited capabilities (it could only fly by, not land), has already provided significant scientific insight into Saturn and its moons. There were 12 instruments on board the Cassini orbiter and 6 on board the Huygens detector for studying the planet. These scientific instruments are like a "Swiss Army Knife" in that they have many different functions and may be used in a variety of situations. They fall generally into three categories: those that measure remote sensing with visible light, those that measure remote sensing with microwaves, and those that examine the immediate vicinity of the spacecraft (ESA, Cassini-Huygens Instruments 2019).

The most well-known are visible-light remote sensing observations, and the majority of the strange Saturn images we see today were captured by Cassini's primary camera. Carolyn Porco, who led the imaging science team on the Cassini mission, created a Captain's Log (Porco 2022) where she posted photos of Saturn alongside personal reflections. In addition, two mapping spectrometers could study Saturn, its moons, and its rings in both the visible and infrared spectrums to learn more about their chemical makeup and temperatures.

In contrast, the second microwave remote sensing instrument required Cassini to "listen" to the feeble echo signal by transmitting its own radio waves or microwaves to the target using a high-gain antenna up to 4 m in diameter. This allowed for further in-depth exploration of Saturn's atmosphere, yielding additional information (Wheeler and Renick 1996) (Fig. 17.4).

The "solo" at the heart of Cassini's lengthy expedition was short but thrilling and fantastic. Although it lasted for only a little over a month and was a relatively minor part of Cassini's 13-year mission, it was the most anticipated by scientists around the world. The landing of the Huygens probe on Titan was undoubtedly the most exciting part of the Cassini mission.

Regrettably, Huygens's battery only had a maximum operating time of about three hours (NASA, Cassini: Unlocking Saturn's Secrets 2004). Exactly why did scientists grant Titan such a golden chance? Among the moons of the solar system, Titan is the second largest. In spite of its rather mundane appearance, it owns an atmosphere thicker than that of the Earth. Also, nitrogen makes up most of the air (98.4%), followed by methane (1.4%) and hydrogen (0.2%) (Coustenis and Taylor, Observations of Titan 1999b). We are all aware that nitrogen and methane can be found in abundance in Earth's atmosphere. Some more creative scientists had speculated that alien life could exist on Titan using liquid methane as a solvent and hiding out in the clouds. Titan, in any case, had piqued scientists' interest due to the possibility of habitability.

The 320-kilogram Huygens spacecraft separated from Cassini on December 25, 2004, and had been orbiting Titan ever since, waiting for a

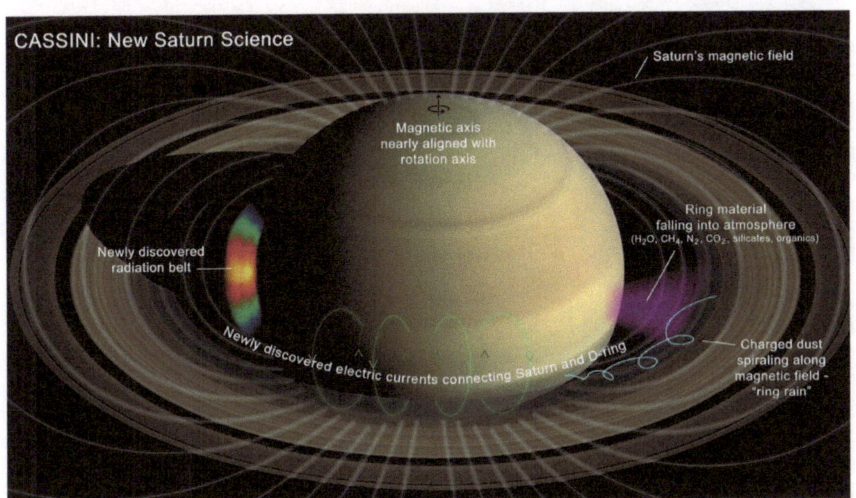

Fig. 17.4 Groundbreaking science emerges from ultra-close orbits of Saturn. (*Credit* NASA/JPL-Caltech)

landing opportunity. On January 14, 2005, the Huygens spacecraft deployed a massive parachute and descended gradually through the thick atmosphere of Titan. Over two and a half hours was spent on the landing process before it was accomplished. During 90 min, far longer than the planned 30 min, scientists clustered around monitors in the mission control center and marveled at the breathtaking barren terrain until the battery went out (Powell et al. 2009) (Fig. 17.5).

Huygens may have had a limited lifespan, but it accomplished its goal of transmitting data from Titan to Earth. These priceless images revealed Titan's surface for the first time to mankind. What would you say it is? If I showed you a picture taken after Huygens landed and asked you to identify it, you might believe it to have depicted a dry lake bed in the western part of China. Entirely accurate, Huygens observed a level "Gobi Desert" strewn with rocks of all sizes, giving the impression that they were pebbles shaped by a moving river (Coustenis and Taylor, Cassini and Huygens: Orbiting Saturn and landing on Titan 1999a). The later released data indicated that these stones are in fact ice, not dry ice but true water ice (Figs. 17.6 and 17.7).

Do aliens inhabit Titan? I can only say that the scientific community has been inconclusive up until this point, but not many scientists are under the

Fig. 17.5 Huygens on Titan (*Credit* NASA)

Fig. 17.6 A panoramic view of Titan taken by Huygens during its descent (*Credit* ESA/NASA/JPL/University of Arizona)

impression that there is life on Titan. Titan is an extraordinarily cold world with a surface temperature of roughly −179 °C, which is all we know about it right now. Titan has no liquid water, and there is no indication of an internal heat source (Raulin and Lunine 2011). To this day, humans have not discovered any life forms that can survive outside of liquid water while maintaining their active state.

People had high hopes that there might be life on Titan before Huygens' visit, but Huygens reminded us that there is just as much disappointment as there is hope regarding this possibility. But, people quickly got over their initial disappointment. Back then, Cassini made a huge discovery, like winning the lottery, that brought everyone out of their despair. This finding ended up being the most fruitful part of the mission to Saturn.

Before Cassini's arrival, Saturn's rings and Titan were the "stars," whereas Enceladus was a "little transparency" mostly ignored by scientists. Enceladus is one-tenth the size of Titan, and in the "eyes" of Voyager, it appeared as a frozen billiard ball reflecting almost 100% of the light from the Sun (NASA, Enceladus: Ocean Moon n.d.). This tiny, cratered moon has no other

Fig. 17.7 Partial view of Titan from Huygens (*Credit* ESA)

redeeming features. Many didn't figure out that they had "clumsy eyes" until more than 30 years after the fact.

After releasing the Huygens lander, Cassini captured a photograph of Enceladus from a distance of 1,167 km on February 17, 2005 (ESA, Cassini finds atmosphere on Enceladus 2005). The shot clearly shows feathery jets erupting from Enceladus' surface. To date, this has been the fourth planet in the solar system confirmed to have volcanic activities. The sophisticated equipment on Cassini has confirmed that the material erupting from Enceladus' surface is ionized water vapor, which is even more of a surprise. In other words, Enceladus likely has water, and perhaps a lot of it (Porco et al. 2006) (Fig. 17.8).

Ever since then, Cassini's primary objective had been to fly over Enceladus. On October 9, 2008, the distance was an amazing 25 km, and the imaging team referred to it as "drifting, (ESA & NASA, ESA Contribution to the Titan Saturn System Mission 2009)" which was fairly suitable given the circumstances. In 2010, a scientist named Candy Hansen working at

Fig. 17.8 Cassini's "drifting" (*Credit* NASA/JPL-Caltech)

the Jet Propulsion Laboratory made a groundbreaking discovery. She examined the Cassini data meticulously and discovered that the speed of some plumes' eruptions reached 2,189 kph, which was faster than Enceladus' escape velocity.

You might be confused by this at first, so let me explain: If Enceladus was made entirely of ice, and these plumes were ice that was expelled and sublimated into water vapor, then it was quite unlikely that they could reach such a high speed. One of the more likely scenarios, in Hansen's opinion, is that Enceladus' ice conceals liquid water and that the volcano acts like a high-pressure water gun, shooting the water to altitudes of tens of thousands of meters (Hansen et al. 2011).

NASA's announcement on April 3, 2014, that calculations suggested there was a subglacial ocean as deep as 10 km under at least 30 km of ice around Enceladus' South Pole, was the culmination of years of diligent observation and research. *Science*, one of the most prestigious scientific publications, also featured this research report (Iess et al. 2014). This has been a monumental discovery, as it confirms the existence of a second extraterrestrial ocean in addition to Europa.

What was found next is much more unexpected. A scientist working on the Cassini mission, Linda Spilker, made the announcement on April 13, 2017, that hydrogen had been discovered in the plumes of Enceladus and could provide energy for potential microorganisms on the ocean floor. Almost every human-identified condition for Earthly living being is satisfied on Enceladus (Seewald 2017) (Fig. 17.9).

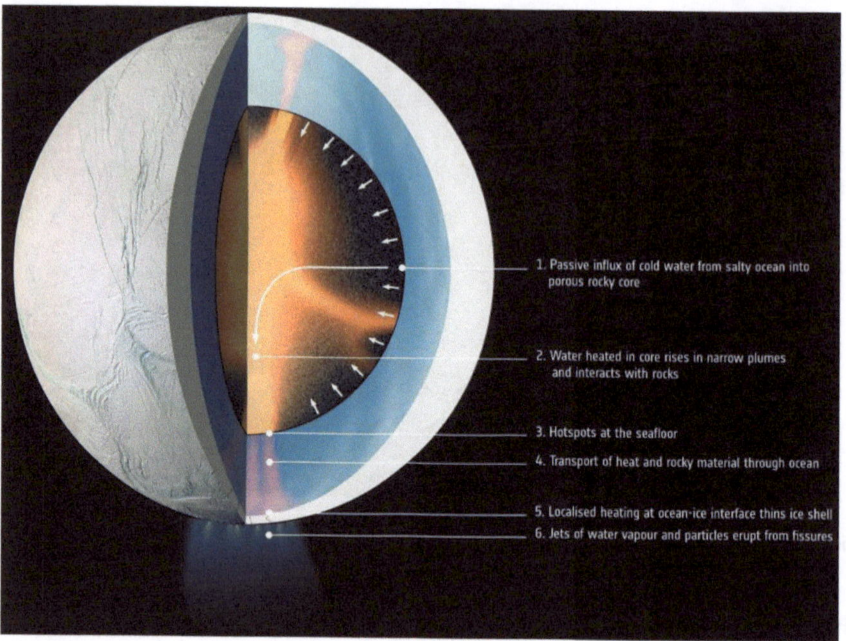

Fig. 17.9 Illustration of how water might be heated inside Enceladus (*Credit* Surface: NASA/JPL-Caltech/Space Science Institute; interior: LPG-CNRS/U. Nantes/U. Angers. Graphic composition: ESA)

This is a fantastic discovery for humanity as a whole, one that paves the way for the possibility of a universe inhabited by intelligent beings from and beyond the Earth. Assume that the moon of a massive planet is adequate for biological existence. If this is the case, then the hunt for Earth-like exoplanets may end up missing numerous possibilities. If a tiny moon orbiting far from the Sun can support life, then billions of planets in the cosmos are suitable for the habitation of organisms. Billions, that's a really thrilling number, right there.

The accomplished Cassini spacecraft "passed away" on September 15, 2017, after a long and fruitful lifetime. Cassini's tragic farewell was necessary in that it could prevent microbial contamination of the alien planet. Like its first indomitable arrival, Cassini left indomitably, rushing into Saturn's dense atmosphere, dissolving and burning in the high temperature, leaving nothing behind, as though the hero's ashes were dispersed on the alien planet he loved (Dougherty 2017) (Fig. 17.10).

Cassini was still contributing to science even in its final days. Using a mass spectrometer, it collected samples of Saturn's atmosphere and transmitted data

Fig. 17.10 Cassini's final trajectory (*Credit* NASA Jet Propulsion Laboratory)

back to Earth in real time, before eventually becoming a permanent part of Saturn.

References

Coustenis, A., Taylor, F.: Cassini and Huygens: orbiting Saturn and landing on Titan. In: Titan, pp. 245–288. World Scientific (1999). https://doi.org/10.1142/9789812815187_0009

Coustenis, A., Taylor, F.: Observations of Titan. In: Titan, pp. 15–52. World Scientific (1999). https://doi.org/10.1142/9789812815187_0002

Dougherty, M.: Cassini the grand finale. New Sci. **235**(3143), 29–35 (2017). https://doi.org/10.1016/s0262-4079(17)31823-7

ESA.: Cassini finds atmosphere on Enceladus (2005). Retrieved from ESA: https://www.esa.int/Science_Exploration/Space_Science/Cassini-Huygens/Cassini_finds_atmosphere_on_Enceladus

ESA.: Cassini-Huygens Instruments (2019). Retrieved from ESA Science & Technology: https://sci.esa.int/web/cassini-huygens/-/34954-instruments?fbodylongid=1612

ESA, & NASA.: ESA Contribution to the Titan Saturn System Mission (2009). Retrieved from ESA: https://sci.esa.int/documents/34334/35976/1567257822200-ESA_ISE_Assessment_report_TandEM.pdf

Falorni, M.: The discovery of the Great Red SPOT of Jupiter. J. Br. Astron. Assoc. **97**(4), 215–219 (1987)

Hansen, C.J., Shemansky, D.E., Esposito, L.W., Stewart, A.I., Lewis, B.R., Colwell, J.E., et al.: The composition and structure of the Enceladus plume. Geophys. Res. Lett. **38**(11) (2011). https://doi.org/10.1029/2011gl047415

Iess, L., Stevenson, D.J., Parisi, M., Hemingway, D., Jacobson, R.A., Lunine, J.I., et al.: The gravity field and interior structure of Enceladus. Science **344**(6179), 78–80 (2014). https://doi.org/10.1126/science.1250551

Lebreton, J.P., Matson, D.L.: The Huygens probe: science, payload and mission overview. In: The Cassini-Huygens Mission, pp. 59–100. Springer Netherlands (2003). https://doi.org/10.1007/978-94-017-3251-2_2

Meltzer, M.: The Cassini-Huygens Visit to Saturn: An Historic Mission to the Ringed Planet. Springer (2015)

NASA.: Cassini: Unlocking Saturn's Secrets (2004). Retrieved from NASA: https://www.nasa.gov/mission_pages/cassini/timeline/huygens_titan.html

NASA. (n.d.). Titan: Facts. Retrieved April 15, 2023, from NASA Solar System Exploration: https://solarsystem.nasa.gov/moons/saturn-moons/titan/in-depth/

NASA. (n.d.). Enceladus: Ocean Moon. Retrieved April 15, 2023, from NASA Solar System Exploration: https://solarsystem.nasa.gov/missions/cassini/science/enceladus/

NASA. (n.d.). Cassini-Huygens. Retrieved April 15, 2023, from NASA Solar System Exploration: https://solarsystem.nasa.gov/missions/cassini/overview/

Peralta, F., Flanagan, S. (n.d.): Cassini interplanetary trajectory design. Control. Eng. Pract. **3**(11), 1603–1610. https://doi.org/10.1016/0967-0661(95)00171-p

Porco, C.C., Helfenstein, P., Thomas, P.C., Ingersoll, A.P., Wisdom, J., West, R., et al.: Cassini observes the active south pole of Enceladus. Science **311**(5766), 1393–1401 (2006). https://doi.org/10.1126/science.1123013

Porco, C.: CICLOPS (2022). Retrieved from Official Source of Cassini Images of Saturn, Its Rings & Moons: https://ciclops.org/

Powell, R.W., Lockwood, M.K., Cruz, J.R., Striepe, S.A., Sutton, K., Fisher, J., et al.: Cassini/Huygens Probe Entry, Descent, and Landing (EDL) at Titan Independent Technical Assessment (2009). Retrieved from NASA Technical Reports Server: https://ntrs.nasa.gov/citations/20090022173

Raulin, F., Lunine, J.: Titan and the Cassini–Huygens mission. In: Origins and Evolution of Life, pp. 489–506. Cambridge University Press (2011). https://doi.org/10.1017/cbo9780511933875.031

Schütte, U.: The Rings of Saturn (1995). In: W.G. Sebald, pp. 72–87. Liverpool University Press (2018). https://doi.org/10.3828/liverpool/9780746312988.003.0005

Seewald, J.S.: Detecting molecular hydrogen on Enceladus. Science **356**(6334), 132–133 (2017). https://doi.org/10.1126/science.aan0444

The Editors of Encyclopaedia Britannica. Gian Domenico Cassini (1998). Retrieved from Encyclopedia Britannica: https://www.britannica.com/biography/Gian-Domenico-Cassini

Wheeler, K., Renick, P.:. Cassini Radar hardware technologies. AIP Conf. Proc. (1996). https://doi.org/10.1063/1.49870

18

Stardust Comet Mission: Capturing Cometary Matter

Stardust Comet Mission

Evidently, the Sun is at least a second-generation star. A second generation means that the present-day location was occupied by another, larger star before the Sun and its system formed. More than 5 billion years ago, that larger star unexpectedly burst as a supernova. In the remnant of the explosion, a fragment of interstellar dust was left behind, which gathered other dust under the influence of gravity like an ever-growing snowball. Eventually, under the influence of gravity, the nuclear fusion process occurred again, giving rise to our present-day Sun (Clery 2018) (Fig. 18.1).

But how did scientists arrive at this conclusion? How did they know that the Sun is not the first-generation star ever created? The rationale has to do with the chemical makeup of the solar system. Heavier elements have larger atomic numbers on the periodic table of chemicals. Iron comes in at number 26. Using theoretical calculations, scientists have determined that the nuclear fusion "furnaces" within stars are unable to generate any elements heavier than iron (copper, silver, gold, etc.) (Lodders 2003). These elements can only be created in the extreme conditions of a supernova explosion, where the temperature and pressure are both extremely high (Foglizzo 2017). As elements with atomic weights greater than iron can be found all around our solar system, we can conclude that they are "second-hand materials" ejected from supernovae.

This is, of course, only conjecture based on scientific theories. What we really need is evidence that supports these claims.

J. Wang, *Eye Beyond the Sky*, Astronomers' Universe,
https://doi.org/10.1007/978-981-99-9818-0_18

Fig. 18.1 Artistic illustration of the early formation of the solar system (*Credit* NASA)

Several minor objects were hurled into the far reaches of the outer solar system early in the process of solar system formation. That world was frigid and gloomy, like a "refrigerator" that preserved the original substance. When a minor celestial body, dragged by the gravity of large planets, deviated from its original orbit and fell into the inner solar system, it would become a comet (Delsemme 1995). Typically, the orbits of these comets were relatively flat ellipses, with one end far out in the reaches of the solar system and the other close to the Sun. When close to the Sun, the frozen material evaporated, leaving behind a long tail (Fig. 18.2).

Hence, we can obtain material from the early solar system by capturing samples either from these comets or their tail ejection. The comet tail ejection material, in comparison, is much less difficult to collect (Mandt et al. 2017). This was exactly what the Stardust probe set out to do.

Stardust intended to fly by Comet Wild 2 (NASA, 81P/Wild (Wild 2) n.d.). What accounted for this choice? Historically, this comet had made near approaches to the inner solar system once every 43 years. In 1974, however, the comet's orbit shifted because of Jupiter's gravity as it passed by. The orbit's radius had been drastically reduced, and its period shortened to six years. At its closest point to Earth, this distance was roughly equivalent to that between Earth and Mars. The comet had then completed five orbits of the Sun. However, it had not been altered significantly from what it was in the past. Comet Wild 2 was, therefore, the ideal celestial body to observe (Fig. 18.3).

Fig. 18.2 Stardust on the expedition (*Credit* NASA/JPL)

How can we gather the dust particles ejected from a comet's tail? The trickiest aspect is maintaining the shape of these dust particles. Aerogel is an amazing solution developed by scientists to collect and store these particles. Some scientists refer to aerogel as "solid smoke," which is essentially what it is: a sponge with an extremely thin mesh. A bulk of aerogel has essentially little gravity considering that it is lighter than the equivalent volume of air. On the other hand, the material is robust enough to bear a pressure of 4,000 times its own weight (Kearsley et al. 2012) (Fig. 18.4).

Particles cannot escape the unique super sponge's fine mesh and become trapped within it. In order for the particles to penetrate without being damaged, aerogel's density must be optimized. Particles entering the aerogel at high velocities will leave trails, making it easy to determine their properties. The impact would be similar to a bullet hitting a bar of soap, leaving behind a distinct imprint of the trajectory of the projectile.

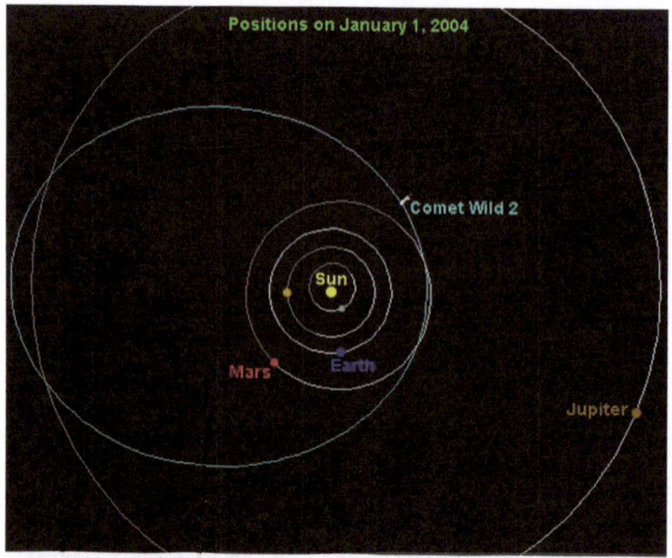

Fig. 18.3 Trajectory of comet Wild 2 after September 1974 (*Credit* NASA/JPL)

Fig. 18.4 Aerogel collector on Stardust (taken before launch) (*Credit* NASA/JPL/Caltech)

The Stardust spacecraft has a highly intricate orbital layout. An enormous ellipse that orbits the Sun best describes its overall shape. In order to encounter Comet Wild 2, however, the ellipse's inclination and the timing must be pinpoint perfect. After collecting samples, it would proceed on an elliptical orbit that crossed the Earth's orbit and arrive at Earth precisely on schedule (Meyer 2000). The aerogel-filled capsule would be dropped back to Earth by Stardust when it flew by, which was another extremely challenging task (Fig. 18.5).

Soon after taking off, Stardust exploited the Earth as a "slingshot" to rapidly close the distance to Comet Wild 2. The smaller the velocity difference between them, the more information could be gleaned from the longer observational period. Nonetheless, the difference in velocity between Stardust and Wild 2 reached 21,000 km per hour (NASA 2005), which is five times the velocity of a bullet fired from a gun. Despite the briefness of the meeting, Stardust managed to snap 72 photos (NASA, NASA spacecraft makes great catch—heads for touchdown n.d.) around the area. Wild 2 had a diameter of around 5 km, was heavily cratered, and was erupting with gas. Stardust was able to collect an abundance of samples throughout its eight-minute adventure through the comet's tail (Fig. 18.6).

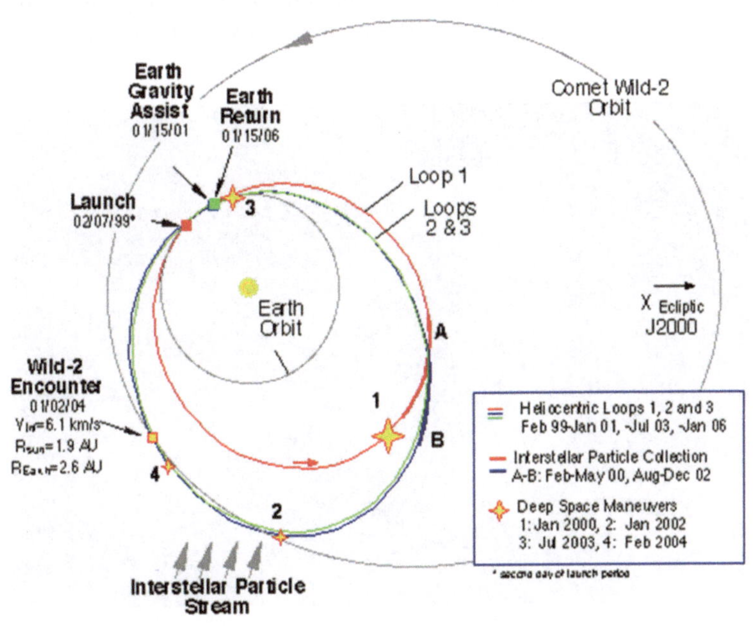

Fig. 18.5 Stardust mission trajectories (1999–2006) (*Credit* NASA/JPL)

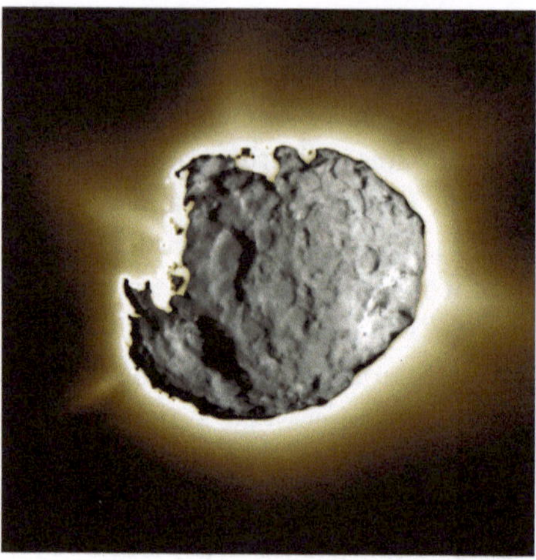

Fig. 18.6 Combinational images of comet Wild 2 photographed by Stardust (*Credit* NASA/JPL)

Stardust performed flawlessly and just as expected in flight. It was exactly 10:12 UTC on January 15, 2006, when the capsule landed in the Utah desert (Muir 2006) (Fig. 18.7).

This re-entry capsule had done a significant orbit around the Sun before returning to Earth. Scientists took the re-entry capsule to a sterile room and

Fig. 18.7 Stardust sample re-entry capsule dropped in the Utah desert. (*Credit* NASA)

opened it. The rare material that erupted from the comet was what the re-entry capsule had been designed to capture, and scientists' greatest concern was that it could be contaminated by the material on Earth (Fig. 18.8).

Many amateur astronomers' computers had been used to analyze the samples carried back. The raw data were collected with a specialized microscope that could take images of the aerogel at varying depths. Working from the surface down, the microscope went through 100 μm of the sample at a time, taking photographs layer by layer. Almost every amateur astronomer's computer participated in analyzing the 7 million photographs uploaded to the Internet (Hand 2010). The software for analyzing data was very much like the screensaver software. It would request the processing resources of a computer to analyze the images as soon as it became available (Fig. 18.9).

As computers could not tell very small particles apart, amateur astronomers had to manually identify some of the images. Particles too small to be recognized disappeared entirely, but larger particles left recognizable tracks. Comet tail particles have such high velocities that they leave a distinct path, much like a bullet does in soap. Particles from comets can be distinguished by their unique "fingerprint" left by the comet's tail. Only a few of the millions of particles were able to leave any sort of imprint. There weren't many trajectories longer than 0.1 mm, and the longest one was barely 1 mm. Images that

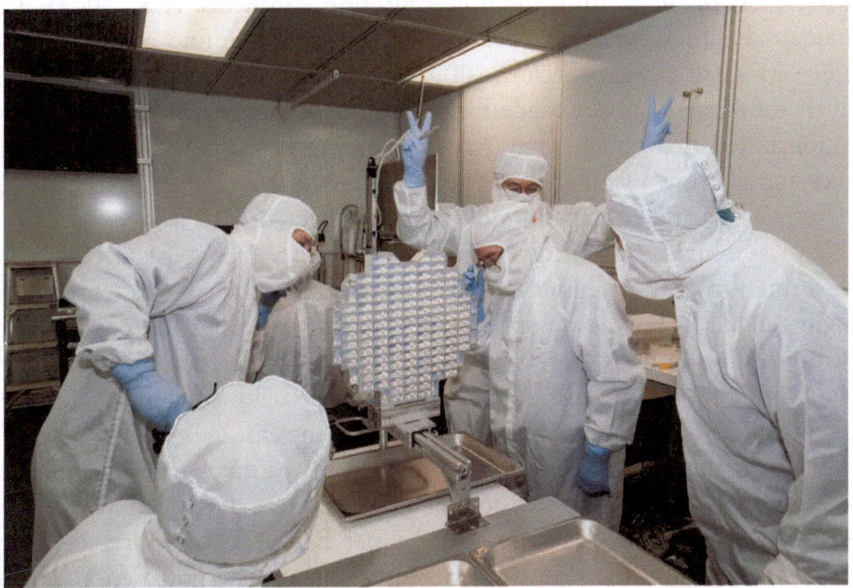

Fig. 18.8 Researchers studied the aerogel collector in the return capsule, and someone excitedly gestured "YEAH." (*Credit* NASA)

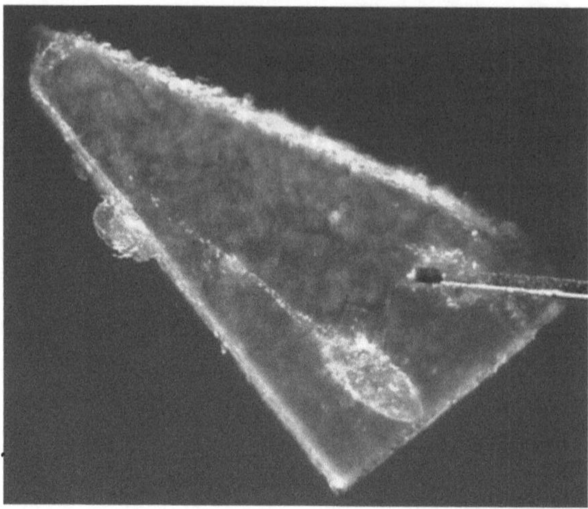

Fig. 18.9 A picture that amateur astronomers needed to identify manually (*Credit* NASA/JPL)

could not be read by a computer needed to be examined using a human eye. Time had come for amateur astronomers to make important contributions to the field.

There have been several breakthroughs in our understanding of the universe thanks to Stardust's sample collection (Brownlee et al. 2006). The Goddard Flight Center team reported detecting the amino acid glycine in some very small particles back in 2009. The glycine was analyzed for isotopes, which showed it was not made from Earth-based ingredients. Research has demonstrated that comets may transport amino acids, a necessary building block for all known forms of life (Elsila et al. 2009).

Iron sulfide and copper sulfide were also detected (Zolensky et al. 2008), which was a highly surprising finding. The presence of water is necessary for the formation of these minerals. This suggested that the comet's temperature might not be as low as was previously thought, opening the possibility of the presence of liquid water. Comets were once considered as dirty snowballs made up of ice, snow, and dust, and in a perpetually cold state. However, recent evidence suggests that this is not always the case.

In short, the Stardust data have provided a wealth of new information, necessitating a significant revision of older models developed with much less data. We once believed that the Kuiper Belt was the source of short-period comets, and that Pluto was its largest member; short-period comets were commonly created when objects from the Kuiper Belt entered the inner Solar system; and those comets with orbital periods of more than 200 years originated in the remote Oort cloud.

Until recently, it was believed that short-period comets and long-period comets originated from entirely different locations and were not related in any way, but the discovery of intermediate extra-Neptunian objects had changed that (Duncan 2008). Disorientation resulting from such a shallow categorizing rule serves to heighten the universe's inherent complexity (Fig. 18.10).

Based on the data gathered by Stardust, it is clear that the comet has a high concentration of ice crystals in the area far beyond Neptune, but a much higher concentration of dust. The minerals that make up the dust appear to be formed at extremely high temperatures, a departure from previous understanding. Generally, we believed that the material in the solar system was stratified, with denser stuff, such as rock, being located closer to the Sun. Only rocks with very high melting points might remain solid in the extremely high temperatures found in the solar system's vicinity. The four largest rocky planets were, therefore, located within the inner ring of the solar system. It

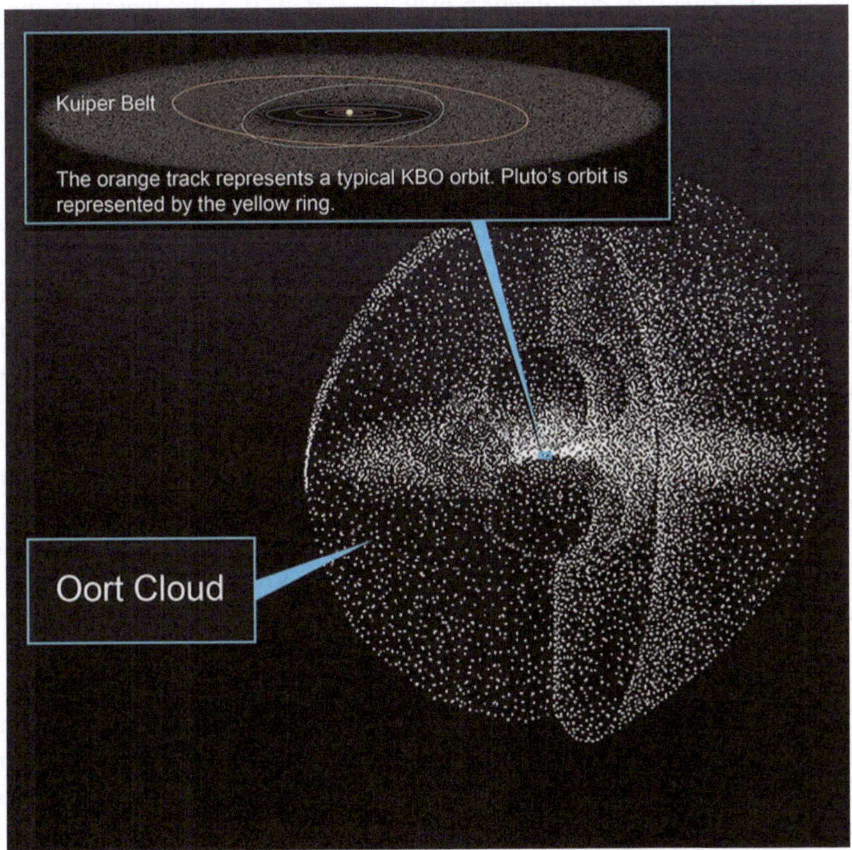

Fig. 18.10 Kuiper Belt and Oort cloud (*Credit* NASA)

would be cooler in the reaches of space. After a certain distance from the Sun, the temperature dropped to a point where water froze into a solid. This is known as the snow line (Ciesla and Sandford 2012; Raymond et al. 2009). Dry ice was formed when carbon dioxide was cooled to a lower temperature. The methane also solidified with the increasing distance. For this reason, scientists believed that the outer regions of the solar system should be covered with ice and snow, and that the majority of the material that was found in comets should also be composed of ice and snow. However, the data Stardust returned contradicted this assessment.

Particles collected in Stardust's aerogel, for instance, include two distinct types of rock. A basic chondrite meteorite was formed from a spherical rock particle, which was melted during its orbit around the Sun and then rapidly cooled. White, irregular particles that are made up of a rare substance called calcium-aluminum-rich inclusion or Ca-Al-rich inclusion (CAI) are another example (Wendel 2014). It is only at very high temperatures that this unusual chemical combination may be created. It defies logic that there would be warm weather in the farthest reaches of our solar system. Thus, scientists believe that these components originated in the inner solar system when the solar system was young and were then expelled to the outskirts. These components, then, date back to the earliest stages of the solar system's development.

However, scientists first built the Stardust probe not to locate the material born in the solar system, but rather to find the material produced in the pre-solar system period, that is, the material of the next generation of stars spread far and wide by the supernova explosion. In other words, the scientists wanted to find pre-solar system material. Stardust's samples did include some pre-solar material, but the amounts were too small to be meaningful. Comets, then, are not made up of leftover debris from other stars but rather of material generated close to the Sun. Thus, they yield crucial information regarding the process of planet and moon formation 4.5 billion years ago.

Before, astronomers thought our solar system was relatively peaceful and well-organized, with rocky planets in the inner solar system and asteroid belts outside of Mars. The asteroid belt formed here due to the gravitational rip caused by the presence of super-giant Jupiter. When you traveled past the asteroids, you would enter the realm of the mega planets. "Snowballs," according to scientific theory, should form the inner core of a massive giant like Jupiter. Snowballs that collided, stuck together, and grew in size. At last, Jupiter surpassed a dozen times the mass of Earth (Chambers 2004). When this snowball's mass hit a critical threshold, it could start sucking in hydrogen and helium. They were the stuff of which our universe was made up mostly.

Hence, Jupiter entered a "winner takes all" mode, and in just 1,000 years, it grew from a dozen Earth masses to 312 Earth masses, becoming the largest planet in the solar system. Saturn had not managed to grab more substances than its "big brother," thus in terms of mass, Jupiter was larger. The third and fourth planets, Uranus and Neptune, were limited to the "leftovers," although their higher methane-to-hydrogen ratio gave them a bluish hue.

This was how scientists once explained the process of the solar system coming into being. The solar system appeared to constantly have a temperature hierarchy, with differing compositions orbiting in relative peace and harmony. It turns out that the early solar system was far more turbulent than previously thought. There was a constant flow of material between the inner and outer rings, with asteroids being ejected and new ones being added. There is growing speculation among scientists over whether the gaseous giants' inner cores are actually massive snowballs. Could the dust be predominating and the giant planets actually be clumps of mud? (Batygin and Laughlin 2015). This theory has some grounding in the findings of the Stardust mission. If gaseous planets were like finished constructions, then comets should be like "bricks and tiles" and "construction trash" that were not put into use. The composition should be the same in both astronomical objects.

Still, a glimpse into comet research, even though Stardust has brought back a wealth of information. Comet Tempel 1 was the target of the next space probe after Stardust, which was launched on July 4, 2005, and was designated Deep Impact. Deep Impact, which had been guided into the Tempel 1's course with the assistance of a modest thruster, slammed into the comet target with a velocity of 37,000 kph (A'Hearn 2006). The force of the impact was roughly equal to 4.7 tons' TNT explosives. A crater 150 m across was blasted out of Comet Tempel 1. Space and ground-based telescopes monitored the comet Tempel 1, gathering a wealth of data and knowledge in a short length of time (Fig. 18.11).

Once Stardust successfully brought its capsule back to Earth, NASA decided to send it on another mission to Comet 103P/Hartley because it still had a sizable amount of fuel remaining.

China is eager to catch up in the exploration of the solar system's outer regions. It plans to launch the "Zheng He" in 2024 to study the near-Earth asteroid 2016HO3 and bring its soil samples back to Earth. "Zheng He" will continue its investigation of the comet's riddles after returning the samples to Earth (Liu 2022).

The questions "Who are we?" "Where are we from?" are perennially relevant to the discipline of philosophy.

Fig. 18.11 Artistic illustration of Deep Impact colliding with Comet Tempel 1 (*Credit* NASA)

Small, far-off celestial bodies are the solar system's living fossils; thus, they may hold the key to answering these two questions. As inquisitive individuals, humans will never stop trying to learn more about how our solar system got started (Fig. 18.12).

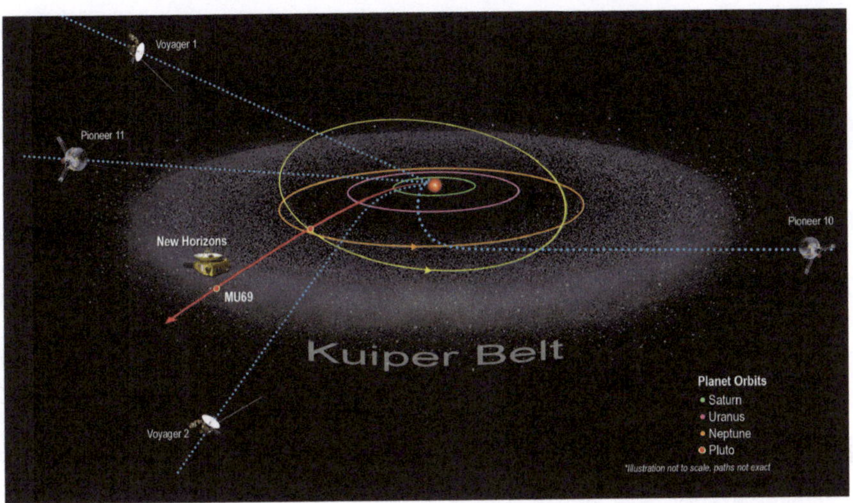

Fig. 18.12 Current positions of the farthest human space probes (*Credit* NASA Solar System Exploration)

References

A'Hearn, M.F.: Deep impact: excavating comet Tempel 1. Proc. Int. Astron. Union **2**(14), 325–326 (2006). https://doi.org/10.1017/s1743921307010861

Batygin, K., Laughlin, G.: Jupiter's decisive role in the inner solar system's early evolution. Proc. Natl. Acad. Sci. **112**(14), 4214–4217 (2015). https://doi.org/10.1073/pnas.1423252112

Brownlee, D., Tsou, P., Aléon, J., Alexander, C.M., Araki, T., Bajt, S., et al.: Comet 81P/Wild 2 under a microscope. Science **314**, 1711–1716 (2006). https://doi.org/10.1126/science.1135840

Chambers, J.E.: Planetary accretion in the inner solar system. Earth Planet. Sci. Lett. **223**(3–4), 241–252 (2004). https://doi.org/10.1016/j.epsl.2004.04.031

Ciesla, F.J., Sandford, S.A.: Organic synthesis via irradiation and warming of ice grains in the solar nebula. Science **336**(6080), 452–454 (2012). https://doi.org/10.1126/science.1217291

Clery, D.: Super-Earth spied in the second-closest star system from the sun (2018). https://doi.org/10.1126/science.aaw0665

Delsemme, A.H.: Cometary origin of the biosphere: a progress report. Adv. Space Res. **15**(3), 49–57 (1995). https://doi.org/10.1016/s0273-1177(99)80063-5

Duncan, M.J.: Dynamical origin of comets and their reservoirs. In: Space Sciences Series of ISSI, pp. 109–126. Springer New York (2008). https://doi.org/10.1007/978-0-387-85455-7_7

Elsila, J.E., Glavin, D.P., Dworkin, J.P.: Cometary glycine detected in samples returned by Stardust. Meteorit. Planet. Sci. **44**(9), 1323–1330 (2009). https://doi.org/10.1111/j.1945-5100.2009.tb01224.x

Foglizzo, T.: Explosion physics of core-collapse supernovae. In: Explosion physics of core-collapse supernovae, pp. 1053–1073. Springer International Publishing (2017). https://doi.org/10.1007/978-3-319-21846-5_52

Hand, E.: Volunteer army catches interstellar dust grains. Nature (2010). https://doi.org/10.1038/news.2010.106

Kearsley, A.T., Burchell, M.J., Price, M.C., Cole, M.J., Wozniakiewicz, P.J., Ishii, H.A., et al.: Experimental impact features in Stardust aerogel: how track morphology reflects particle structure, composition, and density. Meteorit. Planet. Sci. **47**(4), 737–762 (2012). https://doi.org/10.1111/j.1945-5100.2012.01363.x

Liu, X.: "Nature": China sketches out a five-year blueprint for space development (2022). Retrieved from CAS—Science and Technology Daily: https://www.cas.cn/kj/202202/t20220224_4826163.shtml

Lodders, K.: Solar system abundances and condensation temperatures of the elements. Astrophys. J. **591**(2), 1220–1247 (2003). https://doi.org/10.1086/375492

Mandt, K.E., Mousis, O., Bockelée-Morvan, D., Russell, C.T.: Comets as tracers of solar system formation and evolution. In: Comets as Tracers of Solar System Formation and Evolution, pp. 5–7. Springer Netherlands (2017). https://doi.org/10.1007/978-94-024-1103-4_3

Meyer, T.: History of the Comet 81P/Wild-2 and How to Observe it in 1997 (2000). Retrieved from NASA: https://solarsystem.nasa.gov/stardust/comets/wild2.html

Muir, H.: Pinch of comet dust lands safely on Earth (2006). Retrieved from NewScientist: https://www.newscientist.com/article/dn8586-pinch-of-comet-dust-lands-safely-on-earth/

NASA: Stardust Cool Facts (2005). Retrieved from NASA: https://solarsystem.nasa.gov/stardust/cool.html

NASA (n.d.): 81P/Wild (Wild 2). Retrieved April 15, 2023, from NASA Solar System Exploration: https://solarsystem.nasa.gov/asteroids-comets-and-meteors/comets/81p-wild/in-depth/

NASA (n.d.): NASA spacecraft makes great catch—heads for touchdown. Retrieved April 15, 2023, from NASA: https://www.nasa.gov/vision/universe/solarsystem/stardust_update.html

Raymond, S.N., O'Brien, D.P., Morbidelli, A., Kaib, N.A.: Building the terrestrial planets: constrained accretion in the inner solar system. Icarus **203**(2), 644–662 (2009). https://doi.org/10.1016/j.icarus.2009.05.016

Wendel, J.: Seven tiny particles could be first samples of interstellar dust. EOS Trans. Am. Geophys. Union **95**(34), 307–307 (2014). https://doi.org/10.1002/2014eo340003

Zolensky, M., Nakamura-Messenger, K., Rietmeijer, F., Leroux, H., Mikouchi, T., Ohsumi, K., et al.: Comparing Wild 2 particles to chondrites and IDPs. Meteorit. Planet. Sci. **43**(1–2), 261–272 (2008). https://doi.org/10.1111/j.1945-5100.2008.tb00621.x

19

SOlar and Heliospheric Observatory (SOHO): Living with a Restless Star

SOlar and Heliospheric Observatory (SOHO)

Something like this might be announced by the weather forecaster on a program that predicts the space weather: "Greetings, citizens of planet Earth. The date is March 22, 2022. There are currently 30 sunspots in the solar disk, and the solar wind is moving at a speed of 1.11 million kilometers per hour with a density of 4 protons per cubic centimeter. In 24 h, a solar flare of X-ray intensity no higher than C1 is possible."

The solar wind travels at a speed of about a million kilometers per hour, but it shouldn't scare you. When compared to Earth's atmosphere, the solar wind is a mere 10 billionth as dense. There is no need to worry about solar flares of the C1 level. Only flares above the M level produce radio blackouts. If the flare reaches the X level, be on the lookout because it can cause a complete blackout of the Planet and a radiation storm that lasts for a week (Fig. 19.1).

Such space weather forecasts are real, even though there are no local weather forecasters to broadcast this. The data can be found on the front page of Spaceweather.com, the website dedicated to space weather forecasting. Besides, every 10 min, the data are updated on the National Oceanic and Atmospheric Administration (NOAA) website.

It's understandable if you're scratching your head and wondering what relevance the Sun's wind and occasional flares have to our planet, which is around 150,000,000 km distant. Do we need to force a refresh every 10 min? Obviously, that's the tale I will tell in this chapter.

J. Wang, *Eye Beyond the Sky*, Astronomers' Universe,
https://doi.org/10.1007/978-981-99-9818-0_19

Fig. 19.1 Artistic imagination of a solar storm hitting Mars (*Source* NASA)

Little was known about the Sun until the development of astronomical telescopes. Sunspots were unknown to humans for over 2000 years before they were uncovered by chance. It was as late as 1610 that Galileo Galilei made the first recorded observation of sunspots. To him, it appeared as if the Sun had a few black spots, like freckles on a pretty girl's face, except that these "freckles" would move around the Sun as it rotated (Galilei and Scheiner 2010) (Fig. 19.2).

In those days, nobody connected sunspots to Earth. However, subsequent research revealed that the sunspot count decreased to virtually zero for a prolonged period in the seventeenth century, coinciding with a period of severe cold in Europe (so cold that the Dutch could skate on the canals in the summer). The phrase "Maunder Minimum" was given to this time span because of the unusually low number of sunspots. The first inklings of the possibility that solar activity might influence Earth's weather had just been noticed by humans (Soon and Yaskell 2003). The possibility that the Maunder Minimum was caused by a decrease in sunspot activity was merely a scientific hypothesis at that point because there lacked sufficient evidence to support the claim.

By the year 1859, something else momentous had taken place. A white-light flare was seen on the Sun at the moment by two scientists, and two days later, Earth was slammed by a solar magnetic storm that only happened once every thousand years or so (Clauer and Siscoe 2006). Due to this exceptional event, the Aurora was no longer a phenomenon limited to the polar regions. Low latitude dwellers in places like Mexico, Hawaii, and other locations close

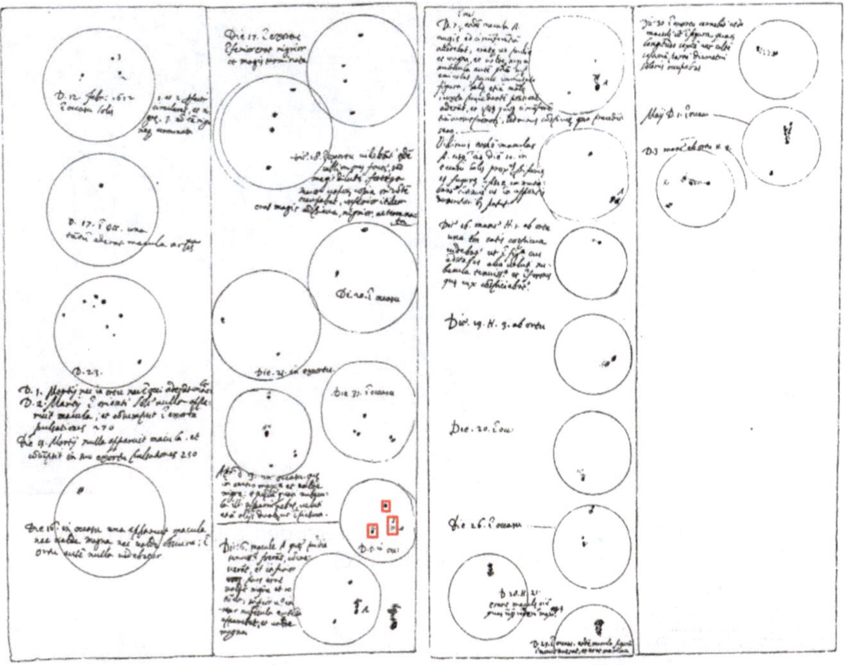

Fig. 19.2 Map of sunspot observations recorded by Galileo Galilei from 12 February to 3 May 1612 (Public Domain)

to the equator had actually witnessed the dazzling Aurora for the first time (Fig. 19.3).

The solar magnetic storm impacted the telegraph systems around the globe. Due to the magnetic storm-induced strong current in the telegraph line, the transmitter continued to function even when the power was cut off. Accordingly, the telegraph operator noted that the current running through the wire was excessively strong even when the electricity was switched off. An electrical spark from the instrument apparently struck them (Clark 2019) (Fig. 19.4).

Was the magnetic storm truly the result of a solar flare? The real culprit seemed to have been uncovered 40 years later when George Ellery Hale (1868–1938) (Carey 2000) constructed the first apparatus to examine solar flares in detail on Mount Wilson in California. This instrument confirmed the existence of a 2-day delay between solar flares and Earth's magnetic storms (Priest 2014) (Fig. 19.5).

An increasing number of researchers have since then focused on how sunspots affect global temperatures and other aspects of Earth's climate. When sunspots were few during the Maunder Minimum, why did global

Fig. 19.3 British astronomer Richard Carrington recorded a blast of white light on the solar surface in 1859. (*Source* Voice of Science)

Fig. 19.4 A telegraph operator encountered an electric spark. (*Source* Voice of Science)

temperatures drop? Sunspots were found to be brighter than the rest of the Sun's surface, which led researchers to conclude that fewer sunspots would lead to a darker Sun. In contrast, digital detectors from the 1970s revealed that the Sun's brightness varies by only a thousandth across its activity cycle. How can a seemingly insignificant shift in brightness cause such a dramatic shift in global temperatures? (Bertaux 2007) In addition, the rise in global temperature during the previous half century could not be explained

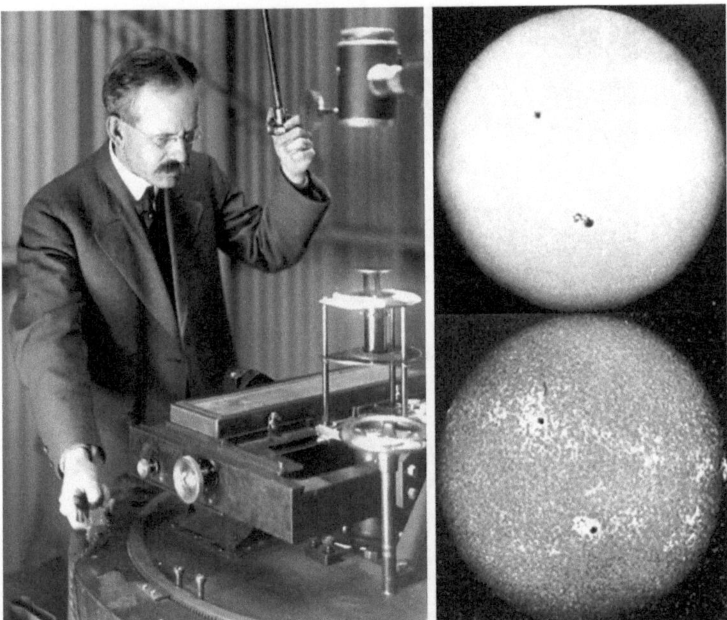

Fig. 19.5 Left: Hale and his instrument for studying solar flares (*Source* California Institute of Technology)
Right: A spectral view of a sunspot taken by the Mount Wilson Observatory in 1906 (*Source* NASA)

by sunspots, and the Maunder Minimum became the only evidence that sunspots affect Earth's temperatures.

In this chapter, SOHO is introduced as the main character to help decipher the enigma of solar activity.

SOHO (SOlar Heliospheric Observatory) is a space-based observatory viewing and investigating the Sun from its deep core (Malinovsky-Arduini and Fröhlich 1984).

It was ESA that came up with the idea for the SOHO program. More than 300 engineers from 14 countries worked on the design and construction, and NASA handled the launch and ground control. This has exemplified the effectiveness of working together across borders (Fig. 19.6).

The Atlas rocket carrying SOHO into space lifted off from Cape Canaveral, Florida, on December 2, 1995. Comparable in size and complexity to Cassini, SOHO made efficient use of its limited volume by housing 12 instruments (ESA & NASA, About the SOHO Mission 2020a) that could produce everything from magnetic fields to X-rays, much like a multifunctional "Swiss Army Knife." The "Super-Thermal and High-Energy Particle Integrated Analyzer," developed at Kiel University in Germany, has

Fig. 19.6 SOHO (*Source* NASA)

a terrifyingly long name but is actually a very useful tool for measuring the location, intensity, and spectrum of high-energy X-rays, UV radiation, and cosmic rays.

In particular, SOHO's position provided the best possible vantage point to watch the Sun's activity. SOHO revolved around the Sun in step with the Earth, while at the same time, it moved slowly around the unusual space point called "the First Sun-Earth Lagrange Point (L1 Point)" (ESA & NASA, SOHO's Orbit 2020b). This orbital position guarantees that the probe will not be obscured by either Earth or the Moon, which makes it the irreplaceable perch of solar observatories such as the SOHO. From then on, SOHO became the envy of many other space probes that also wanted to make use of L1 Point (Fig. 19.7).

SOHO encountered a serious setback three years after its launch. NASA's Goddard Space Center in Maryland recorded the following on the evening of June 24, 1998: SOHO entered emergency attitude control mode (ESA &

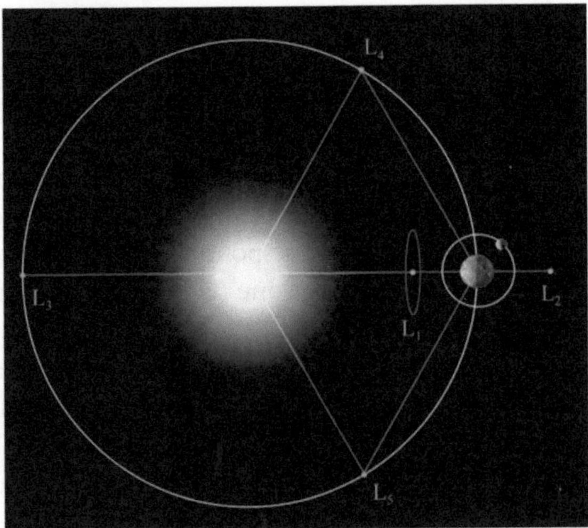

Fig. 19.7 SOHO and L1 (*Source* NASA)

NASA, SOHO Mission Interruption Joint NASA/ESA Investigation Board 1998).

Simply speaking, it meant that the gyroscopes and thrusters were used to realign the telescope's orientation with the Sun. SOHO had entered this emergency mode six times since its inception, so no operators paid enough attention this time. The situation at SOHO, however, deteriorated rapidly over the next few hours. First, a broken gyroscope caused it to spin wildly. At that time, the attitude had gotten out of control, and the solar panels were unable to keep up with the demand for electricity. After losing power, its temperature control system and communication lines stopped functioning. The fate of SOHO hanged in the balance whenever hardware facilities with R&D expenditure of over $100 million were reduced to a pile of scrap iron.

Tensions ran high in the command center. There was an urgent diagnostics conference between NASA and ESA specialists. According to the specialists' evaluations, the spacecraft was still operating in accordance with the preset mode. All the failures up to the point had been caused by human error. The operators accidentally turned off SOHO's safe mode and incorrectly assessed the condition of two gyroscopes. These careless errors are what spiraled the SOHO out of control.

When control is lost, horrible things happen as a result. Since SOHO's solar panels were tilted with their edges facing the Sun, they were unable to provide energy to the spacecraft at this time. If this continued, the temperature would decrease from 100 °C to below zero, and SOHO would reach a frozen state. This could potentially damage the battery and fuel beyond repair.

Losing communication with SOHO, the engineers on the ground had no choice but to wait. They sent a message to SOHO, holding out hope, but received no reply. One month after SOHO last communicated, things began to turn around. After a hesitant second transmission from Arecibo's massive 305-m radio antenna, SOHO finally responded, confirming that it was still alive and spinning at a normal pace of one revolution per minute. An actual miracle had occurred (ESA & NASA, The SOHO Recovery 2020d).

Engineers spent the following two months refilling and warming up the recovered SOHO. Scientists were relieved to discover that their equipment survived the rapid temperature change from 100 above zero to 120 below zero degrees Celsius, with some actually improving in performance as a result.

Now that the storm had passed, SOHO appeared to be more sturdy than ever. Its life expectancy was initially only two years, but it has been prolonged multiple times. Until now it is still active around the Sun, like a lonely "sentry," continuing to diagnose the "pulse" of the Sun (Speed 2020) (Fig. 19.8).

The "buzzing" that sound waves make as they move through the Sun can be detected by SOHO using helioseismology (Domingo 1990), the most essential detection method. This acoustic detection allows researchers

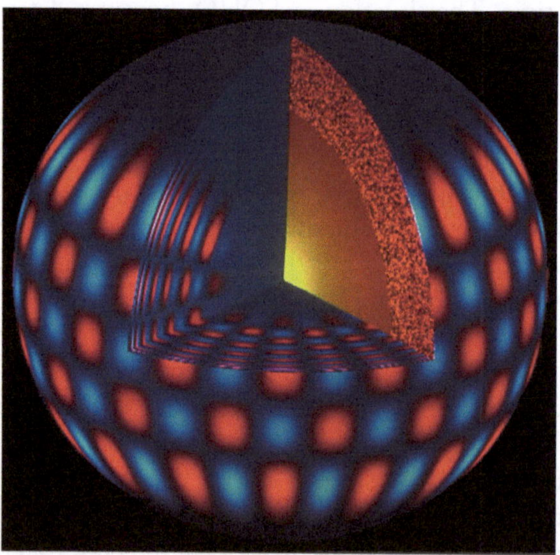

Fig. 19.8 Computer simulation of millions of patterns in the Sun's acoustic oscillations: red for moving away, blue for moving toward us. (*Source* US National Solar Observatory)

to better comprehend the Sun's inner structures, the process by which solar energy is generated, and the occurrence of disruption phenomena on the Sun's surface. Every day, the scientific sensors aboard SOHO send back data that can fill two compact discs. Thanks to the related data analyses, we have learned many fascinating fresh knowledge about the solar system's most massive star. And what exactly are SOHO's scientific accomplishments?

To begin, for the first time, a Solar 3D image was created from high-quality SOHO data. Amazing holographic imaging techniques have been utilized by scientists to rebuild features deep within the Sun, and the results are routinely uploaded to the Internet (ESA & NASA, SOHO's pioneering 25 years in orbit 2020c). Both the depth of the sunspots and the reasons for their presence for weeks remained a mystery. The sunspot, later according to SOHO's findings, is not a shallow-layer event at all. Instead, its entire structure is firmly anchored in the region of high plasma convergence and flow (Fig. 19.9).

To use an analogy, if the emergence of 3D images is like taking an X-ray of the Sun, then the results below are like taking the Sun's temperature and blood pressure. In addition to revealing previously unseen solar activity, such as coronal wave activity and coronal cyclones, SOHO gives the best data to date of the Sun's interior temperature, spin, and gas movement (Lang 1997).

Solar flares, which are sudden outbursts of solar matter, electromagnetic radiation, and high-energy particles, have been shown by SOHO to

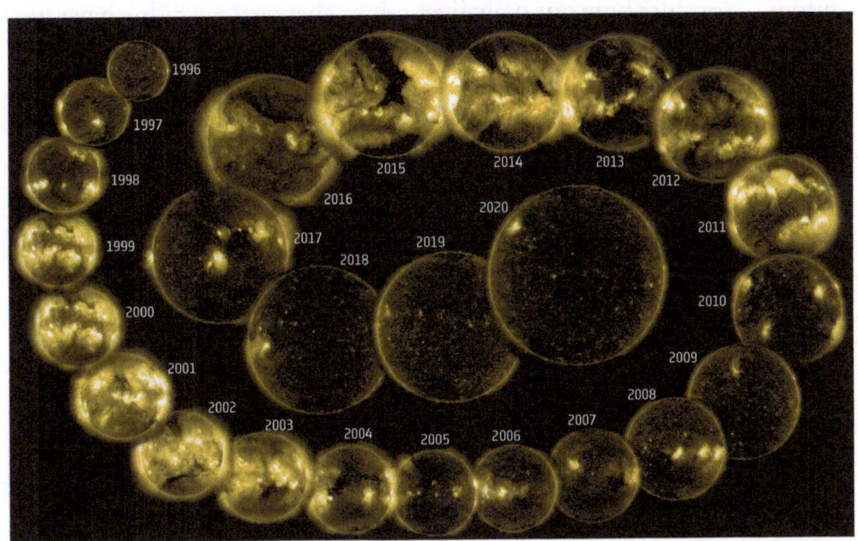

Fig. 19.9 SOHO images of the Sun from 1996 to 2020 (*Source* NASA)

be extremely hazardous. Any significant shift in "space weather" will have far-reaching effects on Earth (Anonymous 2014).

Is this anything you've ever imagined? Green, lush soil in the summer might suddenly become solid ice layers in the winter. Space scientist Mark Moldwin (Moldwin n.d.) has proven that an increase in solar activity during the 350 years between 900 and 1250 AD caused a warm period on Earth. Norse settlers arrived in Greenland when the North Atlantic was far warmer than it is today. They gave the island its lovely current name, Greenland (from the Old Norse word for "green country"). The greatest ice sheet on Earth now covers it, a spectacular transformation (Graf and Pail 2022) (Fig. 19.10).

You might assume that the Earth's warming and freezing periods are part of a very lengthy cycle and that it would be quite unlikely we experienced such dramatic shifts in temperature during lifetime. Is it, however, very close to you if the power goes out and the connection to the network is lost?

You should be well aware that the power grid is extremely vulnerable to power surges brought on by solar storms. Space weather causes significant variations in Earth's electromagnetic field, which can disrupt electrical circuits, damage power lines, and create widespread blackouts (Pirjola 2007).

The situation isn't much better for satellites. According to John W. Freeman, Professor Emeritus and Research Professor of Physics and Astronomy at Rice University (Freeman n.d.): "A solar storm is like a lightning bolt striking the surface of a satellite, which is covered in a cloud of hot electrons, and the order from the ground is comparatively little as a phantom." Consider how our daily lives might change if we didn't have access to the GPS system or weather satellites (Guven 2021).

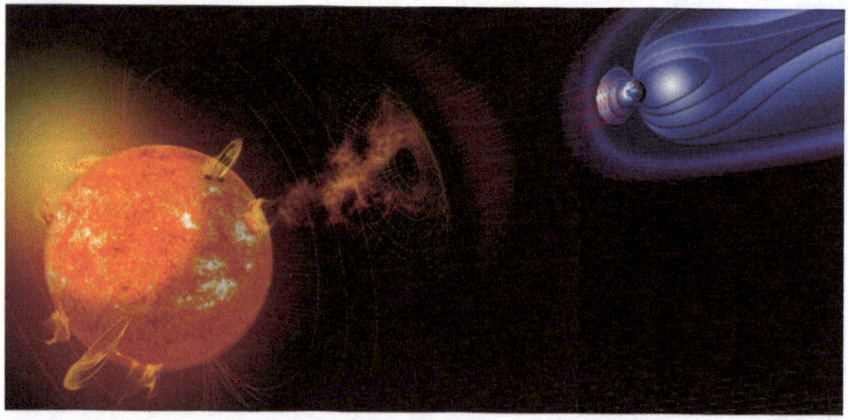

Fig. 19.10 Space weather can have a huge impact on the Earth. (*Source* NASA)

There have been very few solar storms as intense as the one that occurred in 1859. Lucky for us, the influence of power technology was minimal at the time. It may have been less intense, but the ground currents generated by the 1921 solar storm were still enough to cause havoc in the New York subway system (Hapgood 2019). In 1989, after a solar storm, a domino effect of failures in the electrical infrastructure left almost 130 million people in the US without electricity (Nagatsuma et al. 2015). Satellites fared poorly, too, with Sten Odenwald, an American astronomer, author, and NASA scientist-educator (Wikipedian n.d.), claiming that "Satellites in polar orbit were out of control for hours in a jolt, while others nearly toppled over." There were approximately $100 billion worth of satellites in Earth orbits in 2001, and a solar storm might cause billions of dollars in damage to the expensive hardware (Progressive Management 2013) (Fig. 19.11).

In short, we have come to the conclusion that extreme space weather is, like strong earthquakes and tsunamis, a rare but far-reaching natural calamity. It will certainly recur in the following decades, wreaking havoc on the world's electrical infrastructure and satellites each time.

As a result, the concept of space weather forecasting emerged (Moldwin 2008), with SOHO serving as one of the first "weathermen" on the scene. Even though SOHO can't stop natural disasters from happening, it can warn

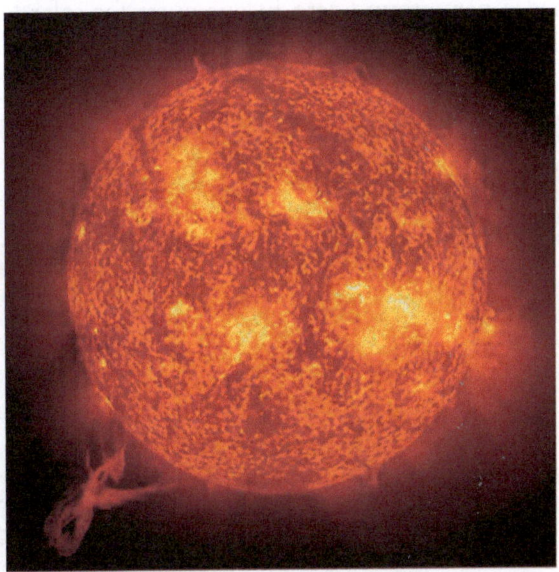

Fig. 19.11 SOHO image of the Sun in January 2000 shows giant prominences escaping the Sun's surface. If it is pinpointed at Earth, communications and power grids would be affected. (*Source* NASA)

us about them two or three days in advance. When our ability to forecast solar storms improves, we will be able to put satellites into protective mode, limit or cut off electrical use, and issue warnings to ships at sea. These precautions can not only save lives but they can also prevent damage to our delicate power infrastructure and global network (Gaunt 2016). Given this information, would you agree that a space weather prediction every 10 min is essential?

The Solar Dynamics Observatory (SDO) (Pesnell et al. 2011) was launched by NASA in 2010. It carried on SOHO's mission of probing the Sun's interior with increasing accuracy. As part of NASA's "Living with a Star" program (NASA n.d.), SDO is sending data back to Earth a thousand times more quickly than SOHO and giving us early notice of electromagnetic storms that could sweep across our vulnerable planet (Fig. 19.12).

Thus far, scientists have not been able to fully explain in detail this very simple-looking, intermediate-aged, intermediate-mass star, despite the fact that SOHO and other satellites have shown that the Sun acts as beautifully in the invisible band as the Byzantine Empire. The future of our race, all life on Earth, and possibly even the planet itself may depend on the Sun and the space weather it generates. Our Sun has been shining for billions of years, its beams traveling more than 100 million kilometers to provide life with energy and warmth.

We've reached a new ground zero in our quest for solar knowledge, and there are untold secrets awaiting the scientists of the future, including you.

Fig. 19.12 SDO (*Source* NASA)

References

Anonymous.: Unexpected impact from medium-sized solar flare. Physics 7 (2014). https://doi.org/10.1103/physics.7.s65

Bertaux, J.L.: Solar variability and climate impact on terrestrial planets. In: Solar Variability and Planetary Climates, pp. 435–444. Springer New York (2007). https://doi.org/10.1007/978-0-387-48341-2_35

Carey Jr., C.W.: Hale, George Ellery. Oxford University Press (2000). https://doi.org/10.1093/anb/9780198606697.article.1300678

Clark, S.: The Sun Kings: The Unexpected Tragedy of Richard Carrington and the tale of how modern astronomy began. Princeton University Press (2019)

Clauer, C.R., Siscoe, G.L.: The great historical geomagnetic storm of 1859: a modern look. Adv. Space Res. **38**(2), 115–388 (2006)

Domingo, V.: The Soho project and helioseismology. In: Progress of Seismology of the Sun and Stars, pp. 257–264. Springer Berlin Heidelberg (1990). https://doi.org/10.1007/3-540-53091-6_90

ESA & NASA.: SOHO Mission Interruption Joint NASA/ESA Investigation Board (1998). Retrieved from Defense Technical Information Center: https://apps.dtic.mil/sti/citations/ADA354771

ESA & NASA.: About the SOHO Mission (2020a). Retrieved from NASA: https://soho.nascom.nasa.gov/about/about.html

ESA & NASA.: SOHO's Orbit (2020b). Retrieved from NASA: https://soho.nascom.nasa.gov/about/orbit.html

ESA & NASA.: SOHO's pioneering 25 years in orbit (2020c). Retrieved from ESA Science & Technology: https://sci.esa.int/web/soho/-/soho-s-pioneering-25-years-in-orbit

ESA & NASA.: The SOHO Recovery (2020d). Retrieved from NASA: https://soho.nascom.nasa.gov/about/Recovery/index.html

Freeman, J.W. (n.d.): Faculty information system—John W. Freeman. Retrieved April 15, 2023, from Rice University: https://fis-archive.rice.edu/faculty7394.html?p=B013E57482467828

Galilei, G., Scheiner, C.: On Sunspots. University of Chicago Press (2010)

Gaunt, C.T.: Why space weather is relevant to electrical power systems. Space Weather **14**(1), 2–9 (2016). https://doi.org/10.1002/2015sw001306

Graf, M., Pail, R.: Combination of geometric and gravimetric data across Greenland for the estimation of changing ice masses. GRACE/GRACE-FO Science Team Meeting, 2022, 18–20. https://doi.org/10.5194/gstm2022-14

Guven, U.: Space Weather: Effects on Ground and Satellites. Independently Published (2021)

Hapgood, M.: The great storm of May 1921: an exemplar of a dangerous space weather event. Space Weather **17**(7), 950–975 (2019). https://doi.org/10.1029/2019sw002195

Lang, K.R.: SOHO reveals the secrets of the sun. Sci. Am. **276**(3), 40–47 (1997). https://doi.org/10.1038/scientificamerican0397-40

Malinovsky-Arduini, M., Fröhlich, C.: The solar and heliospheric observatory, SOHO—A phase-A project of the European space agency. Adv. Space Res. **4**(7), 383–392 (1984). https://doi.org/10.1016/0273-1177(84)90214-X

Progressive Management. 21st Century Complete Guide to Space Weather: Solar Storms, Impacts on Human Activity, Flares and Coronal Mass Ejections, Satellite Sun Observation, Forecasting, Carrington Event (2013)

Moldwin, M.: What is space weather? In: An Introduction to Space Weather, pp. 1–16. Cambridge University Press (2008). https://doi.org/10.1017/cbo9780511801365.002

Moldwin, M. (n.d.): Space—home for prof. Mark Moldwin. Retrieved April 15, 2023, from The University of Michigan: https://space.engin.umich.edu/

Nagatsuma, T., Kataoka, R., Kunitake, M.: Estimating the solar wind conditions during an extreme geomagnetic storm: a case study of the event that occurred on March 13–14, 1989. Earth, Planets and Space **67**(1) (2015). https://doi.org/10.1186/s40623-015-0249-4

NASA. (n.d.): Living with a star. Retrieved April 15, 2023, from NASA Science website: https://science.nasa.gov/heliophysics/programs/living-with-a-star

Pesnell, W.D., Thompson, B.J., Chamberlin, P.C.: The solar dynamics observatory (SDO). In: The Solar Dynamics Observatory, pp. 3–15. Springer US (2011). https://doi.org/10.1007/978-1-4614-3673-7_2

Pirjola, R.: Space weather effects on power grids. In: Space Weather-Physics and Effects, pp. 269–288. Springer Berlin Heidelberg (2007). https://doi.org/10.1007/978-3-540-34578-7_10

Priest, E.: 12—Solar flares and coronal mass ejections. In: Magnetohydrodynamics of the Sun, pp. 416–450. Cambridge University Press (2014). https://doi.org/10.1017/cbo9781139020732.013

Soon, W., Yaskell, S.H.: The maunder minimum and the variable sun-earth connection. World Scientific Publishing Company Incorporated (2003)

Speed, R.: Once considered lost, ESA and NASA's SOHO came back from the brink of death to work even better than it did before (2020). Retrieved from The Register: https://www.theregister.com/2020/07/29/esa_soho_space_extenders/

Wikipedian. (n.d.): Sten Odenwald. Retrieved April 15, 2023, from City of Santa Maria Public Library: https://sm.blackgold.org/Author/Home?author=%22Odenwald%2C%20Sten%22

20

Hipparcos (HIgh Precision PARallax COllecting Satellite): Measuring the Milky Way

Hipparcos (HIgh Precision PARallax COllecting Satellite)

Hipparchus, an ancient Greek astronomer, finished a small chart that contained the results of years of diligent observation in 129 BCE. He looked at it with quivering hands and an elated heart. Hipparchus, the best astronomical observer of his day, saw the sky in a way that no one else did. Most astronomers at the time believed that the positions of stars in the sky were constant relative to each other, but Hipparchus thought otherwise. After many restless nights of staring into the sky, he had finished the "Star Map" (Gysembergh et al. 2022)—a catalog of 850 stars. This is also the earliest extant star catalog (Russo 1994).

A star position catalog could seem like nothing more than a star map to you. It would serve what purpose? While the ancients could utilize the movement of the stars to foretell the weather and navigate their way, this information is of little use to us now. Check the forecast if we want to know what the weather will be like. You would be grossly underestimating the might of the stars if you had such a view. The distances between the stars and the total number of stars visible to the naked eye are actually a coded message. This complicated data conceals insights about the cosmos that are invaluable in ways you can't even begin to imagine (Fig. 20.1).

Humanity has continued the stargazing for thousands of years. The first Western star catalog was compiled by Timorius (Murdin 2001) and Aristius (Wikipedian 2023) of Alexandria in the third century BCE. Almost a century later, the astronomer Hipparchus not only refined the inventory but

J. Wang, *Eye Beyond the Sky*, Astronomers' Universe,
https://doi.org/10.1007/978-981-99-9818-0_20

Fig. 20.1 The starry sky under modern exposure technology. (*Credit* NASA)

also organized the stars by luminosity. This is the foundation upon which the modern magnitude system relies. At a later date, Ptolemy increased the number of stars in the catalog to 1,022 (Encyclopedia Britannica n.d.).

The development of photographic plates in the twentieth century allowed astronomers to finally overcome the limits of the naked eye and make precise measurements of star positions a breeze. Researchers in Europe and the United States had been emboldened to make lofty plans to map the Milky Way by the rapid improvement of observational instruments. Yet, astronomers soon encountered a problem that seemed intractable.

The Earth's dense atmosphere was the main obstacle to locating stars, as you might expect. The Earth's dense atmosphere helps keep the earthly surface at a comfortable temperature by absorbing and reflecting solar radiation, but it also obscures the field view of humans. When viewed through a dense atmosphere, stars appear fuzzy and distorted, as though viewed through a filter. Scientists had long sought a technique to see beyond clouds and other atmospheric obstructions, and with the advent of the contemporary space industry, they finally had a foolproof plan: Simply launch telescopes into orbit (Panek 1999) (Fig. 20.2).

And this chapter's hero, Hipparcos (HIgh Precision PARallax COllecting Satellite), is such a satellite that was designed to measure the Galaxy with extreme accuracy. Undoubtedly, the ancient Greek astronomer Hipparchus

Fig. 20.2 Hipparcos under test (*Credit* ESA/Science & Exploration/Space Science)

inspired its naming. Its sole mission was to take extremely precise measurements of stellar positions and distances (van Leeuwen 2008). Hipparcos' primary telescope had a relatively modest 29-centimeter aperture and 1.4-meter focal length. The telescopes used by amateur astronomers are often significantly larger (Perryman 1983) (Fig. 20.3).

Although putting a satellite into Earth orbit was relatively easy compared to landing on and exploring another planet, the tiny Hipparcos faced a perilous situation and nearly failed throughout its mission.

Hipparcos was successfully launched by an Ariane 4 rocket in France on August 8th, 1989. Despite this promising beginning, an almost devastating tragedy soon befell the satellite. Hipparcos' last engine, which was supposed to fire and propel it into a synchronous orbit from an elliptical transfer orbit, did not fire properly (Kourou 1989).

After three failed efforts to activate the engine, the European Space Agency (ESA), which was in charge of the endeavor, accepted reality. In the present situation, it was crucial to get Hipparcos operational, otherwise the 9-year expenditure of 2,000 technicians and 200 scientists would be for naught (Fig. 20.4).

Fig. 20.3 Crews installed the Hipparcos primary mirror. (*Credit* ESA)

This resulted in Hipparcos operating in an extremely elliptical orbit. Its apogee was 36,000 km from the Earth, whereas its perigee was barely 200. Only about 40% of the time, the satellite could be reached by ESA's ground stations. The Hipparcos satellite "survived the calamity" because all of its instruments were fully functional and scientists were able to adapt to the mounting pressure in time to prolong the satellite's operational lifetime. As time went on, Hipparcos labored doggedly in space until March 1993 (Mignard 1998) (Fig. 20.5).

Hipparcos employed the tried-and-true technique of measuring the parallax of stars to determine stellar distances. During the approximately 6 months when the Earth moves from one side of its orbit around the Sun to the other side, if we keep observing the star from the Earth, it will have a slight displacement, and the angle of this displacement is the stellar parallax. To visualize this, imagine a very narrow triangle drawn in space at this angle. This incredibly narrow triangle, of which one side is the diameter of the Earth's orbit, expands like a sharp awl (van Leeuwen, The HIPPARCOS

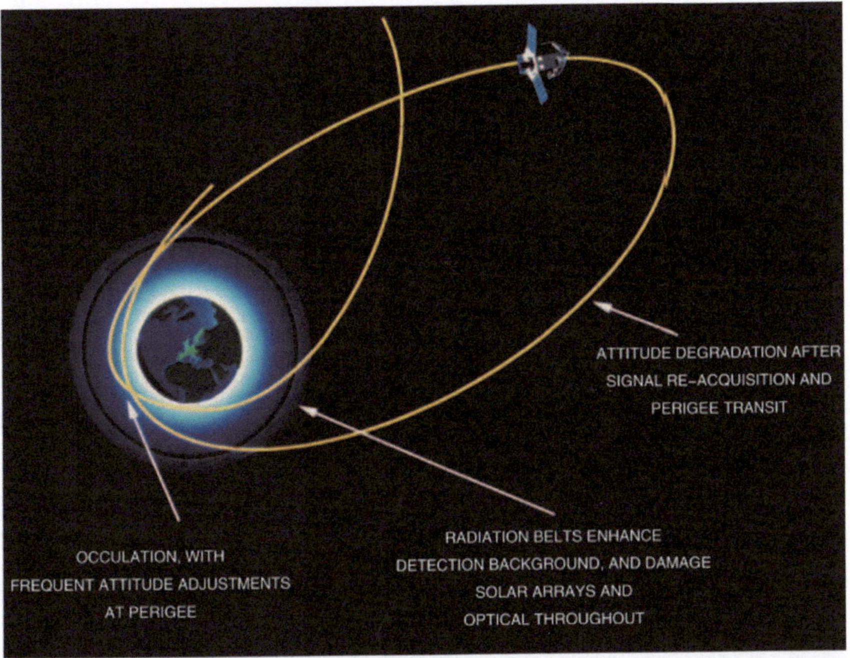

Fig. 20.4 Trajectory of Hipparcos (*Credit* ESA)

parallax for Polaris 2013). The smaller the angle, the narrower the triangle, and the length of the long side can even approach hundreds of thousands of times the length of the short side. From here, we can use trigonometric functions to determine the length of the long side, which is proportional to the distance from the star to the Earth.

Measuring the parallax of distinct stars was the most important aspect of the task that Hipparcos undertook. A more exact angle allows for a more precise measuring distance. The main telescope on Hipparcos had an accuracy of 0.002 arcseconds (Richmond n.d.). What does a precision of 0.002 arcseconds mean? This accuracy is 500 times better than what we get when we look at the stars from Earth through the atmosphere. Hipparcos' two "eyes" surveyed the sky in a wide circle for two hours at a time, and then again after 20 min. Ultimately, by merging over a hundred measurements of the star, a minimal angular error can be derived. Hipparcos has been all about precision and rigor.

I'm going to introduce what Hipparcos has done for science. Before we get started, it's essential that we acknowledge that measuring star positions is the most boring part of the astronomical exploration, but it's also the most fundamental part, and its significance is far larger than you might assume (Chaffee

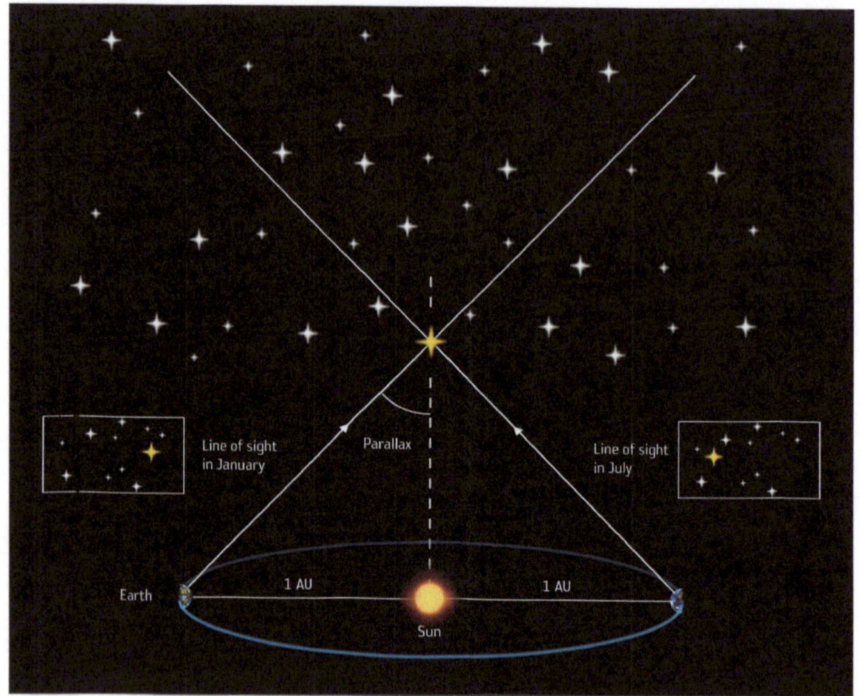

Fig. 20.5 Measuring stellar distances by parallax (*Credit* ESA/ATG medialab)

2018). Measuring the distance between stars relies on accurate measurements of their positions, which is the crucial ingredient in determining many other physical properties of celestial bodies. This is the first domino in the chain, marking the origin from which the magnitude, mass, and luminosity of any planet, star, or galaxy can be calculated. We can only perceive the glimmering look of stars in the sky without knowledge of their precise locations and distances. A star's apparent brightness decreases with distance, so even if it's very bright in reality, it will appear dim if it's very far away, and vice versa. The only way the truth can be reestablished is by precise position measurement. The first piece of the puzzle in understanding the stars has been uncovered (Heifetz 2012).

Throughout the course of its three years of operation, Hipparcos had amassed a substantial amount of information.

As was previously stated, Hipparcos' primary telescope scanned the positions of stars with an accuracy of 0.002 arcseconds. With this information, researchers had created the most precise Hipparcos catalog to date, which contained data on 118,218 stars (ESA 2019).

With a beam splitter, Hipparcos was able to scan the sky with an accuracy of 0.03 arcseconds, or one-fifteenth of the accuracy of its primary telescope, yet still captured images of 1,058,332 stars (ESA, The HIPPARCOS Space Astrometry Mission n.d.), putting it on par with telescopes capable of capturing images of millions of stars. The name "Tycho Catalogue" was given to this collection. The "Tycho-2 Catalogue," which scientists revised and expanded upon seven years after Hipparcos stopped functioning, contains an astounding 2,539,913 stars when it was first released in 2000. Included are 99% of all stars down to magnitude 11, which means almost 100,000 times fainter than the brightest star, Sirius (Urban and Wycoff 2000) (Fig. 20.6).

Hipparcos has not only boosted the total number of stars known but also increased the precision with which their distances are measured by an order of magnitude. Prior to Hipparcos, only about a hundred stars had their distances to Earth measured within 5% accuracy. Hipparcos contributed to the increase to over 7,000 (Turon 1997). A distance of 500 light years from the Sun was then within the measurement range of such precision. While this is still a very small fraction of the Milky Way, it has a wealth of stars of nearly every known type.

The Milky Way is a barred spiral galaxy, which we learned about thanks to Hipparcos. There are three common shapes for galaxies, and they are elliptical, spiral, and irregular. There are two distinct classes of spiral galaxies: barred and unbarred. Our Milky Way is a barred spiral, a type of galaxy that makes up around two-thirds of all spirals in the cosmos. The center of a barred spiral galaxy is typically a huge, fast rotator in the form of a

Fig. 20.6 Tycho-2 Catalog (*Credit* NASA/Goddard Space Flight Center Scientific Visualization Studio)

short bar, with spiral arms stretching outwards from either side of the bar, as suggested by the name. Yet the Milky Way is not a flat "pie," but rather has a curved edge, with one end pointing upward and the other downward, forming a tiny "S" shape, with the degree of curvature being more pronounced with the increasing distance from the galactic center (Science Update 2021) (Fig. 20.7).

Do you now have an impression of what Hipparcos was like? Essentially a little telescope, Hipparcos relied on straightforward observation techniques to glean even the most fundamental information about the stars. Compared to other space exploration missions, like Voyager or Cassini, the work done by Hipparcos might not seem all that significant. You would be crudely dismissing Hipparcos and the significance of astrometry if you thought so (Fig. 20.8).

This seemingly basic and dull work done by Hipparcos actually has implications for all branches of astrophysics, from planets to the cosmos. If you imagine the field of modern astronomy as a pyramid, then astrometry would be the base upon which all other theories rest. Many situations call for it than you might think.

Fig. 20.7 The Milky Way galaxy (*Credit* NASA JPL)

Fig. 20.8 A dark sky over Sequoia National Park (*Credit* NASA/A.Fujii)

Hipparcos is a potent tool for astronomers in their search for potential extraterrestrial life habitats and has helped mankind narrow down our search for terrestrial planets (Sozzetti and de Bruijne 2018). Although there are still a significant number of stars that are beyond our grasp, it has been established that hundreds of reasonably close stars contain planets, and Hipparcos has determined the most fundamental distance data for these planets. Distances to these attractive, potentially inhabited worlds range from a few to hundreds of light years. 8.8 billion Earth-like habitable worlds (Petigura et al. 2013), each with the potential to sustain life, have been estimated for the entire galaxy by astronomers.

Galactic archaeology can also benefit from Hipparcos' research. Galactic archaeology is the practice of using modern data to learn about the distant past. Using the accurate stellar motion data provided by Hipparcos, astronomers performed a magical "time travel" back 500 million years to track the Sun's trajectories as it orbited the galactic center and traveled up and down the galactic disk. Hipparcos data led some researchers to conclude that four times in the past 500 million years, the Sun had passed through the Milky Way's spiral arm, and each time almost coincided with an extraordinarily cold phase in the history of the Earth's climate. That was a fascinating discovery. Cosmic rays are thought by some scientists to have been more powerful in the spiral arms, possibly causing denser cloud covers and extended ice ages on Earth.

The Hipparcos data are also a stunning confirmation of general relativity. In 1919, during a total solar eclipse, the deflection of starlight near the edge of the Sun provided the first observational confirmation of Einstein's theory that light beams are bent by mass. The greater the distance from the Sun, the less noticeable the deflection, and the more challenging it will be to identify. That minuscule angle was picked up by Hipparcos (Fig. 20.9).

As of right now, the fundamental job of pinpointing star locations and determining their distances continues. Gaia, the next generation of astrometric satellite missions launched by ESA in late 2013, significantly extended the scope of Hipparcos' research. Gaia has performed more thorough and precise star surveys, monitoring the location, distance, brightness, and motion of one billion stars across the Milky Way (Cacciari 2014) (Fig. 20.10).

About half of the global population currently resides in urban areas. They no longer have unobstructed views of the night sky as a direct result of the expansion of light pollution. Science missions like Hipparcos and Gaia, however, can serve as their "eyes" and help them reconnect with the stars, learn about the history and destiny of the galaxy, and locate Earth within it. Our efforts to crack the galaxy's mystery continue unabated.

Fig. 20.9 Gaia Space Observatory (*Credit* ESA/ATG medialab)

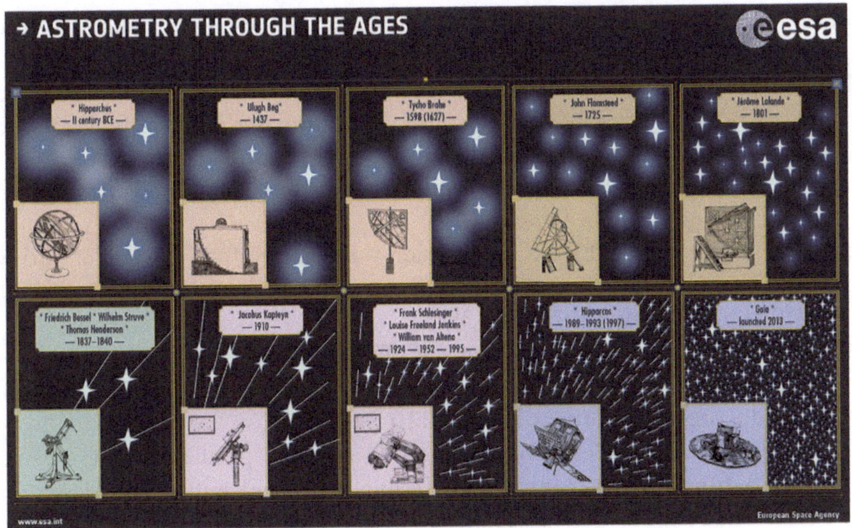

Fig. 20.10 From Hipparchus to Hipparcos (*Credit* ESA)

References

Cacciari, C.: Mapping the stellar populations of the Milky Way with gaia. In: Asteroseismology of Stellar Populations in the Milky Way, pp. 155–163. Springer International Publishing (2014). https://doi.org/10.1007/978-3-319-10993-0_18

Chaffee, K.: Astrometry measuring the stars in the sky (2018). Retrieved from Online Star Register: https://osr.org/blog/kids/astrometry-astronomy/

Encyclopedia Britannica. (n.d.): Ptolemy. Retrieved April 15, 2023, from University of Oregon: https://pages.uoregon.edu/jschombe/glossary/ptolemy.html

ESA.: The Hipparcos all-sky map (2019). Retrieved from ESA Science & Technology: https://sci.esa.int/web/hipparcos/-/52887-the-hipparcos-all-sky-map

ESA. (n.d.): The Hipparcos Space Astrometry Mission. Retrieved April 15, 2023, from ESA: https://www.cosmos.esa.int/web/hipparcos

Gysembergh, V., Williams, P.J., Zingg, E.: New evidence for Hipparchus' Star Catalogue revealed by multispectral imaging. J. Hist. Astron. **53**(4), 383–393 (2022). https://doi.org/10.1177/00218286221128289

Heifetz, M.: Measuring distances in the sky. In: A Walk through the Southern Sky, pp. 3–16. Cambridge University Press (2012). https://doi.org/10.1017/CBO9781139344531.002

Kourou, J.D.: Europe's satellite looks doomed as power fails (1989). Retrieved from NewScientist: https://www.newscientist.com/article/mg12316780-600-europes-satellite-looks-doomed-as-power-fails/

Mignard, F.: The Hipparcos double and multiple star solutions. Highlights Astron. **11**(1), 539–541 (1998). https://doi.org/10.1017/s1539299600021985

Murdin, P.: Timocharis (c. 280 BC). In: Encyclopedia of Astronomy & Astrophysics. Boca Raton: CRC Press (2001). https://doi.org/10.1888/0333750888/4057

Panek, R.: Seeing and Believing: How the Telescope Opened Our Eyes and Minds to the Heavens. Penguin Group (1999)

Perryman, M.A.: The HIPPARCOS mission—astrometry from space. In: Surveys of the Southern Galaxy, pp. 281–284. Springer Netherlands (1983). https://doi.org/10.1007/978-94-009-7217-9_39

Petigura, E.A., Howard, A.W., Marcy, G.W.: Prevalence of Earth-size planets orbiting Sun-like stars. Proc. Natl. Acad. Sci. **110**(48), 19273–19278 (2013). https://doi.org/10.1073/pnas.1319909110

Richmond, M. (n.d.): Measuring distances to stars via parallax. Retrieved April 15, 2023, from Rochester Institute of Technology: http://spiff.rit.edu/classes/phys301/lectures/parallax/parallax.html

Russo, L.: The astronomy of Hipparchus and his time: a study based on pre-ptolemaic sources. Vistas Astron. **38**, 207–248 (1994). https://doi.org/10.1016/0083-6656(94)90030-2

Science Update.: A barred galaxy's massive molecular inflow (2021). Retrieved from Center for Astrophysics: https://www.cfa.harvard.edu/news/barred-galaxys-massive-molecular-inflow

Sozzetti, A., de Bruijne, J.: Space astrometry missions for exoplanet science: Gaia and the legacy of Hipparcos. In: Handbook of Exoplanets, pp. 1205–1228. Springer International Publishing (2018). https://doi.org/10.1007/978-3-319-55333-7_81

Turon, C.: From Hipparchus to Hipparcos: Measuring the Universe (1997). Retrieved April 15, 2023, from Sky & Telescope: https://wwwhip.obspm.fr/heritage/hipparcos/SandT/hip-SandT.html

Urban, S.E., Wycoff, G.L.: Densifying the optical reference frame: the tycho-2 catalog of 2.5 million stars. In: International Astronomical Union Colloquium, vol. 180, pp. 75–79 (2000). https://doi.org/10.1017/s0252921100000130

van Leeuwen, F.: The Hipparcos mission. In: Hipparcos, The New Reduction of the Raw Data, pp. 3–37. Springer Netherlands (2008). https://doi.org/10.1007/978-1-4020-6342-8_1

van Leeuwen, F.: The HIPPARCOS parallax for Polaris. Astron. Astrophys. **550**, L3 (2013). https://doi.org/10.1051/0004-6361/201220871

Wikipedian.: Aristarchus of Samos (2023). Retrieved from Wikipedia: https://en.wikipedia.org/wiki/Aristarchus_of_Samos

21

Spitzer Space Telescope: Revealing the Galaxy's Chilling Nature

Spitzer Space Telescope

By October 4, 2003, 40 days after its launch (Caltech n.d.), the Spitzer Space Telescope had been in an Earth-trailing orbit around the Sun for a month. The 4-m-long telescope stood out against the blackness of space, cloaked in a silvery blackness like a dumb rifle barrel. It had lived up to the label "cold" by diligently performing its chilling duty over the past 40 days, reaching as low as minus 268 °C (Martin 2012) at the moment. At such a temperature, even single atoms and molecules stop moving. Furthermore, Spitzer would take great care to keep this aberrant "cool" going until its final moment (Fig. 21.1).

Spitzer is unquestionably unique among space telescopes. For what purpose did it require such a frigid setting? Knowledge of space is essential for comprehension.

Obviously, space is not completely empty. Robert Trumpler (1886–1956) (The Editors of Encyclopaedia Britannica 1998), an American astronomer, demonstrated in the 1930s that the space between stars is permeated by very thin gas and dust called the interstellar medium. The interstellar medium absorbs and scatters light from distant stars, diluting and reddening the light as it travels through (van de Kamp 1932). This causes the starlight to appear redder than it actually is. For a long time, it has been a major challenge for humanity to figure out how to peer through interstellar medium

Supplementary Information The online version contains supplementary material available at https://doi.org/10.1007/978-981-99-9818-0_21.

J. Wang, *Eye Beyond the Sky*, Astronomers' Universe,
https://doi.org/10.1007/978-981-99-9818-0_21

Fig. 21.1 Spitzer Space Telescope (*Credits* NASA/JPLCaltech)

and investigate the enormous clouds responsible for the creation of stars (Fig. 21.2).

Electromagnetic waves are necessary to resolve this issue. The electromagnetic waves in our material universe range in wavelength by a factor of trillions. They can be classified into radio waves, microwaves, infrared light, visible light, ultraviolet light, X-rays, and gamma rays, depending on their wavelength. Only a tiny fraction of it, called visible light, can be seen by the human eye (NASA n.d.). Using an analogy of an 88-key piano keyboard to represent all electromagnetic wavelengths, the human eye can only play two of those keys (Fig. 21.3).

Fig. 21.2 Interstellar cloud (*Credit* NASA/JPL-CALTECH)

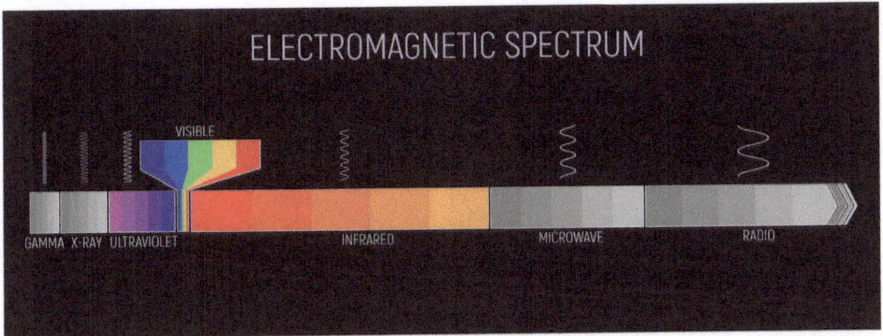

Fig. 21.3 Electromagnetic spectrum (*Credit* NASA and J. Olmsted)

It's likely that you're well-versed in different electromagnetic waves that have become ubiquitous in modern society. Radio waves allow for the broadcast of music and news; microwaves allow for the use of mobile phones; ultraviolet rays can cause skin damage and produce vitamin D at the same time; X-rays are so potent in hospitals that many patients wait in line to get checked.

In the family of electromagnetic waves, the infrared band is the one that provides the solution to the challenge of observing star clouds. While the word "infrared" may conjure images of remote controls, physiotherapy instruments, and other niche uses, its real-world applications are much broader. This is due to the fact that infrared radiation is emitted by all objects in space, including the Sun, the Moon, asteroids, and even carbon particles with a diameter of 1/10000–1/100 mm. Astronomical objects can appear very different when viewed in the infrared spectrum. In the optical image of Orion Nebula, there is a dark spot that appears to have nearly no stars; but in infrared photographs that can penetrate the dust, you can observe a dense cluster of stars in the same area that cannot be seen even with the most powerful optical telescope (NASA, ESA, & Bacon, The Orion Nebula: Visible and Infrared Views 2018) (Fig. 21.4).

In this chapter, we follow Spitzer, a space telescope designed for infrared observations. It was conceived as a tool for infrared astronomy studies and developed accordingly. It has the capability of sensitively detecting infrared signals from stars and galaxies billions of light-years away and can even see through the dust clouds that envelop most newborn stars. It's like discovering a whole new world of sensation, allowing you to see the cold, invisible cosmos for what it really is.

It was first suggested in 1970, but the Spitzer program had endured a rocky history ever since, with setbacks caused by the Space Shuttle Challenger disaster, near cancellation, political impasse, funding cuts, and more.

Fig. 21.4 Comparison of optical (left) and infrared (right) images of the Orion Nebula (Credits for near-infrared image: NASA; K.L. Luhman (Harvard-Smithsonian Center for Astrophysics, Cambridge, Mass.); and G. Schneider, E. Young, G. Rieke, A. Cotera, H. Chen, M. Rieke, R. Thompson (Steward Observatory, University of Arizona, Tucson, Ariz.) Credits for visible-light picture: NASA, C.R. O'Dell and S.K. Wong (Rice University))

After more than 30 years of planning and preparation, the $800 million (Kwok 2006) Spitzer program finally got underway in 2012. It was named after Lyman Spitzer (1914–1997), a pioneer in the field who initially recognized the potential of putting a telescope in orbit (Soter and Tyson 2000) (Fig. 21.5).

The Spitzer infrared telescope was sent into space by a Delta II rocket on August 25, 2003, from Cape Canaveral in Florida. Spitzer was 4.5 m tall, 2.1 m in diameter (NASA, Spitzer Space Telescope 2021), and had a mass on par with a compact truck. A lengthy mask, similar to a cloak, might be seen hanging behind the telescope. The black shield on the side facing away from the Sun helped dissipate heat into space while the other side of the mask facing the Sun absorbed solar energy and sheltered the telescope from radiation.

After 40 days, Spitzer was in its designated orbit and had cooled down to an astounding minus 268 °C for operation. Normally, a temperature this low would freeze a space telescope, but for Spitzer, it had been absolutely essential. This is due to the fact that Spitzer, being an infrared telescope, also generates infrared rays, severely interfering with and polluting the infrared rays emitted by other celestial objects. The pollutants can be cleaned up only

Fig. 21.5 Lyman Spitzer (1914–1997) (*Source* NASA)

if the temperature drops far enough. As a result, Spitzer had to remain "cool until the finish" to ensure reliable observational data.

The liquid helium is the Spitzer Space Telescope's secret weapon for keeping the temperature down to these extremes. Like how evaporation cools your skin, releasing liquid helium into space's vacuum effectively reduces the temperature. This is why Spitzer came prepared with a thermostat able to store 350 L of liquid nitrogen (ipac, Caltech, JPL, & NASA n.d.), roughly the size of a large truck's gas tank. Three Coke bottle caps' volume of liquid helium was what Spitzer required to maintain its temperature each day. To get the job done before the depletion of the liquid helium supply, Spitzer faced a race against time (Fig. 21.6).

Spitzer has a background radiation level of one millionth (NASA, Spitzer View of the Galactic Center) that of a ground-based infrared telescope. As a result of working in such a favorable environment, Spitzer was able to produce many astounding and significant results across the three different scales, including cosmos, galaxies, and stars.

Fig. 21.6 Illustration of a black hole discovered by Spitzer (*Credit* NASA/JPL-Caltech)

Spitzer reveals the universe's earliest stages of star creation on a galactic scale. Stars and galaxies formed rapidly in the first three billion years after the Big Bang, and Spitzer has two distinct advantages in reconstructing this thrilling era in our universe's past. The first advantage has to do with the redshift resulting from the expansion of space. Over the 13 billion light-year journey to the Earth, a star's light will have elongated into the infrared spectrum. The second advantage is that the infrared spectrum, which is the working band of the Spitzer telescope, is where the light from newly formed stars is emitted after being wrapped and absorbed by the dust of the early cosmos. Spitzer is now one of the most reliable tools to gauge the cosmic history of star formation, thanks to the "gift" of infrared light (Fazio 2021).

Spitzer's intriguing work reveals the first generation of celestial bodies formed under gravity's force at the conclusion of the "dark age" of the universe, and this is where the story of the "elders" of the celestial world begins. All first-generation stories have the potential to pique readers' interest. The scientists first removed information about foreground stars and neighboring interstellar objects from Spitzer's infrared photos, leaving only the earliest diffuse radiation, which they then studied for oscillations (Stapelfeldt 2005). Alexander Kashlinsky (Mohr 2023), an astrophysicist, clearly illustrates the process by using the following metaphor: "Imagine attempting to view fireworks over a highly populated city at night, and if you can totally turn off the city lights, you can see the fireworks. Well, scientists did dim the city lights so that we could make out the fireworks' original silhouette" (Fig. 21.7).

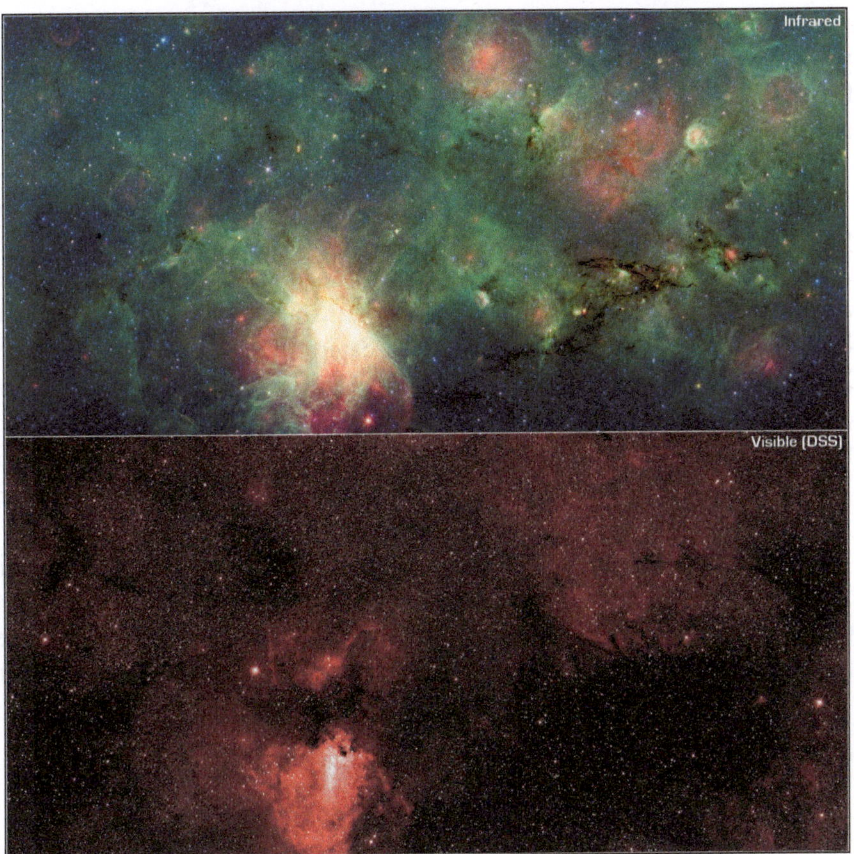

Fig. 21.7 Spitzer's images show star formation in the Omega Nebula (M17) (*Credit* NASA/JPL-Caltech/M. Povich (Penn State Univ.)

The Spitzer Space Telescope has conducted an aggressive-sounding study of the entire Milky Way galaxy. The program's official term is the "Galactic Legacy Infrared Mid-Plane Survey Extraordinaire," or GLIMPSE (Churchwell et al. 2009) for short. Spitzer's sensitivity is 100 times higher, and its resolution is 10 times higher than that of the earlier sky surveyors. The GLIMPSE project's painstaking stitching of 110,000 pointing locations and 840,000 photos yielded a priceless seamless "jigsaw" brimming with vibrant hues like blue, red, and green. Please understand that I am referring to Spitzer's data in the infrared spectrum, and that the colors shown in the "jigsaw" are not the same as those seen in the visible spectrum (Churchwell et al. 2009).

Radiation from elderly stars is responsible for the blue in this "jigsaw," while radiation from young stars that are still in the cold dust "womb" is responsible for the red. The green comes from stellar "embryos" and is mostly caused by radiation from complex chemicals known as polycyclic aromatic hydrocarbons. This molecule, which results from the incomplete combustion of carbon, can be found, for example, in-vehicle tailpipes and barbecue grills. A "guide" like this will help you piece together the jigsaw properly (Fig. 21.8).

A new perspective on the Milky Way's shape has also been revealed thanks to the GLIMPSE survey. More than 100 million stars were counted and used by the GLIMPSE imaging team to create a drawing of the Milky Way's spiral arms and central bars. This image is significant because it provides compelling evidence for the theory that the Milky Way is a barred spiral galaxy. The Norma Arm, the Perseus Arm, the Carina-Sagittarius Arm, and the Scutum-Centaurus Arm were all assumed to be part of the Milky Way's spiral structure before the surveys of Spitzer and Hipparcos. The GLIMPSE findings reveal that the Milky Way's only two major spiral arms are the Scutum-Centaurus Arm and the Perseus arm, which branch off the bar in the galactic center like the left and right hands of a human being (Vorontsov-Velyaminov 1970).

Spitzer can be used to uncover the origins of the solar system on a galactic scale. Dusty, chilly nebulae are not only stellar "delivery rooms" but also potential planet formation sites. Little particles aggregated to form larger dust grains, observed through Spitzer, illustrating the disk's evolution in its

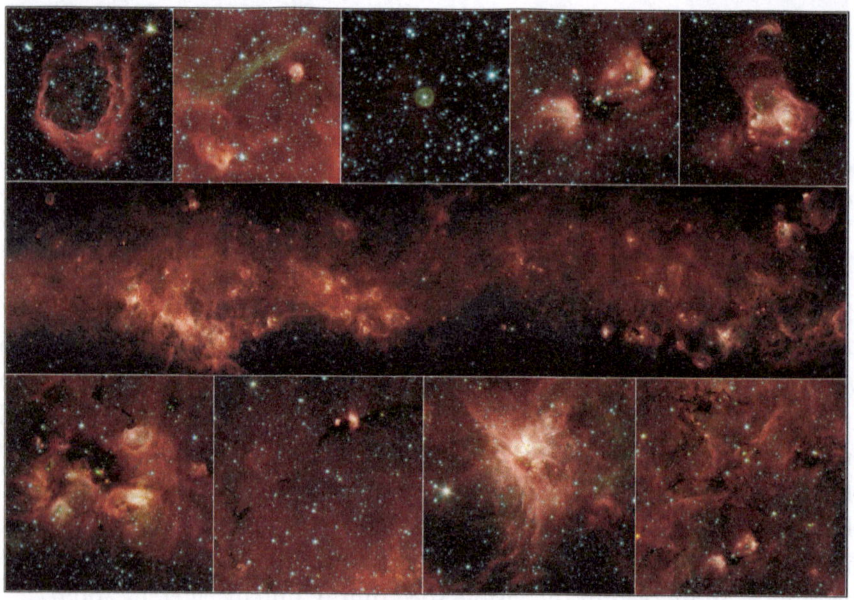

Fig. 21.8 Seamless "jigsaw" (mid-level) and detailed view of the Milky Way produced by the Spitzer GLIMPSE project (*Credit* NASA/JPL-Caltech/E. Churchwell (University of Wisconsin-Madison) and the GLIMPSE Team)

Fig. 21.9 Spitzer images of the Eagle, Omega, Trefoil, and Lagoon Nebulae, all located in the Sagittarius Arm of the Milky Way (*Credit* NASA/JPL-Caltech)

early stages. For rocky planets like Earth and Mars, the process of accretion involved the gradual buildup of material from gravel to a rock, to a hill, and finally to a planet. It's possible that this will take 10,000,000 years. The birth of the solar system and its subsequent evolution can be pieced together by comparing dust disks at various times in the process (Fig. 21.10).

Fig. 21.10 Spitzer obtained evidence of several rocky collisions in the distant solar system. (*Credit* NASA/JPL-Caltech)

Upon hearing Spitzer's "work resume," how could you not be concerned for its future? Wasn't its total capacity for liquid helium only 350 L? What would happen to Spitzer once the liquid helium was depleted? The liquid helium in Spitzer was finally depleted on May 15, 2009, after the spacecraft had been in service for six years (Michael W. Werner et al. 2022). Spitzer's long-wavelength observing instruments were rendered useless without liquid helium (Fig. 21.11).

The scientific work done by Spitzer, however, continued. It was an unexpected "highlight moment" in Spitzer's life when he was at his lowest. Its late-stage accomplishments are so unexpectedly surprising that they considerably surpassed scientists' initial predictions.

While Spitzer's long-wave equipment had lost its sensitivity, the spacecraft's two short-wave channels remained active and could be transferred to the "warm mission" (Gardner et al. 2007) phase. Spitzer was reborn a second time due to this warm mission. Its "warmth" is, however, simply relative; in absolute terms, the temperature was still quite cold (Fig. 21.12).

During Spitzer's "warm mission," the emphasis turned to "hot" exoplanets whose orbital planes aligned with the line of sight. Because the infrared radiation from both the star and the planet is reduced when the planet transits in front of its parent star, the size and temperature of an exoplanet can be determined from spectral measurements taken in different wavelength ranges. Nevertheless, it is remarkable that Spitzer can also identify the atmospheric and even geological components of exoplanets. According to Spitzer's project scientist Michael Werner (Werner n.d.): "Spitzer data can be used

Fig. 21.11 Twelve images of the universe taken by Spitzer (*Credit* NASA/JPL-Caltech)

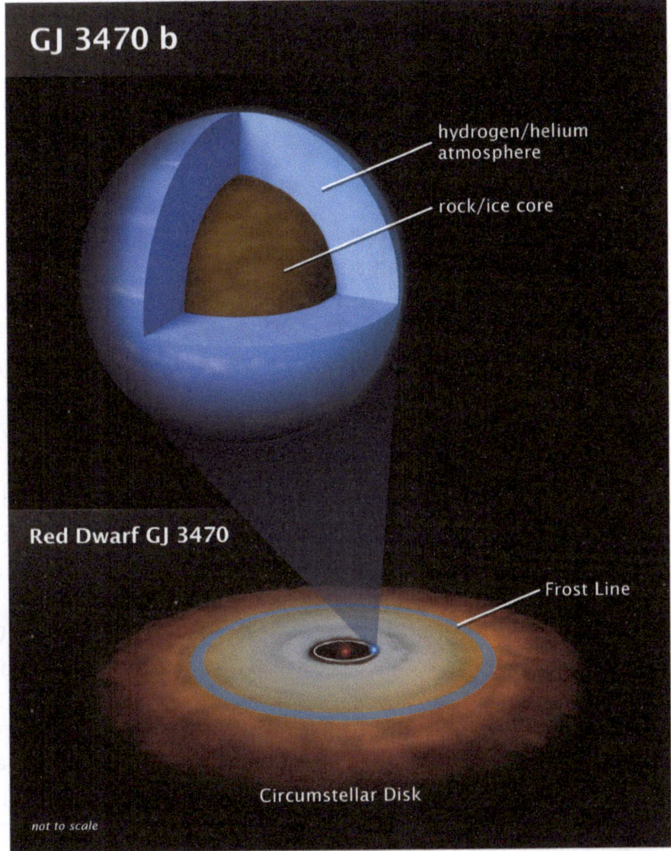

Fig. 21.12 Spitzer teamed up with Hubble to determine the structure of exoplanet GJ3470b (*Credit* NASA, ESA, and L. Hustak (STScI))

to determine the temperature of the planet and give constraints on atmospheric structure, chemical composition, and so on because the wavelengths produced by areas of different chemical composition at different levels in the atmosphere are different." While Spitzer's primary function is to uncover objects shaded by interstellar dust, it may prove most useful in analyzing the peculiarities of exoplanets (Kenney 2017).

After 11 years of "warm mission," the exhausted Spitzer Space Telescope officially completed its mission on January 30, 2020. It began aimlessly floating through space (ipac, Caltech, JPL, & NASA, Spitzer Space Telescope Mission overview n.d.).

I believe that our cosmos must be full of new stars and exoplanets and it must be full of life. But not all of them are so easily accessible. That's why we keep sending fresh space-based "eyes" into orbit. These frigid instruments are heroes in the name of humanity. As long as they keep moving forward, who knows what incredible feats they will accomplish?

References

Caltech. (n.d.): Spitzer Space Telescope Mission. Retrieved 15 April 2023, from Caltech: https://www.spitzer.caltech.edu/mission

Churchwell, E., Babler, B.L., Meade, M.R., Whitney, B.A., Benjamin, R., Indebetouw, R., Bracker, S.: The Spitzer/GLIMPSE surveys: a new view of the Milky Way. Publ. Astron. Soc. Pac. **121**(877), 213–230 (2009). https://doi.org/10.1086/597811

Fazio, G.: The Spitzer space telescope: exploring the infrared universe. Scientia (2021). https://doi.org/10.33548/scientia688

Gardner, J.P., Fan, X., Wilson, G., Stiavelli, M., Storrie-Lombardi, L.J., Silbermann, N.A.: A Spitzer Warm Mission Ultra-Wide Survey as A Target Finder for The James Webb Space Telescope. AIP Conference Proceedings (2007). https://doi.org/10.1063/1.2806782

ipac, Caltech, JPL, & NASA. (n.d.): Spitzer Space Telescope Mission Overview. Retrieved 23 April 2023, from Caltech: https://www.spitzer.caltech.edu/mission/mission-overview

ipac, Caltech, JPL, & NASA. (n.d.): Spitzer Telescope handbook. Retrieved 15 April 2023, from IRSA—Spitzer Documentation & Tools: https://irsa.ipac.caltech.edu/data/SPITZER/docs/spitzermission/missionoverview/spitzertelescopehandbook/4/

Kenney, K.L.: Exoplanets: Worlds Beyond Our Solar System. Twenty-First Century Books TM (2017)

Kwok, J.: Finding a Way: The Spitzer Space Telescope Story (2006). Retrieved from NASA APPEL Knowledge Services: https://appel.nasa.gov/2006/10/01/finding-a-way-the-spitzer-space-telescope-story/

Martin, J.: A Thousand Days in the Life of the Infrared Universe (Photos) (2012). Retrieved from CNET: https://www.cnet.com/pictures/a-thousand-days-in-the-life-of-the-infrared-universe-photos/2/

Mohr, K.: Alexander Kashlinsky Maniac Lecture (2023). Retrieved from NASA Earth Sciences: https://earth.gsfc.nasa.gov/maniac/kashlinsky

NASA: Spitzer Space Telescope (2021). Retrieved from eoPortal: https://www.eoportal.org/satellite-missions/spitzer

NASA. (n.d.): Introduction to the Electromagnetic Spectrum. Retrieved 15 April 2023, from NASA Science website: https://science.nasa.gov/ems/01_intro

NASA. (n.d.): Spitzer View of the Galactic Center. Retrieved 15 April 2023, from Caltech: https://coolcosmos.ipac.caltech.edu/system/media_files/binaries/73/original/galcen_back_panels.pdf?1375834871

NASA, ESA, & Bacon, G.: The Orion Nebula: Visible and Infrared Views (2018). Retrieved from NASA Scientific Visualization Studio: https://svs.gsfc.nasa.gov/30959

Soter, S., Tyson, N.D.: Lyman Spitzer and the Space Telescope (2000). Retrieved 15 April 2023, from American Museum of Natural History: https://www.amnh.org/learn-teach/curriculum-collections/cosmic-horizons-book/lyman-spitzer-hubble-telescope

Stapelfeldt, K.: Planet formation studies with the Spitzer Space Telescope. Space **2005** (2005). https://doi.org/10.2514/6.2005-6670

The Editors of Encyclopaedia Britannica: Robert Julius Trumpler (1998). Retrieved from Encyclopedia Britannica: https://www.britannica.com/biography/Robert-Julius-Trumpler

van de Kamp, P.: On the absorption of light in the galactic system. Astron. J. **42**, 97 (1932). https://doi.org/10.1086/105123

Vorontsov-Velyaminov, B.A.: Spiral structure of our galaxy and of other galaxies. In: The Spiral Structure of Our Galaxy, pp. 15–17. Springer Netherlands (1970). https://doi.org/10.1007/978-94-010-3275-9_3

Werner, M.W., Lowrance, P.J., Roellig, T., Gorjian, V., Hunt, J., Bradford, C.M., Krick, J. (2022). On-orbit performance of the Spitzer space telescope: science meets engineering. Instrum. Methods Astrophys. **8**(1)

Werner, M. (n.d.): JPL Science: Michael Werner. Retrieved 15 April 2023, from NASA: https://science.jpl.nasa.gov/people/mwerner/

22

Chandra X-ray Observatory: Probing into a Violent Universe

Chandra X-ray Observatory

An artillery officer in his forties was resting on a bed in a field hospital in Germany at the end of 1915, surrounded by the dense smoke of the First World War and aware that his days were numbered because of a terminal illness. In light of this, he spent countless hours laboring over the rough draft, even though no one in the hospital could decipher the mathematical symbols he used.

Anyone familiar with the artillery lieutenant's last name of Schwarzschild would be shocked by his background. Karl Schwarzschild (1873–1916) (Karl schwarzschild—Important scientists, 2009) was the head of the Leibniz Institute for Astrophysics Potsdam and a respected member of the Royal Prussian Academy of Sciences before he was called to serve in the war. Schwarzschild solved Einstein's field equation on shaky hospital drafting paper while confined to his bed. He died the following year having missed out on the significance of his calculation (Schwarzschild, 1992). Yet, his findings would go on to form the basis of a significant subfield of astronomy (Goenner, 2017), and the solution he calculated on the rough copy would be given the name "Schwarzschild black hole" (Journey into a Schwarzschild black hole, n.d.) decades later, becoming a significant scientific term that still influences our understanding today (Fig. 22.1).

When it comes to black hole, I think everyone has heard of it. American theoretical physicist John Archibald Wheeler (1911–2008) coined the term in 1967 (American Institute of Physics, 2008). It describes one of the most peculiar objects we know of in the universe. The Nobel Prize-winning

J. Wang, *Eye Beyond the Sky*, Astronomers' Universe,
https://doi.org/10.1007/978-981-99-9818-0_22

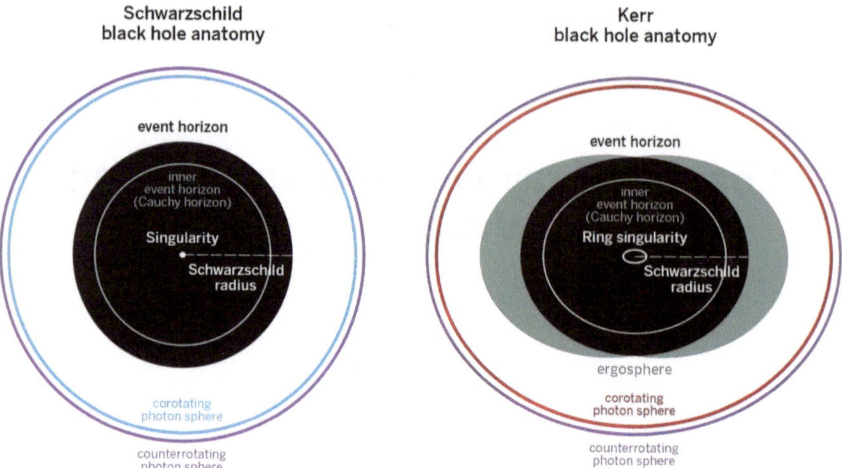

Fig. 22.1 Schematic diagram of Schwarzschild and Kerr black hole (Source: Voice of Science)

theoretical physicist Kip Thorne (Gregersen, 2017) has claimed: "Of all the conceptions of the human mind from unicorns to gargoyles to the hydrogen bomb, perhaps the most fantastic is the black hole: a hole in space with a definite edge over which anything can fall and nothing can escape; a hole with a gravitational field so strong that even light is caught and held in its grip; a hole that curves space and warps time" (Thorne, 1977).

Although the concept of a black hole was well-known, it had for a long time only existed in the theoretical calculations of scientists, and it was unknown whether such a peculiar celestial object actually existed in the real cosmos. Some argued that black holes might never exist outside of mathematical speculation and could not be proven through observation. Their reason lay in that a black hole is an entirely opaque object (not even light can escape its gravitational pull), and studying one turned impossible.

X-ray observation, on the other hand, came into being in the 1960s and marked the beginning of a new technique era. Theoretically, this technique can be used to observe black holes during astronomical observations. X-ray astronomy is the study of the universe through the use of X-rays, which are emitted by celestial bodies (Tucker & Giacconi, 1985) (Fig. 22.2).

The X-ray telescope throne goes to the Chandra X-ray Observatory (CXC, n.d.), the protagonist in this chapter. With its help, a violent cosmos with great collisions, explosions, and radiation all the time has been revealed to humanity for the first time, showing mankind the other side of the deep and tranquil universe.

Fig. 22.2 A typical black hole (Credit: NASA's Goddard Space Flight Center background, ESA/Gaia/DPAC)

At the Kennedy Space Center in Florida, the US, on July 23, 1999, the Chandra X-ray Observatory was successfully launched into space by the Space Shuttle Columbia (Mohon, 2015). For X-ray astronomy, Chandra's launch marked the beginning of the spectroscopic period, which followed the photometric era (Fig. 22.3).

NASA has a long-standing tradition of honoring scientists by naming telescopes and probes after them, usually for achievements directly connected to

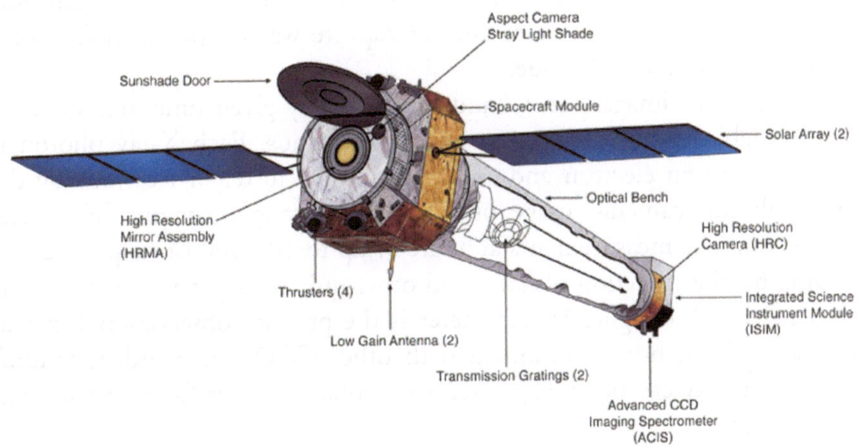

Fig. 22.3 Chandra X-ray observatory component map (Credit: NASA/CXC Program)

the equipment's scientific purpose. NASA's series of Great Observatories satellites are four large, powerful space-based astronomical telescopes launched between 1990 and 2003, each following this tradition. They are the Hubble Space Telescope launched in 1990, the Compton Gamma Ray Observatory launched in 1991, the Chandra X-ray Observatory launched in 1999, and the Spitzer Infrared Space Telescope launched in 2003 (Nakra, 2020).

Subrahmanyan Chandrasekhar (1910–1995), popularly known as Chandra (Wali, 2010), was a well-known Indian-American theoretical physicist who received the Nobel Prize in Physics in 1983. Chandrasekhar limit, the theoretical maximum mass of a white dwarf, is around 1.4 times the mass of the Sun, and is his most notable piece of work (Chandrasekhar, 1931). The stellar debris will collapse into an extremely dark celestial body, a neutron star, or, if the mass of the celestial body is very great, a black hole, as long as the gravitational pull of the celestial body exceeds the resistance of any interaction between particles. Chandrasekhar proposed this hypothesis at an early age, but it was harshly criticized by the renowned British scientist Arthur Eddington (1882–1944) (Douglas, 1998), almost causing him to give up (Wali, Chandrasekhar vs. Eddington—an unanticipated confrontation, 1982). Fortunately, evidence will always play a deciding role in determining whether a scientific conclusion is valid. The legend of Chandra serves as undeniable proof of this, and it has been recorded for posterity in the annals of science (Fig. 22.4).

With very small fields of view and low angular resolution, the best X-ray telescopes before the launch of the Chandra X-ray Observatory were on par with the best optical telescopes of Galileo's day. As soon as astronomers began operating Chandra, the sensitivity increased by several orders of magnitude, and the image detail they were able to capture was on par with that of a medium-sized optical telescope.

There are two imagers on Chandra, and at any given time, one of them would be able to pick up X-rays from the cosmos. Each X-ray photon is converted into an electron and amplified into a cluster of electrons by the high-resolution cameras' usage of vacua and strong magnetic fields. The cameras can take measurements at a rate of up to 100 000 times per second, allowing for the detection of flares and other quick cosmic occurrences. The Advanced CCD Imaging Spectrometer is the primary observational instrument aboard Chandra. Combined with other CCDs on board, Chandra's imaging capacity can be one hundred times that of earlier X-ray instruments (Arcand, n.d.).

In reality, the universe's depths are not as tranquil as summer nights. The X-ray band is a valuable tool for studying the violent cosmos and provides

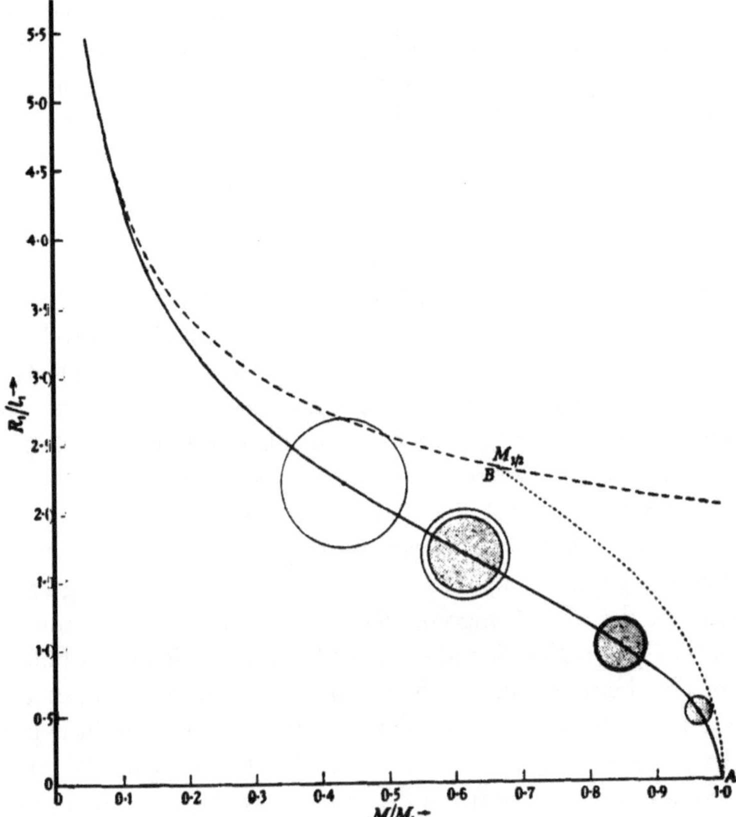

Fig. 2. The full-line curve represents the exact (mass-radius)-relation (l_1 is defined in equation (46) and M_3 denotes the limiting mass). This curve tends asymptotically to the ---- curve appropriate to the low-mass degenerate configurations, approximated by polytropes of index 3/2. The regions of the configurations which may be considered as relativistic ($\varrho > (K_1/K_2)^3$) are shown shaded. (From Chandrasekhar, S., Mon. *Not. Roy. Astr. Soc., 95, 207* (1935).)

Fig. 22.4 In his paper published in 1931, Chandrasekhar first proposed the concept of "Chandrasekhar limit" (Public Domain)

evidence of some cosmic events of unfathomable intensity. To what end is that? The primary cause is that the X-rays emitted by typical stars like the Sun are quite faint in comparison to the visible light they mainly generate. It is only through highly extreme physical processes that X-rays can be created at temperatures of hundreds of thousands or millions of degrees Celsius or through the acceleration of particles to extremely high energies (NASA, 2013). Consequently, it is reasonable to assume that the discovery of X-rays in the cosmos indicates the occurrence of a very "violent" event. You can probably guess how high expectations the scientific community had for the Chandra X-ray Observatory.

Each X-ray photon is extremely valuable since the number of X-ray photons produced by an astronomical burst is significantly lower than that of visible light beams produced. X-ray astronomy is focused on capturing these individual photons. It can take up to 2 weeks for Chandra to collect just two or three photons, but that's more than enough to verify the existence of a source. Many scientific articles have more authors than photons actually measured.

Everyone's high hopes for the Chandra X-ray Observatory were finally realized. As more information was gathered, a picture of a violent, high-energy universe emerged, and the mystery of black holes began to unravel.

In 1964, a team led by scientist Riccardo Giacconi (1931–2018) (Chandra, 2018) utilized ground-based X-ray telescopes to discover a unique X-ray source they dubbed Cygnus X-1 (Tananbaum, Gursky, Kellogg, Giacconi, & Jones, 1972). After some time, it was speculated that Cygnus X-1 was a black hole, and the X-rays had originated from its accretion disk. However, ground-based X-ray telescopes were too feeble to provide compelling evidence. Stephen Hawking (1942–2018), a betting enthusiast, had fueled the fire of this ongoing debate in the scientific community over the nature of Cygnus X-1. Whether or not Cygnus X-1 is a black hole was the subject of a bet between him and Kip Thorne, and Hawking bet that it is not (Brooks, 2017). The object's rising fame can be traced back to its appearances in the American TV series Star Trek, a song by the rock band RUSH (Peart, 1977), and a number of works of literature. Hawking conceded publicly in 1990 that he had lost the bet, but the key factor in his concession had been his intuition. In fact, the absence of conclusive evidence that Cygnus X-1 is a black hole remained. Cygnus X-1 had risen to a top priority for Chandra's observational efforts (Fig. 22.5).

Fig. 22.5 Cygnus X-1's optical image (left) and its artistic illustration (right) (Credit: X-ray: NASA/CXC; Optical: Digitized Sky Survey)

Cygnus X-1, as seen in an X-ray image taken by the Chandra X-ray Observatory, appears as a diffuse, fan-shaped luminous object stimulated by the jet as it plows into the interstellar medium (NASA & CXC, The Amazing Universe, n.d.). Wallace Tucker, an American astrophysicist who specializes in high-energy astrophysics (Tucker, 1984), after analyzing the history of Chandra's discovery, concluded that Chandra had provided substantial evidence for the existence of black holes (Fig. 22.6).

The black hole, found in the nearby spiral galaxy M33 in 2007 with the help of Chandra observations, has a mass 16 times that of the Sun, making it the then-heaviest stellar black hole ever found. The fact that a huge star 70 times the mass of the Sun enters a binary orbit with this black hole is even more intriguing. There has been no revelation yet of the mechanism responsible for this phenomenon's creation. For the first time, astronomers have been able to precisely determine the star's masses and other features, thanks to Chandra's successful observation of an occultation-displaying binary black hole system. Like its smaller companion, the enormous one will end up as a black hole (CXC, Heaviest stellar black hole discovered in nearby galaxy, 2007) (Fig. 22.7).

Two black holes were also spotted by Chandra in the galaxy merger. There is just a distance of about 3 000 light-years between the two enormous black holes. In the next 100 million years, they will most likely collide and form

Fig. 22.6 Image of Cygnus X-1 taken by Chandra (Credit: NASA/CXC)

Fig. 22.7 M33's X-ray optical image (bottom left) and its artistic illustration (Credit: Illustration: NASA/CXC/M.Weiss; X-ray: NASA/CXC/CfA/P.Plucinsky et al.; Optical: NASA/STScI/SDSU/J.Orosz et al.)

a single entity, which will cause a catastrophic gravitational wave. A single, larger black hole will dwell in the center of the merged galaxy. This is the first direct evidence that mergers and accretion can cause a galaxy's central supermassive black hole to develop over hundreds of millions of years.

Astronomers at the time used these observations to predict that in the future, they would be able to witness an unfathomably violent event in the universe: the merger of two black holes into a single "monster" in a spiral of death, releasing energy equivalent to several solar masses in an instant. The forecast came true in the year 2016. As a major breakthrough for science, gravitational waves produced by the merger of two black holes were detected by the Laser Interferometer Gravitational-Wave Observatory (LIGO) in the United States (Chandra, EXCITING NEWS: Direct Detection of Gravitational Radiation, 2016) (Fig. 22.8).

We live in a universe where cosmic phenomena like these occur constantly, and by the time their signals reach Earth, they are incredibly diluted. Nevertheless, once such occurrences happen in our galaxy, our perception of them will be as intuitive as witnessing a nuclear bomb explosion for the first time. We can't even begin to fathom the "violence" of the cosmos.

Thousands of times denser than the stars near the solar system in the "urban–rural intersection" of the Milky Way, the star clusters surrounding

Fig. 22.8 The spiral galaxy M51 is merging with its smaller companion. (Credit: NASA/CXC/M.Weiss; X-ray: NASA/CXC/CfA/P.Plucinsky et al.; Optical: NASA/STScI/SDSU/ J.Orosz et al.)

the supermassive black hole Sagittarius A* with about 4 million times of the Sun (Boen, 2015) at the center of our galaxy are a good indicator of the cosmic violence. The galaxy's supermassive black hole is as calm as a freshly fed predator stroking its fur after devouring a large meal. Scientists theorize that a black hole's activity is stimulated when stuff falls into it, causing it to emit high-energy particles that clear the area around it and calm it down until the black hole's next "prey" is grabbed by gravity. Because of this, supermassive black holes are typically dormant. This explanation is supported by data from the Chandra X-ray Observatory. Several X-ray lobes and plasma blocks were spotted in close proximity to the galactic core black hole, suggesting the black hole is active on a quasi-periodic basis. More locally, it found what may be the "recoil" of supermassive black holes in the form of huge bubble-like cavities and wave ripples in the centers of some galaxies (Fig. 22.9).

Deep sky surveys are required to learn more about the evolution of black holes and galaxies. The initiative was taken by Chandra. By surveying the entire sky over the course of 3 weeks, the Chandra Deep Field South observation captured the deepest and most distant X-ray image to date. How much of the sky was actually being observed? Only two square degrees in the area. To a lesser extent than the angle formed by holding a stamp at shoulder height,

Fig. 22.9 Images taken by Chandra (Credit: NASA)

Chandra has found over 2,000 supermassive black holes in such a limited scan, giving astronomers a wealth of information with which to examine the history of black holes and galaxies (Mohon, Deepest x-ray image ever reveals black hole treasure trove, 2017) (Fig. 22.10).

Scientists have discovered a connection between star creation and the emergence of black holes in galaxies by analyzing data from Chandra's sky surveying mission. For the first few billion years following the Big Bang, the most gigantic black holes, which were hundreds of millions of times more massive than the Sun, were in a feeding frenzy. As the first to develop, they devoured the "youth" of the huge galaxies that accompanied them. They were then compelled to cut back on, or abstain from, "eating," and consequently clam up.

These black holes, with masses between 10 and 100 million times that of the Sun, are more "gentle" than the earlier. They consume matter more slowly and over longer periods of time. This is why these black holes continue to release radiation even now. Furthermore, the survey solves a long-standing puzzle about the X-ray sky. Ever since the first X-ray telescope gathered data, scientists had noticed a faint layer of X-ray background radiation throughout the whole X-ray sky. The extensive surveys performed by Chandra and XMM-Newton (ESA, n.d.) confirmed that this background radiation was the X-rays produced by a large number of active galaxies located between three and eight billion light-years away (Fig. 22.11).

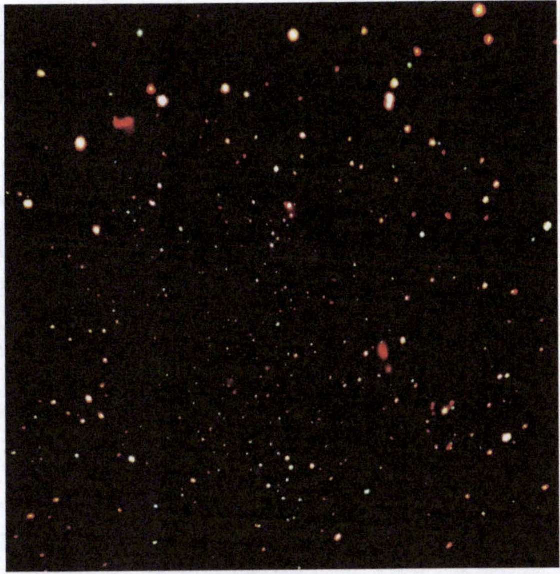

Fig. 22.10 The Chandra Deep Field North (CDF-N) (Credit: NASA/ESA/CXC/Penn State/D.M. Alexander, F.E. Bauer, W.N. Brandt et al.)

The Chandra X-ray Observatory is part of a series of instruments that have provided a new window through which to study the cosmos, revealing for the first time the essential nature of stellar black holes and neutron stars and the widespread presence of supernova remnants in galaxies. They also mark a new stage in the evolution of human astronomy by providing evidence for the existence of supermassive black holes at the centers of galaxies and advancing the search for dark matter and dark energy.

Fig. 22.11 Image of a supermassive black hole based on The Chandra Deep Field South (CDF-S) (Credit: NASA/CXC/Penn. State/G. Yang et al. and NASA/CXC/ICE/M. Mezcua et al.; Optical: NASA/STScI; Illustration: NASA/CXC/A. Jubett)

References

American Institute of Physics. John Archibald Wheeler 1911–2008. (2008). https://doi.org/10.1063/PT.5.022125

Arcand, K. Image processing. (n.d.). Retrieved April 16, 2023, from Chandra X-ray Center: https://chandra.harvard.edu/imaging/

Boen, B. Supermassive black hole sagittarius A*. (2015). Retrieved from NASA: https://www.nasa.gov/mission_pages/chandra/multimedia/black-hole-SagittariusA.html

Brooks, M. Odds on: Five scientific theories decided by wager. (2017). Retrieved from New Scientist: https://www.newscientist.com/article/2130956-odds-on-five-scientific-theories-decided-by-wager/

Chandra. EXCITING NEWS: Direct Detection of Gravitational Radiation. (2016). Retrieved from Fresh Chandra News: https://chandra.si.edu/blog/node/587

Chandra. A hero of the heroic age of astronomy. (2018). Retrieved from Fresh Chandra News: https://chandra.harvard.edu/blog/node/704

Chandrasekhar, S.: The maximum mass of ideal white dwarfs. Astrophys J **74**, 81 (1931). https://doi.org/10.1086/143324

CXC. Heaviest stellar black hole discovered in nearby galaxy. (2007). Retrieved from Phys.Org: https://phys.org/news/2007-10-heaviest-stellar-black-hole-nearby.html

CXC. Chandra X-ray Observatory – NASA's flagship X-ray telescope. (n.d.). Retrieved April 16, 2023, from Chandra X-ray Center: https://chandra.harvard.edu/

Douglas, A.V. Arthur Eddington. (1998). Retrieved from Encyclopedia Britannica: https://www.britannica.com/biography/Arthur-Eddington

ESA. SOC home page–XMM-Newton. (n.d.). Retrieved April 16, 2023, from Cosmos: https://www.cosmos.esa.int/web/xmm-newton

Goenner, H.: General relativity and the growth of a sub-discipline "gravitation" in Germany. Eur. Phys. J. H **42**(3), 395–430 (2017). https://doi.org/10.1140/epjh/e2017-70057-4

Gregersen, E. Kip thorne. (2017). Retrieved from Encyclopedia Britannica: https://www.britannica.com/biography/Kip-Thorne

Journey into a Schwarzschild black hole. (n.d.). Retrieved April 16, 2023, from University of Colorado Boulder: https://jila.colorado.edu/~ajsh/insidebh/schw.html

Karl schwarzschild - Important scientists. (2009). Retrieved April 16, 2023, from The Physics of the Universe: https://www.physicsoftheuniverse.com/scientists_schwarzschild.html

Mohon, L. Launch of Chandra X-ray Observatory. (2015). Retrieved from NASA: https://www.nasa.gov/centers/marshall/history/chandra_140723.html

Mohon, L. Deepest x-ray image ever reveals black hole treasure trove. (2017). Retrieved from NASA: https://www.nasa.gov/mission_pages/chandra/chandra-deep-field-south-deepest-x-ray-image-ever-reveals-black-hole-treasure-trove.html

Nakra, R. Our eyes in the universe—These are the nasa's 4 great observatories in space. (2020). Retrieved from The Secrets Of The Universe: https://www.secretsofuniverse.in/nasa-4-great-observatories/

NASA. Observatories Across the Electromagnetic Spectrum. (2013). Retrieved April 16, 2023, from NASA Goddard Space Flight Center: https://imagine.gsfc.nasa.gov/science/toolbox/emspectrum_observatories1.html

NASA, & CXC. The Amazing Universe. (n.d.). Retrieved April 16, 2023, from Multiwavelength Astronomy: https://ecuip.lib.uchicago.edu/multiwavelength-astronomy/x-ray/history/12.html

Peart, N. Cygnus X-1: Book One: The Voyage. (1977). Retrieved April 16, 2023, from RUSH: https://www.rush.com/songs/cygnus-x-1-book-one-the-voyage/

Schwarzschild, K. Biography of Karl Schwarzschild (1873–1916). In: Gesammelte Werke Collected Works, pp. 1–28. Springer Berlin Heidelberg (1992). https://doi.org/10.1007/978-3-642-58086-4_1

Tananbaum, H., Gursky, H., Kellogg, E., Giacconi, R., Jones, C.: Observation of a correlated x-ray transition in Cygnus X-1. Astrophys J **177**, L5 (1972). https://doi.org/10.1086/181042

Thorne, K.S. Kip Stephen Thorne quotes - 1 science quotes. (1977). Retrieved April 16, 2023, from Dictionary of Science Quotations and Scientist Quotes: https://todayinsci.com/T/Thorne_Kip/ThorneKip-Quotations.htm

Tucker , W.H. [182] About The Author. (1984). Retrieved from SP-466 The Star Splitters: https://history.nasa.gov/SP-466/author.htm

Tucker, W.H., Giacconi, R.: The x-ray universe. Harvard University Press (1985)

Wali, K.C. Chandrasekhar vs. Eddington—an unanticipated confrontation. Chandrasekhar Vs. Eddington—an Unanticipated Con-Front. **35**(10), 33–40 (1982). https://doi.org/10.1063/1.2914790

Wali, K.C.: Chandra: A biographical portrait. Phys. Today **63**(12), 38–43 (2010). https://doi.org/10.1063/1.3529401

23

Hubble Space Telescope: Permanently Changing the Way We See the Universe

Hubble Space Telescope

The Hubble Space Telescope was placed into orbit by the Space Shuttle Discovery on April 24, 1990. The $2.5 billion telescope, which measures 13 m in length, carries the high hopes of scientists and the general public. But a few weeks later, the first image from Hubble shocked everyone: The starlight did not concentrate on a focus plane but instead formed a wide and unsightly halo that was muddled in a way not much different from the large telescope on the ground (Fig.(Scoles, 2020) (Fig. 23.1).

It was discovered that the Hubble telescope's primary lens was flawed. Specifically, its outermost edge was somewhat thicker than it should have been. Hubble's severe "myopia" and subsequent descent from heaven to hell were caused by this little imperfection in thickness of just one-fifth of a hair. When the news became public, NASA came under immediate fire from all walks of life. It was a nationwide complaint that commentators and talk show hosts quipped that NASA had launched the largest piece of space garbage in history and that some business schools used it as a negative example (Fig. 23.2).

Perhaps you're having some doubts as you read this: "No. I've heard that the Hubble Space Telescope is a very successful project. There are innumerable scientific study outcomes. I've seen Hubble's sky images, and they're not blurry at all." Your perspective is right, and it doesn't contradict with the "myopic" Hubble I mentioned. For the explanation, keep reading further.

The "first life" of the Hubble Space Telescope is where we should begin our story. There were difficulties with the Hubble project right from the

J. Wang, *Eye Beyond the Sky*, Astronomers' Universe,
https://doi.org/10.1007/978-981-99-9818-0_23

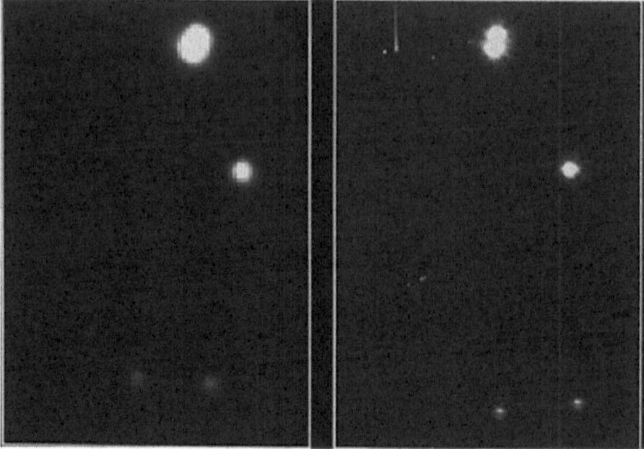

Fig. 23.1 Hubble's first image (right) is compared to a ground-based observatory in Chile that photographed the same celestial object. (*Credit* NASA, ESA, and STScl; Ground Image: E. Persson (Las Campanas Observatory, Chile)/Observatories of the Carnegie Institution of Washington)

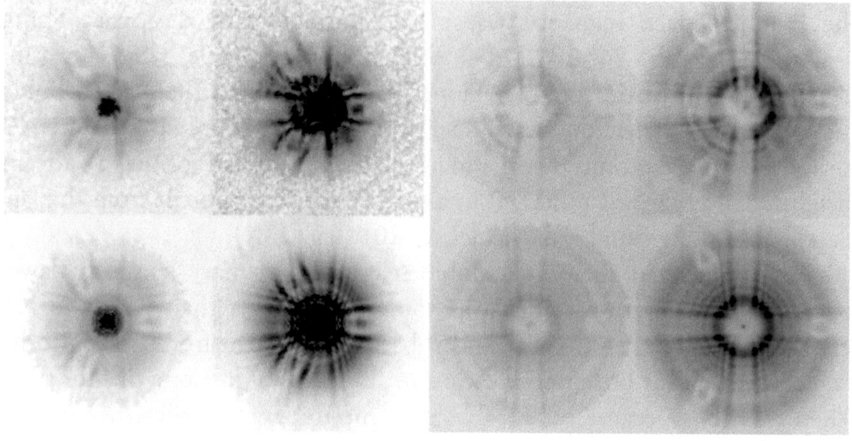

Fig. 23.2 Experts analyzed the difference between normal imaging (left) and "near-sighted" Hubble (right). (*Credit*: NASA)

start. The idea of putting telescopes into Earth orbit to make deep surveys of the universe was first proposed in 1946 by Lyman Spitzer, a professor of astronomy at Yale University and the man after whom the Spitzer Space Telescope is named. After nearly two decades of planning, the US National Academy of Sciences gave their blessing to the installation of a 3-m-long telescope in space. Nevertheless, in 1975, Congress flatly rejected the budget because it was too exorbitant (between $400 and $500 million). NASA and the ESA did all they could to get the project approved by reducing the telescope's size from 3 m to 2.4 m and the budget to $200 million.

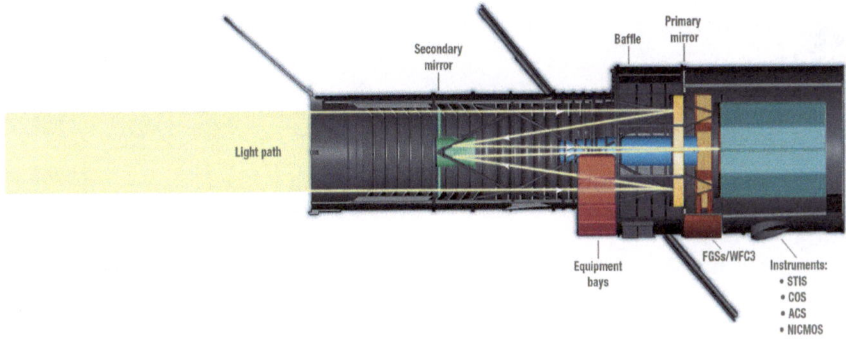

Fig. 23.3 Hubble's structural design (*Credit* NASA/STScI)

For once, Congress gave in, passing a budget in 1977 and deciding to launch it in 1983. As its initial launch attempt in 1983 was canceled for unknown reasons, and the Space Shuttle Challenger disaster in 1984 caused the grounding of all space shuttles, the project's completion was pushed back by seven years. The Hubble Space Telescope wasn't put into orbit until April 24, 1990. It had been 44 years since the idea was introduced and the spacecraft was launched (NASA, 2018) (Fig. 23.3).

It was Sod's law. Hubble finally reached orbit after numerous obstacles, but it "poured cold water" on the heads of scientists who had been feverishly anticipating its launch. Actually, Hubble had suffered terrible "myopia" before it ultimately ascended skyward. The investigation team concluded that poor management rather than faulty engineering was to blame for the tragic error. The Hubble project was a collaborative effort involving numerous institutions and individuals. The optical telescope components, for instance, were built by PerkinElmer, while Lockheed Martin was in charge of the telescope's support system. In addition, there had been about 20 secondary contracts with various space businesses. This calamity was caused by systemic flaws, including a lack of strict and transparent management systems and a failure to effectively communicate with technical experts (Tatarewicz, 2001).

When the primary mirror was finally finished, PerkinElmer's optical equipment used to detect its shape had two components that were off by 1.3 mm, resulting in the erroneous "poisonous seedling" being planted (NASA, The Hubble Space Telescope optical systems failure report, 1990) (Figs. 23.4 and 23.5).

Later, after conducting two additional inspections, PerkinElmer (NASA, Hubble's Mirror Flaw, n.d.) discovered the "poisonous seedling" but elected to keep the results secret (NASA, The Hubble Space Telescope optical systems failure report, 1990). NASA didn't require a third party to verify the primary

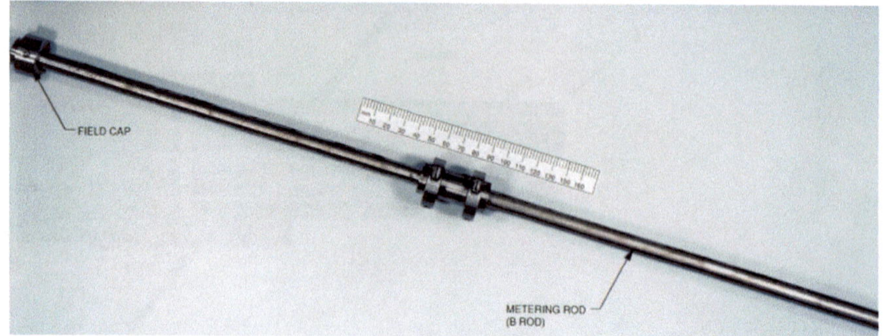

Fig. 23.4 Metering rod for testing mirror shape (note field cap) (*Credit* NASA)

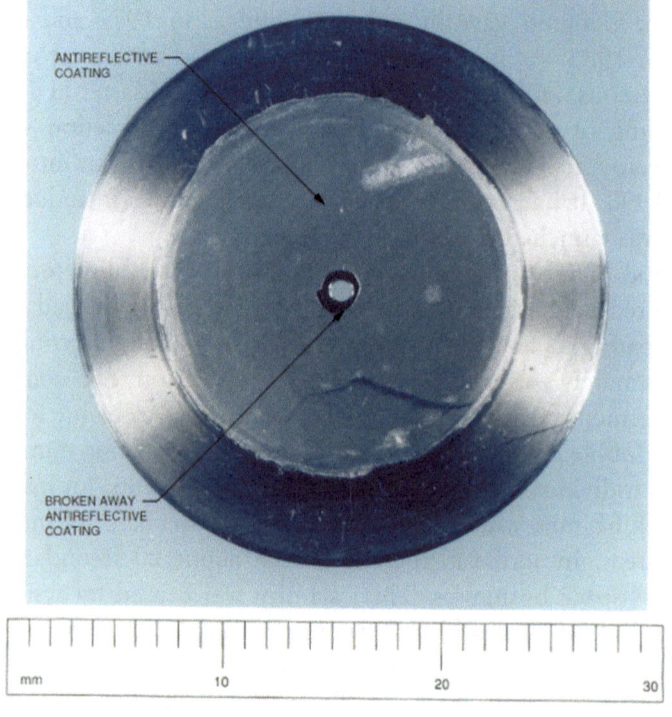

Fig. 23.5 Field cap on the metering rod causing Hubble's "myopia" (*Credit* NASA)

mirror's performance. Due to financial constraints, the pre-launch testing of parts was omitted, and the issue was instead taken into orbit. A war can be lost with the hammering of just one nail (Fig. 23.6).

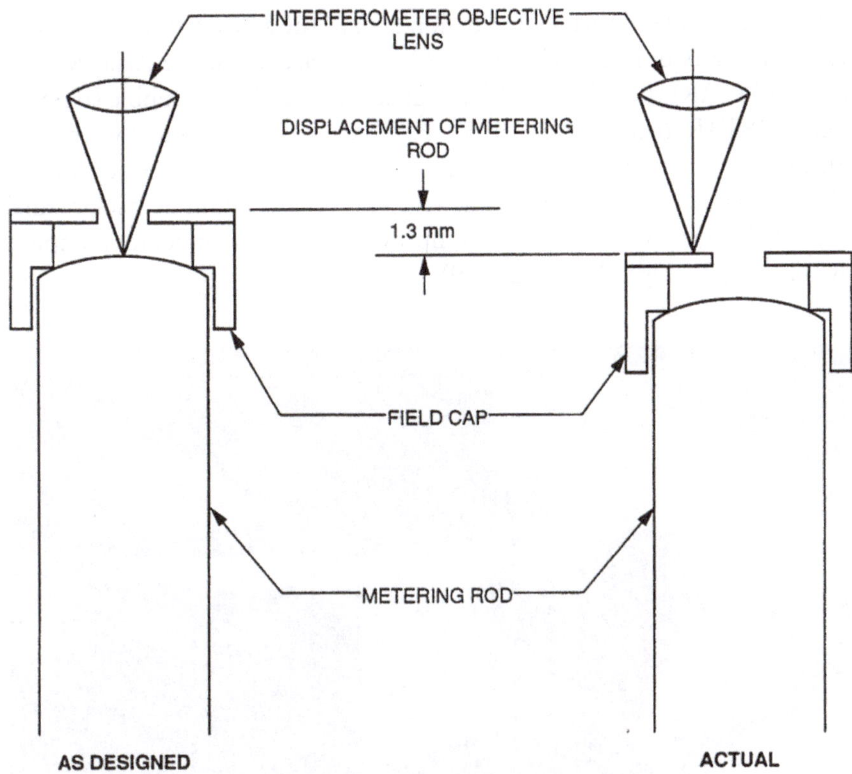

Fig. 23.6 Schematic diagram of 1.3 mm displacement (*Credit* NASA)

Thankfully, Hubble did not have such a sad ending. If you believe that when God shuts a door, he opens a window, then God has blindfolded Hubble but sent multiple "angels" to it (Fig. 23.7).

Fig. 23.7 COSTAR production equipment (*Credit*RadioFan (talk)) and its mechanism (*Credit* LouScheffer)

During their December 1993 mission aboard the Space Shuttle Endeavour, seven astronauts donned "Corrective Optics Space Telescope Axial Replacement (COSTAR) (NASM, 2012)" designed specifically for Hubble (NASA & ESA, COSTAR Installation, 1993). Hubble's vision was greatly improved with the help of COSTAR that compensated for the optical deviation caused by its original lenses. In addition, the seven astronauts put in new solar panels, overhauled four gyroscopes, and improved Hubble's onboard computers (HOTZ, 1993) (Figs. 23.8 and 23.9).

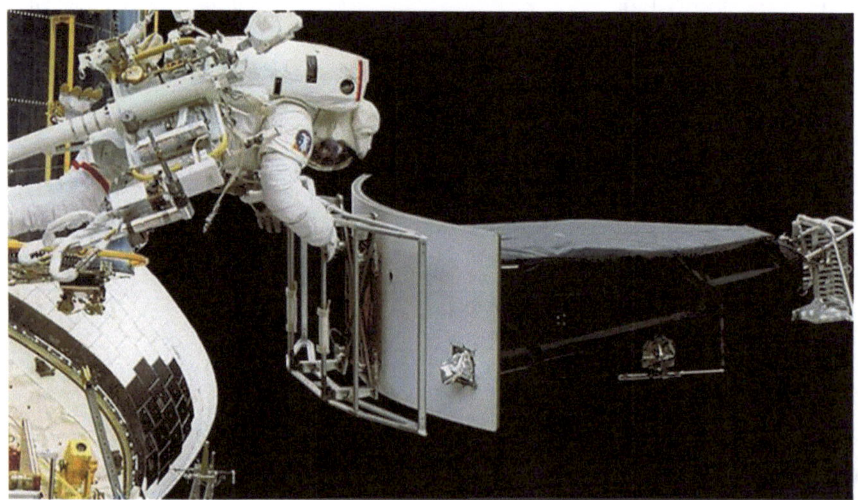

Fig. 23.8 Astronauts left the capsule to repair Hubble. (*Credit* NASA)

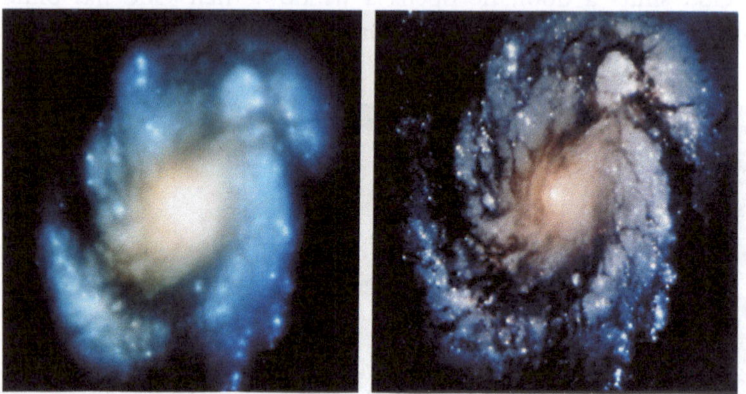

Fig. 23.9 Comparison of image quality before and after Hubble's installation of COSTAR (*Credit* NASA)

Over the course of 11 days, astronauts performed five spacewalks totaling more than 35 h of extravehicular activity as part of the maintenance effort. The complexity of this mission is nearly unparalleled in space travel.

As a result of the astronauts' "space medical service," Hubble was functioning again. In addition to removing the "myopia" spell and regaining its due imaging capabilities, the astronauts' installation of more sophisticated cameras substantially improved Hubble's imaging effectiveness. NASA heaped accolades from the "space medical service," and the public's perception of humanity and technology had undergone a dramatic about-face as a result of the confidence brought about by the mission's success.

After tasting the fruit of success, NASA went on to perform four more such space services within the following decade. Each human "doctor" helped rejuvenate Hubble's instruments. In 1997, 4 years after the first mission, astronauts returned to Hubble with various upgrades, including a new spectrometer and the first equipment optimized for infrared light. In 1999, astronauts assisted Hubble in slowly addressing the issue that was brought on by the malfunction of the gyroscopes. In 2002, astronauts upgraded the Hubble Telescope by replacing its solar panels, power units, spectrometers, and infrared cameras, and then re-positioned it in a higher orbit with the aid of the space shuttle (NASA, About—Hubble servicing missions, 2015). Hubble always kept on a fresh appearance thanks to this series of work (Fig. 23.10).

Fig. 23.10 Astronauts repaired the Hubble Space Telescope in 1999. (*Credit* NASA/ESA)

Tragic mishaps occurred, however, before the fifth space service, nearly causing the program to be scrapped. The space industry suffered a substantial blow after the Space Shuttle Columbia disaster in February of 2003. Against this backdrop, NASA's then-administrator Sean O'Keefe assessed the situation and decided in 2004 that he would no longer attempt to manually repair the telescope. This meant Hubble might peacefully "pass away" when its gyroscope or data transmission equipment failed (NASA, Hubble Space Telescope: NASA's plans for a servicing mission, 2005).

O'Keefe was almost swept away by the tidal surge of public protest as his choice sparked widespread outrage. Over 400 emails of protest were sent to NASA offices every day, and dozens of articles and op-eds expressed objections in a simple and clear manner. Hubble had been dubbed "the people's telescope" by numerous publications, all of which had voiced their support for the spacecraft's continued operation. The astronauts signed on to show that they were committed to the telescope despite the risks involved (Hartnett, 2015). The new NASA administrator, Michael Griffin, was swayed by public pressure to reverse O'Keefe's decision, and the agency's fifth space service mission was successfully completed in May 2009 (Gainor, 2020).

This time around, the astronauts upgraded and replaced a lot of Hubble's main components, as well as installed a bunch of the newest cutting-edge instruments, some of which were one hundred times more powerful than the original design. The "myopic glasses" COSTAR were no longer necessary because the newly fitted equipment could automatically correct the mistakes caused by the primary lens, hence the astronauts disposed of it (NASM, 2012).

The Hubble telescope is similar to the ship of Theseus. Except for the main mirror, every piece of components had been replaced during Hubble's space travel.

These five upgrades kept the Hubble Space Telescope at the cutting edge of astronomical research. Hubble has a massive price tag, but its benefits are enormous. Its area of study does not fit neatly into any astronomical subfield, but it has benefited the entire discipline. Even if your interests lie elsewhere, you have undoubtedly heard of the Hubble Space Telescope. Because of space constraints, we can only highlight a small subset of Hubble's many scientific contributions, notwithstanding how outstanding it is (Fig. 23.11).

To begin, Hubble determined the expansion rate of the cosmos. At what rate does the universe grow larger? The value of the Hubble constant establishes this speed. Measured in kilometers/second per million parsecs, the Hubble constant is a measure of the accelerating expansion of the universe (White, 2015). If the Hubble constant is 70, then the retrograde velocity

Fig. 23.11 The classic image of the Pillars of Creation inside of the Eagle Nebula was the first image taken by Hubble to cause public attention. (*Credit* NASA, Jeff Hester, and Paul Scowen (Arizona State University))

will rise by 70 km per second for every million parsecs, or 3.3 million light-years. To put it another way, galaxies located a million light-years away from the Milky Way will move backwards at a rate of around 21 km per second. More than 30 galaxies' distances were measured by the Hubble Space Telescope, and from those measurements, we are able to calculate the current value of the Hubble constant of the universe: 73 (with a random error of 6% and a systematic error of 8%) (Freedman & Feng, 1999). The Hubble Space Telescope has also contributed to the verification of the acceleration of the universe's expansion (Tyndall, 2020) (Fig. 23.12).

To continue, Hubble is able to determine how much of the mysterious dark matter permeates clusters of galaxies. Hubble is able to quantify dark matter through the study of the "lensed arcs" phenomenon. Massive clusters of galaxies often display "lensed arcs," which look like ripples in the water caused by a dropped pebble. These arcs exist because of the fact that

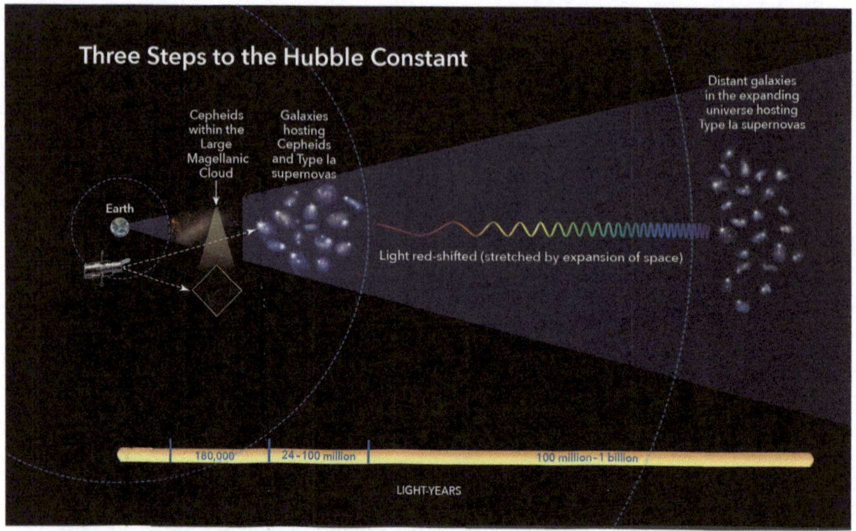

Fig. 23.12 Steps to measure the Hubble constant using the Hubble Telescope (*Credit* NASA, ESA, and A. Feild (STScI))

mass causes a deflection in the path of light, which may be used to accurately "count" dark matter alongside regular matter. Hubble's high-resolution photos allow for the investigation of lensing arcs in dozens of clusters in galaxies. One can observe hundreds of tiny arcs in certain clusters of galaxies, as if all these light rays were aiding in optical experiments. Their radians allow us to learn about the distribution of dark matter in clusters of galaxies. Hubble's measurements reveal that the universe has about five times more dark matter than regular matter and is organized around an immense network of dark matter filaments that have grown over time (NASA, Discoveries - Highlights I Shining a Light on Dark Matter, 2017) (Fig. 23.13).

Finally, black holes were found to be ubiquitous thanks to Hubble's research. Supermassive black holes in the galaxies orbiting around quasars were regarded to be exceptional by astronomers for a long time. These black holes are roughly a billion times the mass of the Sun. Yet, during the past 15 years, Hubble has oriented its spectrograph on a small sample of ordinary galaxies in our "local neighborhood," and discovered that nearly all of them exhibit a rapid rise in velocity in their central regions. An extremely potent "gravitational engine," as calculated by the data, can only be a black hole. To put it another way, black holes are not rare but rather "typical" of the galactic core (NASA, Hubble's exciting universe: Finding supermassive black holes, n.d.) (Fig. 23.14).

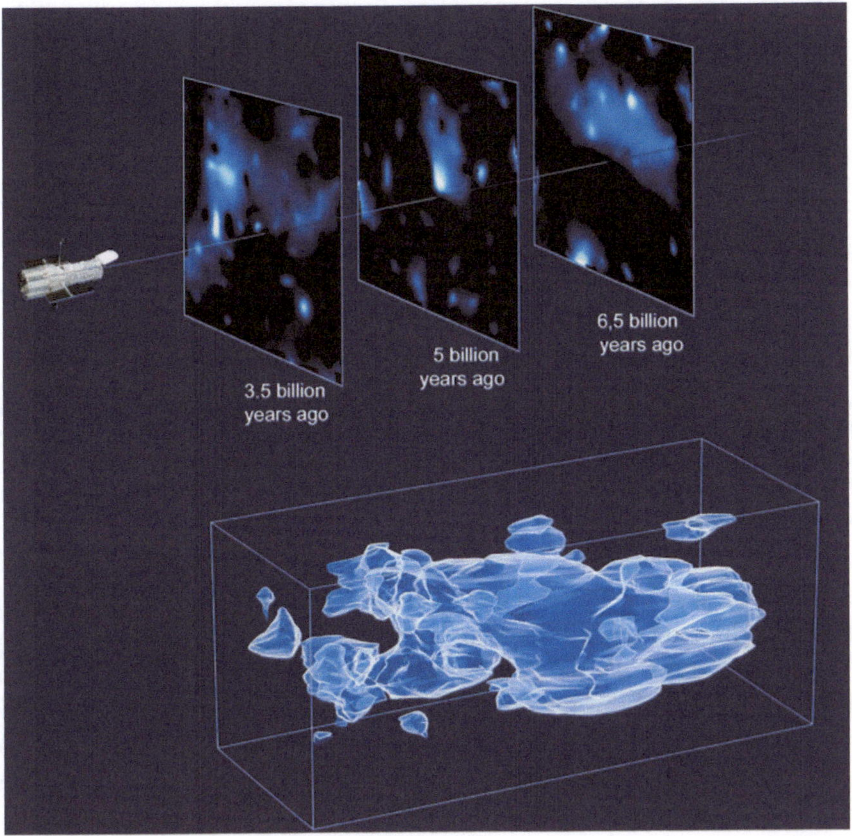

Fig. 23.13 A 3D map of dark matter created by scientists using Hubble data (*Credit* NASA, ESA, and R. Massey (California Institute of Technology))

The list of comparable accomplishments is long. Next, I want to present not a groundbreaking scientific finding but rather an image: the Hubble Deep Field. While the importance of this image is now widely acknowledged, more than 20 years ago there was significantly less enthusiasm for this observational plan. That was a radical move, one made out of desperation.

Scientists from all across the world submitted thousands of observation requests to Hubble each year, but only about 15% were accepted. They counted down the seconds until they might fully take advantage of that precious observing time. Robert Williams, who led the Space Telescope Science Institute back in 1995, "willfully sacrificed" 10 days of Hubble's observation time, from December 18 to 28, to picture a tiny patch of sky in the constellation Ursa Major, measuring about 2.6 arc minutes across, accounting for only one twenty-four-millionth of the whole sky area, whose effect is equivalent to observing a tennis ball from 100 m away on the

Fig. 23.14 Hubble image of galaxy NGC 3147 (left) and artist's rendering of the black hole nebula at the center of the galaxy (*Credit* Hubble Image: NASA, ESA, S. Bianchi (Università degli Studi Roma Tre University), A. Laor (Technion-Israel Institute of Technology), and M. Chiaberge (ESA, STScI, and JHU); illustration: NASA, ESA, and A. Feild and L. Hustak (STScI))

Earth. What you see in the Hubble Deep Field currently is the result of superimposing 342 separate exposures (Nardo, 2017).

More than three thousand sparkling "diamonds" dot the otherwise pitch-black background of this photograph. Those "diamonds" you see are actually galaxies, some of which are the oldest and most distant in the universe as far as we know and contain hundreds of billions of stars just like our own galaxy. To tell the truth, they dwarf our Milky Way in comparison (Fig. 23.15).

As one of NASA's most valued photos, the Hubble Deep Field image is not only stunningly gorgeous and mysterious but also laden with profound scientific undertones. There had been more than a thousand citations to relevant scholarly papers before January 2019. Meanwhile, it prompted a swift response from astronomers working in the infrared, radio, and X-ray aspects, whose findings supplemented Hubble's results in optical wavelength. Following then, deep space fields of view began to materialize.

A smaller and deeper field of view was captured by Hubble in 2003 and 2012, yielding the Hubble Ultra Deep Field (NASA, ESA, Beckwith, & HUDF Team, 2006) and the Hubble eXtreme Deep Field (NASA, et al., 2012). Over 10 000 galaxies, the weakest of which is one fifty-billionth of the faintest light visible to the human eye, have been imaged by Hubble in the Hubble Ultra Deep Field (Fig. 23.16 and 23.17).

Hubble has completed its primary mission of monitoring practically all visible-universe celestial bodies in more than 30 years since its launch in 1990. It has investigated solar system astronomy and exoplanets, scanned the birth and death of stars with unprecedentedly high accuracy, and contributed

Fig. 23.15 Hubble Deep Field (*Credit* Robert Williams and the Hubble Deep Field Team (STScI) and NASA/ESA)

Fig. 23.16 Hubble Ultra Deep Field (*Credit* NASA and the European Space Agency. Edited by Noodle Snacks)

Fig. 23.17 Hubble eXtreme Deep Field (*Credit* NASA; ESA; G. Illingworth, D. Magee, and P. Oesch, University of California, Santa Cruz; R. Bouwens, Leiden University; and the HUDF09 Team)

many magnificent photographs of galaxies far and near. It also provided a crucial quantitative basis for cosmology, including the measurements of the size, age, and expansion rate of the universe (Bell, 2020) (Fig. 23.18).

The Hubble Space Telescope represents the greatest concentration of human effort to date in its pursuit of knowledge about the universe's remote past and distant future. Even now, Hubble is making its way across space, snapping and sending back breathtaking photographs of distant galaxies and nebulae. Exactly what will occur to Hubble in the future remains unclear. Why not just let Hubble continue its journey for a while longer? (Fig. 23.19)

Fig. 23.18 "Space Reef" released by NASA to commemorate the 30th anniversary of Hubble's launch (*Credit* NASA, ESA, and STScI)

Fig. 23.19 Hubble Space Telescope (*Credit* NASA)

References

Bell, J. Hubble legacy: 30 years of discoveries and images. Union Square & Company (2020)

Freedman, W.L., Feng, L.L.: Determination of the Hubble constant. PNAS **96**(20), 11063–11064 (1999). https://doi.org/10.1073/pnas.96.20.11063

Gainor, C.: Not Yet Imagined: A Study of Hubble Space Telescope Operations. National Aeronautics and Space Administration, Washington (2020)

Hartnett, K. *Hubble: An Overview of the Space Telescope.* National Aeronautics and Space Administration, Goddard Space Flight Center, Office of Communications (2015)

HOTZ, R.L. Prescription lenses inserted to focus hubble. (1993). Retrieved from Los Angeles Times: https://www.latimes.com/archives/la-xpm-1993-12-08-mn-65094-story.html

Nardo, D. Hubble Deep Field: How a photo revolutionized our understanding of the universe. Capstone (2017)

NASA. The Hubble Space Telescope optical systems failure report. (1990). Retrieved from NASA: https://ntrs.nasa.gov/api/citations/19910003124/downloads/19910003124.pdf

NASA. Hubble Space Telescope: NASA's plans for a servicing mission. (2005). Retrieved from Congressional Research Service: https://www.everycrsreport.com/reports/RS21767.html

NASA. About—Hubble servicing missions. (2015). Retrieved from NASA: https://www.nasa.gov/mission_pages/hubble/servicing/index.html

NASA. Discoveries - Highlights|Shining a Light on Dark Matter. (2017). Retrieved from NASA: https://www.nasa.gov/content/discoveries-highlights-shining-a-light-on-dark-matter

NASA. About–Hubble history timeline. (2018). Retrieved from NASA: https://www.nasa.gov/content/goddard/hubble-history-timeline

NASA. Hubble's exciting universe: Finding supermassive black holes. (n.d.). Retrieved April 16, 2023, from HubbleSite.Org: https://hubblesite.org/hubble-30th-anniversary/hubbles-exciting-universe/finding-supermassive-black-holes

NASA. Hubble's Mirror Flaw. (n.d.). Retrieved April 16, 2023, from NASA Science website: https://science.nasa.gov/mission/hubble/overview/hubbles-mirror-flaw

NASA, & ESA. COSTAR Installation. (1993). Retrieved from esahubble: https://esahubble.org/images/costar/

NASA, ESA, Beckwith, S., & HUDF Team. Hubble Ultra Deep Field. (2006). Retrieved from esahubble: https://esahubble.org/images/heic0611b/

NASA, ESA, Illingworth, G., Magee, D., Oesch, P., Bouwens, R., & HUDF09 Team. Hubble eXtreme Deep Field (XDF). (2012). Retrieved from Hubblesite: https://hubblesite.org/contents/media/images/2012/37/3098-Image.html

NASM. COSTAR. (2012). Retrieved April 16, 2023, from National Air and Space Museum: https://airandspace.si.edu/collection-objects/costar-corrective-optics-space-telescope-axial-replacementhubble-flown/nasm_A20120157000

Scoles, S. Hubble's blurry years. (2020). Retrieved from Physics Today: https://physicstoday.scitation.org/do/https://doi.org/10.1063/pt.6.4.20200401d/full/

Tatarewicz, J.N. The Hubble Space Telescope servicing mission. (2001). Retrieved April 16, 2023, from NASA: https://history.nasa.gov/SP-4219/Chapter16.html

Tyndall, A. Hubble's greatest discoveries: Accelerating expansion of the Universe. (2020). Retrieved from BBC Science Focus Magazine: https://www.sciencefocus.com/space/hubble-space-telescope-accelerating-expansion-of-the-universe/

White, M. The Hubble Expansion. (2015). Retrieved April 16, 2023, from UC Berkeley: https://w.astro.berkeley.edu/~mwhite/darkmatter/hubble.html

24

WMAP (Wilkinson Microwave Anisotropy Probe): Revealing Our Universe in Its Infancy

WMAP (Wilkinson Microwave Anisotropy Probe)

The first cosmic background explorer, or COBE, was launched into sun-synchronous orbit on November 18, 1989 from Vandenberg Space Force Base in the United States. This probe cost around $160 million (Sawyer, 1989) (Fig. 24.1).

More than a thousand researchers, led by physicists George Smoot and John Mather, toiled on the detector. More than two years later, they made a big announcement: Evidence has indicated that the temperature differences among different sky areas are about one part in hundreds of thousands of the average temperature. As soon as the news broke, it was front page material in all the major newspapers across the United States, and it generated widespread enthusiasm among scientists. Delighted scientists dubbed this extremely minute deviation "God's fingerprint" (Singh, 2005). To honor their work, George Smoot and John Mather were awarded the Nobel Prize in Physics in 2006 (Nobel Prize Outreach AB 2023, 2023).

You may be wondering as you read this: How can this be dubbed God's fingerprint and be worthy of a Nobel Prize if it is merely a slight difference in the microwave radiation of the universe? To grasp its full import, we must return to the year 1929.

In 1929, American astronomer Edwin Hubble (1889–1953) and his assistant made a discovery that would change the world. They found that faraway galaxies are receding from our own at a rate of 1,100 km per second or more, indicating that the universe is expanding (Hubble, 1929). The scientific community as a whole was taken aback by this revelation, and the concept

J. Wang, *Eye Beyond the Sky*, Astronomers' Universe,
https://doi.org/10.1007/978-981-99-9818-0_24

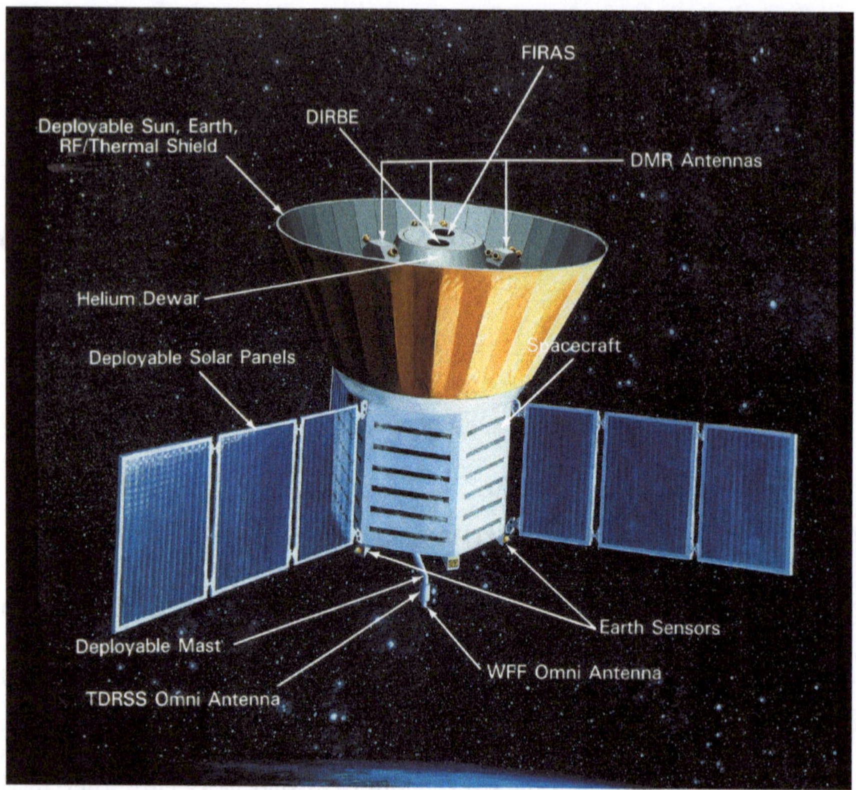

Fig. 24.1 Schematic of the components of COBE (*Credit:* NASA/COBE Science Team)

seemed so preposterous that not even Einstein could at first fully embrace it. However, the Belgian astronomer Georges Lemaître (1894–1966) (Clery, 2018) did not disagree with this theory but instead built a more audacious hypothesis upon it. A beginning existed, and the universe extended outward from there, in his view. To be more precise, it means that the entire universe, including over a billion galaxies, was instantly produced from an extremely compact spacetime. The renowned Big Bang theory can trace its roots back to this (Weinberg, 1977).

If you were a regular person living a century ago and heard this idea for the first time, would you think it was ridiculous? The Big Bang theory was still only a wild speculation until it could be backed up by hard data from observations.

The possibility of an observable feature supporting the Big Bang theory was first proposed in 1949 by George Gamow (1904–1968) (AIP, 2023) and Robert Hermann (1931–2020) (Boston Globe, 2020). They hypothesized that the afterglow of the Big Bang, after cooling over billions of years, could produce the pervasive microwave photons known as the cosmic microwave background radiation (Alpher & Herman, 1948) (Gamow, 1948). Not only did Gamow provide a qualitative description of what the microwave background radiation was, but he also determined quantitatively that this radiation should have a temperature of 5 Kevin. With today's measured values for the parameters, this equation yields a solution of 2.7 Kevin. If you convert it to the Celsius scale, you get a temperature of minus 270.45 degrees. Being able to make such an accurate prediction is both incredible and admirable. Is there, then, such a phenomenon as microwave background radiation? Can we observe it through experiments?

Now that the mystery has been solved, thank goodness. The cosmic microwave background radiation predicted by Gamow was first detected in 1965, only three years before his death. Surprisingly, it was not astronomers who first noticed it; rather, it was two engineers at Bell Labs who detected it by chance. Then the researchers calculated the temperature of the microwave radiation signal to be 3.5 K, which was just slightly off from Gamow's forecast (Penzias & Wilson, 1978) (Fig. 24.2).

It is important to note that the temperature of about 3 K observed at present is equivalent to that any place of one square centimeter in the universe can receive about 10 photons per second, which is why the cosmic microwave

Fig. 24.2 Robert W. Wilson and Arno Penzias with their antenna where they detected the Big Bang's ripples (*Credit:* NASA)

background radiation can become the key evidence of the Big Bang theory. Any radiation source would have to be enormous to produce such a large amount of energy, given the vastness of the cosmos (Partridge, 1995). Just like the remaining heat of a gigantic inferno after 13.8 billion years of expansion, these photons can only have been produced at the same time as the universe's birth (Fig. 24.3).

The Big Bang theory, with the "help" of the cosmic microwave background radiation, was a smashing success and aced the first test. However, supporters of the Big Bang theory began to perspire and fidget as scientists around the world became increasingly focused on and precise in their measurements. They were shocked to learn that these measurements were moving away from supporting the Big Bang theory, casting doubt on the veracity of the initial "angel" that inspired the theory. What was happening?

When scientists took precise measurements of the cosmic microwave background radiation, they were shocked to discover that the radiation was perfectly constant throughout the whole universe, with a temperature of accurate 2.725 Kevin no matter which way the antenna was pointed (Robitaille, 2009). There's a chance that you won't see the significance of the problem or the meaning of this constant temperature right away. Don't fret. As I explain further, you'll see what I mean.

Truth be told, the early universe expanded at a tremendous velocity. In the early stages of microwave radiation, the distance between two points in space could reach 60 times the speed of light. With the rate of expansion, it

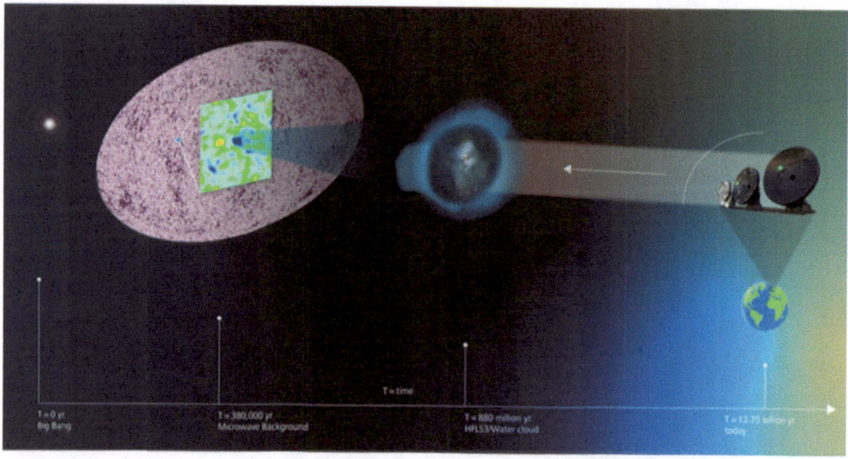

Fig. 24.3 Big Bang's remnant traveling to Earth (*Credit:* ESA and the Planck collaboration; zoom-in panel: Dominik Riechers, University of Cologne; image composition: Martina Markus, University of Cologne)

is impossible for the entire cosmos to cool down to the same precise temperature at the same time. The phenomenon is analogous to fireworks erupting in the sky; however, the measurement reveals that each individual spark of fireworks has the same temperature.

Subsequently, a scientist by the name of Alan Guth proposed (Guth, 1991) the now-famous theory of inflation to account for this occurrence. According to this explanation, the "cosmic fireball" had ample time to cool to the same temperature everywhere because the speed of spatial separation at the beginning of the universe was slower than that predicted by the conventional model. As the cosmos reached thermal equilibrium, it had a brief explosive expansion that was utterly insane. In just a billion billion billionth of a second, the universe's size rose by 10 to the 26th power, or 10 billion billion times. Fundamental to the model is the fact that everywhere in the cosmos eventually heats up to the same temperature before expanding eternally (Peacock, 2001) (Fig. 24.4).

The "ludicrous" theory of inflation was even more unsettling than the Big Bang theory to skeptics at first, so it required stronger evidence before it could gain acceptance in the scientific community. Extraordinary claims require extraordinary evidence.

So, can inflation theory provide any testable predictions? In a word, yes. A large group of theoretical physicists (including the well-known Stephen Hawking (Hawking, 1975)) worked together and discovered that the rapid expansion of space did not result in the absolute homogeneity of radiation. Instead, quantum fluctuations imply that the seemingly uniform microwave background radiation of the universe is subject to tiny temperature variations,

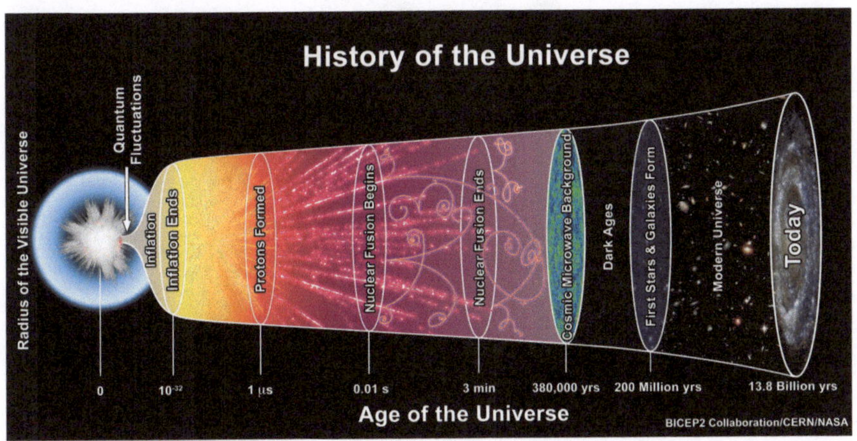

Fig. 24.4 Alan Guth's theory of inflation (*Credit:* The AstronomyBum)

similar to a tiny ripple on a still body of water. The most crucial evidence for the inflation theory would be gathered as long as this faint "ripple" was detected.

The COBE probe, introduced at the beginning of this chapter, set out in 1989 to investigate this claim. Key evidence supporting the theory of inflation has been found in the temperature fluctuations of the cosmic microwave background radiation, which was discovered by COBE, to be compatible with theoretical predictions with a divergence of only one part in a few hundred thousand. Scientists' enthusiasm in this field has been tremendously stoked by COBE's discovery. The Wilkinson Microwave Anisotropy Probe (WMAP), serving as COBE's successor and our finale protagonist in this chapter, was launched from Cape Canaveral on June 30, 2001 (Smith, 2016). WMAP has taken more accurate readings of the cosmic microwave background radiation, presenting a rigorous test of the validation of inflation theory (Fig. 24.5).

Even though WMAP appears to be considerably larger and more intricate than COBE, it actually only weighs about a third as much. The collection of microwave radiation is accomplished by the utilization of a pair of dish antennas with a diameter of 1.5 m, and the receiver is able to function on

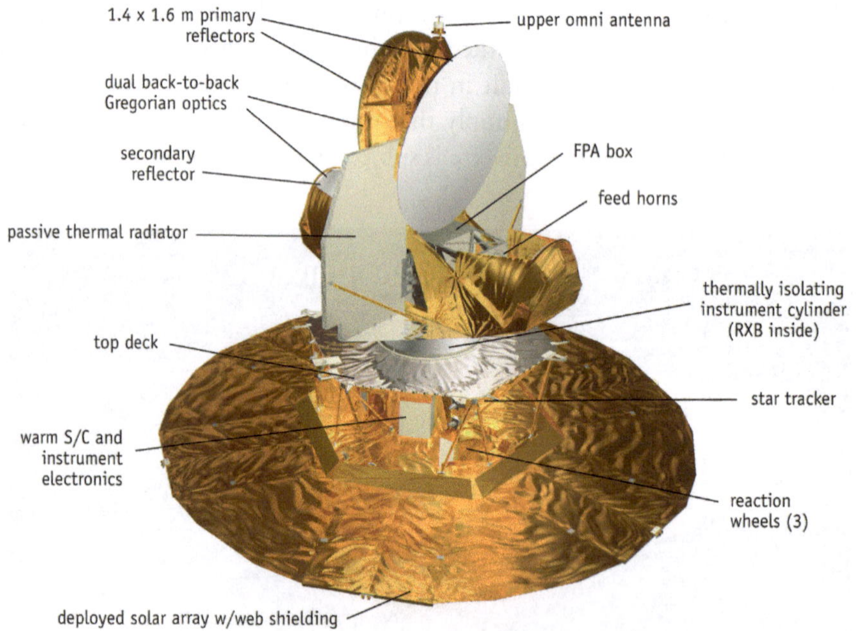

Fig. 24.5 Schematic of the components of WMAP (*Credit:* NASA/WMAP Science Team)

five different frequencies. WMAP records minute fluctuations in temperature throughout the entire sky at a rate of once every 130 s over the course of six months (NASA & WMAP Science Team, 2010).

In light of the fact that COBE has confirmed the inflation theory, you might be wondering why WMAP, rather than COBE, is being presented as the protagonist here.

The fact that WMAP can improve the precision with which the Big Bang theory is tested is the main reason for its significance. WMAP's placement at the 2nd Sun-Earth Lagrange point (L2), 1.5 million kilometers from Earth, where radiation pollution is considerably less than that of lower Earth orbits, gives it a substantial advantage over COBE. Because of its exceptional "location," WMAP is 45 times more sensitive, with 33 times the angular resolution of its COBE satellite predecessor (NASA & GSFC, Wilkinson Microwave Anisotropy Probe DR5, n.d.). When comparing WMAP to COBE, it's like comparing the Hubble Space Telescope to a 30-cm telescope on Earth's ground. Dimensionality reduction wins big. In other words, if there is a discrepancy between WMAP and COBE observations, the latter is more likely to be correct (Mandolesi & Vittorio, 2011).

Do the WMAP measurements back up the inflation theory or disprove it? First, let me state the obvious: It validates the theory. Inflation theory continues to hold up in the face of more stringent examinations (Komatsu, 2006).

Over the course of more than a decade of observations, WMAP has refined fresh data on numerous crucial physical constants into finely calibrated cosmological parameters, including temperatures measured to a thousandth of a degree. Is there any use for such pinpoint accuracy? Perhaps it hasn't occurred to you that it can be used to determine if the universe is flat or curved. Curving spacetime into a massive lens is a theoretical possibility in general relativity. So, to what degree does this lens point? Microwaves with a journey time of over 10 billion years may be magnified or diminished by the universe's curvature, exactly like light passing through a convex or concave lens, depending on the direction of the lens's optical axis. What WMAP actually saw was surprising, though: The universe is flat as a mirror and microwaves don't have curvature. This means that the universe might be flat to within one percent of its apparent size and that there is no curvature anywhere in it (Hinshaw, et al., 2013). This uniform smoothness is consistent with predictions of inflation theory.

Inflation theory with a major revision of the Big Bang theory, now called the "Standard Cosmological Model", has been the dominant model in cosmology since the release of the WMAP data. And what happened at the inflationary instant, a billion billion billionth of a second after the Big Bang, is the cutting edge of cosmology right now (CTC, n.d.).

In addition, WMAP has accomplished a great deal more. WMAP provides extremely accurate measurements of the cosmological mass and energy. With an inaccuracy of only 0.1% for ordinary matter and 1% for dark matter, it measures that they make up only 4.6% and 24% (NASA & WMAP Science Team, WMAP's Universe - What is the Universe Made Of?, 2014), respectively, of the total mass energy of the universe. Just what do these two numbers represent? They reveal an inconvenient truth: The vast majority of the universe is composed of unknown forms of dark matter and dark energy, while only a tiny fraction is made up of familiar forms of stuff like stars, planets, and humans. Based on general relativity and the Hubble-Lemaitre law, we can determine how old the universe is by measuring its total mass and energy composition. The final calculation based on the WMAP measurements and theoretical models, the universe is estimated to be 13.82 ± 0.12 billion years old (Banerjee, 2021) (Fig. 24.6).

In addition, WMAP's data enables us to piece together how the earliest stars and galaxies were formed. It is plausible that a huge number of photons

Fig. 24.6 Comparison of sensitivities for detecting microwave background radiation in recent 40 years (*Credit:* NASA)

would have been scattered by electrons in the early universe if stars had emerged soon after the Big Bang, ionizing the diffuse gas. Microwave radiation would be marked with polarization due to this back-scattering effect. WMAP's polarization data reveals that hundreds of millions of years after the Big Bang, the fog of ionized gas scattered around 20% of the photons (Luisi, Anderson, Balser, Wenger, & Bania, 2017). To put it simply, astronomers were taken aback by this finding, as they had not anticipated the first generation of stars to form so quickly (Fig. 24.7).

To WMAP, the cosmic microwave background radiation represents a snapshot of the universe in its infancy. Imagine the incredible scenario of viewing your infant self for the first time as an adult. The universe, stars, and planets are full of unfathomable space just waiting to be explored and listened to; perhaps electromagnetic waves are the "language" through which they convey messages. In 2009, the European Space Agency (ESA) launched the Planck Space Observatory (ESA, n.d.), the third generation of microwave background detection satellites, which has permitted humanity to further probe the mysteries of the early cosmos with increased sensitivity and angular resolution, considerably solidifying the evidence of the Big Bang gained by COBE

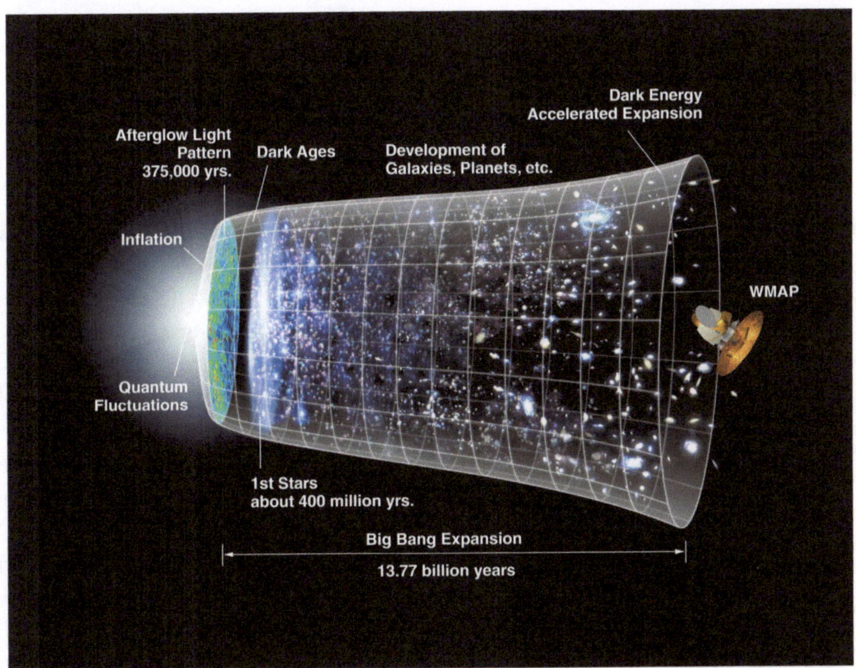

Fig. 24.7 WMAP helps us restore the timeline of the birth of the universe. (*Credit:* NASA/WMAP Science Team)

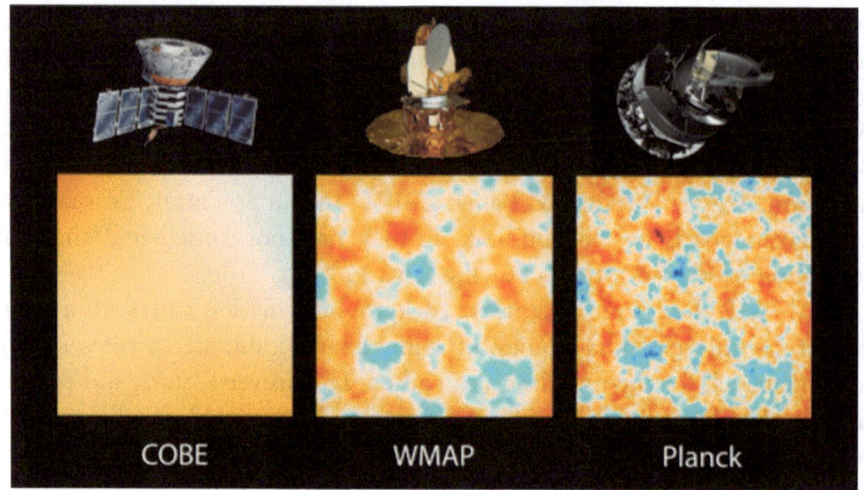

Fig. 24.8 Sensitivity comparison of three generations of probes (*Credit:* NASA/JPL-Caltech/ESA)

and WMAP. Although our future is impossible to foresee, scientists will continue their tireless search for the cosmic origin (Fig. 24.8).

References

AIP: George Gamow 1904–1968. (2023). Retrieved April 16, 2023, from AIP Physics History Network. https://history.aip.org/phn/11510011.html

Alpher, R.A., Herman, R.C.: On the relative abundance of the elements. Phys. Rev. **74**(12), 1737–1742 (1948). https://doi.org/10.1103/physrev.74.1737

Banerjee, G.: Measuring the age of the Universe: An overview. (2021). Retrieved from Cosmic Varta: https://cosmicvarta.in/measurig_age_uiverse_gourav

Boston Globe: DR. ROBERT C. HERMANN. (2020). Retrieved from Legacy—The Boston Globe: https://www.legacy.com/us/obituaries/bostonglobe/name/robert-hermann-obituary?id=2241298

Clery, D.: Move over, Hubble: Discovery of expanding cosmos assigned to little-known Belgian astronomer-priest. (2018). https://doi.org/10.1126/science.aav8870

CTC: The origins of the universe: Inflation introduction. (n.d.). Retrieved April 16, 2023, from Centre for Theoretical Cosmology: https://www.ctc.cam.ac.uk/outreach/origins/inflation_zero.php

ESA: ENABLING & SUPPORT – Planck. (n.d.). Retrieved April 16, 2023, from THE EUROPEAN SPACE AGENCY: https://www.esa.int/Enabling_Support/Operations/Planck

Gamow, G.: The Evolution of the Universe. Nature **162**(4122), 680–682 (1948). https://doi.org/10.1038/162680a0

Guth, A.H: Fundamental arguments for inflation. In: Observational Tests of Cosmological Inflation, (pp. 1–21). Springer Netherlands. (1991). https://doi.org/10.1007/978-94-011-3510-8_1

Hawking, S.W.: Particle creation by black holes. Commun. Math. Phys. **43**(3), 199–220 (1975). https://doi.org/10.1007/bf02345020

Hinshaw, G., Larson, D., Komatsu, E., Spergel, D.N., Bennett, C.L., Dunkley, J., ...Wright, E. L.: NINE-YEAR WILKINSON MICROWAVE ANISOTROPY PROBE (WMAP) OBSERVATIONS: COSMOLOGICAL PARAMETER RESULTS. Astrophys. J. Suppl. Ser. **208**(2), 19 (2013). https://doi.org/10.1088/0067-0049/208/2/19

Hubble, E.: A relation between distance and radial velocity among extra-galactic nebulae. Proc. Natl. Acad. Sci. **15**(3), 168–173 (1929). https://doi.org/10.1073/pnas.15.3.168

Komatsu, E.: Recent results from the wilkinson microwave anisotropy probe. Prog. Theor. Phys. Suppl. **163**, 185–203 (2006). https://doi.org/10.1143/ptps.163.185

Luisi, M., Anderson, L.D., Balser, D.S., Wenger, T.V., Bania, T.M.: Diffuse ionized gas in the Milky Way disk. Astrophys J **849**(2), 117 (2017). https://doi.org/10.3847/1538-4357/aa8fd2

Mandolesi, N., Vittorio, N.: The Cosmic Microwave Background: 25 years later: Proceedings of a meeting on 'The Cosmic Microwave Background: 25 years later', held in L'Aquila, Italy, June 19–23, 1989. Springer Science & Business Media (2011)

NASA, & GSFC: Wilkinson Microwave Anisotropy Probe DR5. (n.d.). Retrieved April 16, 2023, from NASA Goddard Space Flight Center: https://lambda.gsfc.nasa.gov/product/wmap/current/

NASA, & WMAP Science Team: WMAP Facts. (2010). Retrieved from NASA: https://wmap.gsfc.nasa.gov/news/facts.html

NASA, & WMAP Science Team: WMAP's Universe - What is the Universe Made Of? (2014). Retrieved from NASA: https://wmap.gsfc.nasa.gov/universe/uni_matter.html

Nobel Prize Outreach AB 2023. The Nobel Prize in Physics 2006. (2023). Retrieved from NobelPrize.org: https://www.nobelprize.org/prizes/physics/2006/9545-the-nobel-prize-in-physics-2006-2006-2/

Partridge, R.B. 3K: The cosmic microwave background radiation. Cambridge University Press. (1995)

Peacock, J.A.: An introduction to the physics of cosmology. CRC Press, In Modern Cosmology (2001). https://doi.org/10.1887/0750308109/b1051c2

Penzias, A.A., Wilson, R.W.: The cosmic microwave background. Phys. Bull. **29**(12), 564–564 (1978). https://doi.org/10.1088/0031-9112/29/12/026

Robitaille, P.M.: COBE: A Radiological Analysis. Progress in Physics, 4. (2009). Retrieved from https://www.ptep-online.com/2009/PP-19-03.PDF

Sawyer, K.: Searching Skies for Faint Glow From Dawn of Time. (1989). Retrieved from The Washington Post: https://www.washingtonpost.com/archive/politics/1989/11/19/searching-skies-for-faint-glow-from-dawn-of-time/6fb3ee9f-7dad-47e8-9a41-8de25a43bb15/

Singh, S.: Big bang: The origin of the universe. (2005). Harper Collins.

Smith, Y.: Wilkinson microwave anisotropy probe. (2016). Retrieved from NASA: https://www.nasa.gov/feature/making-sense-of-the-big-bang-wilkinson-microwave-anisotropy-probe/

Weinberg, S.: The First Three Minutes: A Modern View of the Origin of the Universe. (1977)

25

DAMPE: DArk Matter Particle Explorer

DAMPE

Those of you who are familiar with the history of physics should be aware that physicists at the turn of the 20th century faced two formidable challenges. These are the renowned "two dark clouds." As Lord Kelvin put it: the relative motion of the ether with respect to massive objects and Maxwell–Boltzmann's theorem on the equipartition of energy (Passon 2021). As time went on, the two ominous clouds solidified into the two cornerstones of physics. Since then, the phrase "black clouds" has come to symbolize physics' many mysteries that have yet to be explained.

Once again, physicists found themselves in peril as humans embrace the twenty-first century. They confront two new "dark clouds", of which one is known as "dark matter" and the other as "dark energy". Physicists do not even have a comprehensive understanding of their most fundamental properties, and all they can do is proposing some hypotheses based on the astronomical events that have been witnessed.

The investigation that led to the discovery of dark matter has been fascinating. When observing the Milky Way's rotation, American astronomer Vera Rubin (1928–2016) (NSF n.d.) noticed something a bit off. With increasing distance from the inner ring, velocity decreases as predicted by Newtonian mechanics. Her actual observations contradicted this, though. Outlying stars in the Milky Way rotate at a much higher rate than their central counterparts.

Supplementary Information The online version contains supplementary material available at https://doi.org/10.1007/978-981-99-9818-0_25.

J. Wang, *Eye Beyond the Sky*, Astronomers' Universe,
https://doi.org/10.1007/978-981-99-9818-0_25

This peculiar phenomenon could be explained only if the bulk of the Milky Way's mass was not located at its center. With binoculars, however, we can make out that the core is the brightest and the cantilevers around it are quite faint. Doesn't it imply that the core of the galaxy contains the greatest number of stars? There appears to be a discrepancy between gravitational calculations and optical observations (Yeager 2008).

To our good fortune, there exist a variety of approaches to determining a galaxy's total mass. Observing the rotation rate of the galaxy is one method, while using gravitational lenses is another. Light from distant stars will be distorted when it travels close to a galaxy because, according to Einstein's theory of relativity, the enormous mass of galaxies acts like a magnifying glass. We can observe this distortion of the image. In addition to deducing the gravitational mass of galaxies, this phenomenon can also be utilized to determine galaxy's optical mass by measuring its brightness. These two values almost never correspond, and there is a significant gap between them (Schechter 2001). In most cases, the gravitational mass exceeds the optical mass by a significant margin (Fig. 25.1).

In point of fact, Vera Rubin is not the first astronomer to recognize the peculiar occurrence in question. Her predecessors in this field, Jan Hendrik Oort (1900–1992) (NASA 2015) and Fritz Zwicky (1898–1974) (Swart 2019), made similar observations but didn't investigate further. It was Rubin's meticulous observational data that caught the attention of astronomers everywhere (Yeager 2021).

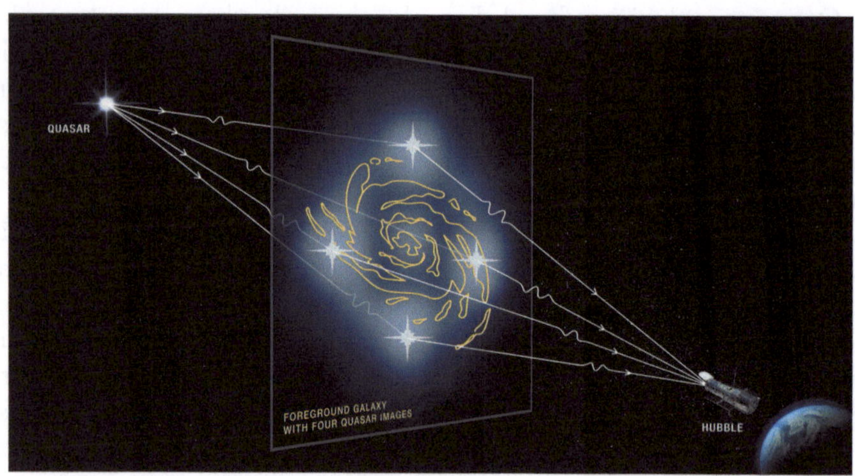

Fig. 25.1 Principle of gravitational lensing (*Credit* NASA, ESA and D. Player (STScI))

Without a clear explanation, scientists could only presume that the observable matter in the galaxy is only a fraction of the total amount of matter in the universe. Immediately, a buzz of excitement spread throughout the scientific community. As it turned out, the portion of the universe we could observe was only the tip of the iceberg. Dark matter is a term used to describe invisible yet gravitationally active mass.

There is no doubt that dark matter possesses gravity, and there is no question that it does not interact electromagnetically (Randall 2018). Electromagnetic interaction permeates every facet of human existence. Light can be thought of as both a type of electromagnetic wave and an example of electromagnetic phenomena. The reason why different chemical components can be organized and combined to form a wide variety of molecules is due to the interaction between electromagnetic fields. For the same reason, all human sense organs and humans' highly developed sensors are dependent on electromagnetic interactions (Fig. 25.2).

Since direct detection of dark matter is theoretically impossible if it lacks electromagnetic interaction, astronomers have to make do with indirect evidence. Accordingly, the idea of WIMPs (short for "Weakly Interacting Massive Particles") has been put forward by scientists as a reasonable assumption. These hypothetical, microscopic particles have been proposed as possible candidates for dark matter (Randall 2016).

Fig. 25.2 Black hole and gravitational lensing (*Credit* Depositphotos/nltoshevikova)

However, the formation of hypotheses is merely the first step in the process of scientific investigation, and the most essential phase is the search for evidence to prove the existence of such particles. One option is to crash into the world's largest and most powerful particle accelerator, the Large Hadron Collider (LHC). When particles traveling at high speeds hit with one another, they release a large amount of energy and can even recreate certain aspects of the Big Bang. If the energy is powerful enough, everything will be viable as an option. Nobody can predict what particles will be ejected. Perhaps among them are the mysterious dark matter particles. The current situation, however, is disheartening. The Higgs boson has been the most recently discovered particle by LHC (Voir en 2022) for more than a decade (Fig. 25.3).

Our solar system's orbit around the Milky Way's galactic core takes 230 million years (NASA 2000). It's safe to presume that there's a huge number of dark matter since maintaining the solar system's velocity requires 10^{11} solar masses. According to this hypothesis, there should be enormous quantities of dark matter surrounding the Earth. Then the second option is to simply wait for dark matter particles to collide with the nuclei of regular matter. We will be able to detect this collision if it does occur. Indeed, there are far too many particles in the world's natural settings. To prevent uninvited interference from cosmic rays, researchers should bury the detector as deep as they can to shield it from the interference of other particles. Neutrinos and dark matter

Fig. 25.3 CMS detector of large hadron collider (*Credit* SimonWaldherr)

particles are the only ones that can travel to the depths of the earth. In spite of having waited for a number of years under the ground, the researchers came up empty-handed (Bernabei and Incicchitti 1996) (Figs. 25.4 and 25.5).

Can dark matter be detected in any other way? Current theory suggests that the collision of two dark matter particles could lead to annihilation and result in the creation of standard observable particles. Indicators include the production of positron and negatron pairs or the release of gamma rays. It

Fig. 25.4 Alpha magnetic spectrometer (AMS) (*Credit* NASA)

Fig. 25.5 Fermi Gamma-ray satellite telescope (*Credit* NASA)

is possible to identify such traces with the help of the Alpha Magnetic Spectrometer (AMS) (Ghose 2013) and the Fermi Gamma-ray Satellite Telescope (FGST) (Howell 2018). Due to unforeseen technical challenges, the total cost of the AMS project increased from an initial estimate of $33 million to $2 billion. The largest component of the 7.5-ton AMS is a 2.6-ton Chinese-made permanent magnet. China is among the best in the world when it comes to the production of super-strong magnets (Chen 2011). AMS was deployed aboard the International Space Station (ISS) to ensure its 2,500 watts (AMS n.d.) of power consumption. In addition, the only place that can supply AMS with electricity is ISS. While FGST is less expensive than AMS, it still costs close to $700 million. Both of these detectors come with a hefty price tag.

At a cost of only $100 million, China's DAMPE (DArk Matter Particle Explorer) is among the most cost-effective options. As the first professional astronomical exploration satellite to be launched by our nation, DAMPE was successfully placed into a sun-synchronous orbit on December 17, 2015 (AIP 2015). About a week after liftoff, DAMPE's onboard instruments were switched on in sequence and embarked on their regular detecting duties. Whenever it travels over China, it will make contact with the ground in order to send back an average of 16 GB of data each day. The data from this detector has laid a solid foundation for all scientific inquiries afterward.

Even though DAMPE is not particularly large, only a big box that measures 1.5 m on each side, it has an expert level of operation. It is equipped with plastic scintillator detector, silicon tungsten tracker, BGO calorimeter, and neutron detector. As a top dark matter particle space detector, it possesses the greatest energy range and the finest energy resolution (Vitillo 2018) (Fig. 25.6).

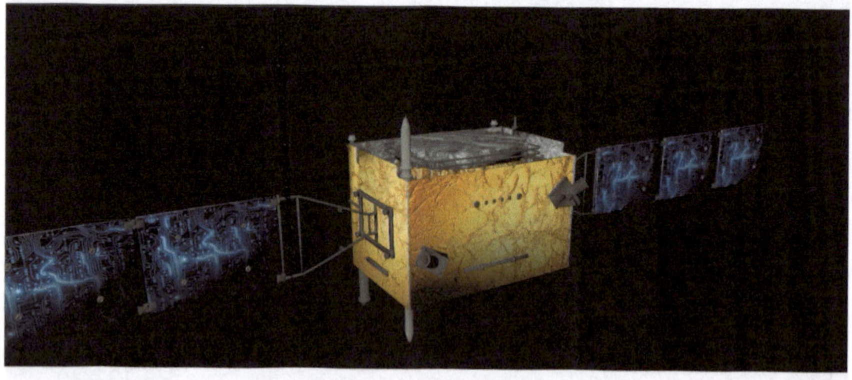

Fig. 25.6 DAMPE (*Credit* Chinese Virtual Observatory)

When high-energy particles accelerated by stellar objects like stars, neutron stars, supernova remnants, accretion black holes, and other astronomical things collide with interstellar matter, they create antiparticles like positrons. However, the number of positrons created by these processes is negligible compared to the number of positrons released by dark matter. If we filter out everything but positrons, the signal will be much clearer. If you want to tell the difference between positive and negative electrons, you'll need a magnetic spectrometer. When positrons and negatrons enter a powerful magnetic field, their deflection will be totally reversed, making it simple to tell them apart (Mazziotta et al. 2019). Because of this, AMS needs to take a super magnet that weighs a significant amount into orbit.

With a total mass of only 1.8 tons, DAMPE is sufficiently light to be launched by the Long March-2D carrier rocket (NASA 2020). Engineers always strive to lower one gram of weight whenever possible, because the heavier the satellite, the larger the rocket, and the higher the cost. However, this means that DAMPE cannot take a heavy magnet into space with it, and its data on positrons and negatrons isn't as clean as that from a magnetic spectrometer. But theoretical calculations show that there are only ten times as many electrons as positrons in space (Griffiths 2008), and if there is a sudden rise in the number of positrons, we may easily identify anomalies in the total data of positrons and negatrons, therefore this method is viable in general (Figs. 25.7 and 25.8).

Fig. 25.7 Fast-rotating neutron star (Pulsar) (*Credit* Depositphotos/Juric.P)

Fig. 25.8 The Nebula produced by a supernova explosion (*Credit* Depositphotos/Juric.P)

Eliminating needless interference is another significant challenge that must be overcome. There are a lot of high-energy protons in cosmic rays. At 100 giga-electron volts, protons are 300 times as numerous as electrons, and at 1 tera-electron volt, they are 800 times as many. Electrons are the target, while protons are just interference. One needs to cultivate a pair of golden eyes in order to differentiate electrons from a vast number of protons without making mistakes (Close 2004). According to some estimates, the probability of making an error in the calculation should not be higher than one in tens of thousands, which is an unacceptably high standard.

Fortunately, in 2008, Chang Jin, then a researcher at the Purple Mountain Observatory, presented a method to efficiently differentiate between protons and electrons by making use of the complete information provided by a variety of detectors. This process eliminates the need for cumbersome and costly apparatus. The Americans were planning to travel high into the air using balloons they would soon release at the South Pole in an effort to detect cosmic rays. During their conversation, Chang Jin assured the Americans that their detector was sensitive enough to pick up the electrons and gamma rays present in cosmic rays (Xinhua 2018). After much persuasion on Chang Jin's part, the Americans agreed to send over their data to him for processing, and the results were promising. Chang Jin is the first author of this study published in *Nature* (Chang et al. 2008) (Fig. 25.9).

Fig. 25.9 Chang Jin (*Credit* Yuan Jing, a photojournalist from Wenhui Bao)

It should come as no surprise that this technology is utilized in the DAMPE probe. This innovation allows us to search for proof of dark matter using lighter and less expensive methods.

The BGO (Bismuth Germanium Oxide) calorimeter plays a pivotal part in DAMPE's operation, while the other instruments play supportive functions. As a crucial feature of the detector, BGO crystals will glow when exposed to X-rays and gamma rays. The Shanghai Silicate Research Institute is an authority in BGO crystals and supplies them to other institutions, including CERN. China now possesses the largest space probe in the world thanks to the BGO calorimeter on DAMPE, whose crystal is 60 centimeters on a side, more than doubling the previous world record (Chang 2014) (Fig. 25.10).

The plastic scintillator detector, the silicon tungsten tracker, and the neutron detector make up the remaining three sub-detectors. The Chinese Academy of Sciences created a plastic scintillator detector that can tell the difference between gamma rays and charged particles and also measure the charge of charged particles. Together with the University of Perugia in Italy and the University of Geneva in Switzerland, the Chinese Academy of Sciences' Institute of High Energy Physics produced the silicon tungsten tracker, whose primary role is to detect the particle's direction of origin and transform some of the gamma rays into pairs of positrons and negatrons. Purple Mountain Observatory's neutron detector was designed to better distinguish between electrons and protons.

Fig. 25.10 BGO crystal (*Credit* Materialscientist)

DAMPE's first data results were compiled after 530 days' operation. DAMPE amassed a total of 2.8 billion high-energy cosmic rays' data during this period. Around 1.5 million (Chen 2017) cases of 25 giga-electron-volt cosmic rays were randomly chosen. This information allowed Chinese researchers to acquire the most accurate results for the detection of electron cosmic rays at a level of tera-electron volts at that time. These findings show that DAMPE owns a detection range that is greater than that of either AMS or FGST. This is similar to taking photographs that capture both the radiance of the sun and the flickering of candle flames at the same time. And in the 1 tera-electron volt range, the data captured by DAMPE was the purest. All of these factors created excellent conditions for the investigation of dark matter (Fig. 25.11).

Theoretical calculations suggest that there should be a relatively smooth curve describing the distribution of the various types of energy particles found in the cosmos. However, DAMPE probe's graph displays an anomalous "bulge" in 1 tera-electron volt range, which may be the result of dark matter particles' mutual annihilation. In addition, several fine structures can also be observed in the energy level region of 1.4 tera-electron volts, which is also worthy of consideration (Okada and Seto 2018).

We can only claim that we have identified some indicators at this point and cannot come to the conclusion that these signs point to the emission of dark matter particles. This is because we have not yet been able to observe dark matter particles directly. Even if these electrons are not released by dark matter, they are nonetheless unexpected findings. The mechanism that causes 1 tera-electron volt energy level of electrons to burst is unknown. Is that a

Fig. 25.11 DAMPE in space (*Credit* Chinese Virtual Observatory)

capability of the neutron star? Another possibility is that it's a black hole. All of these possibilities are worth devoting time and effort to researching.

In this context, we'd also like to stress a key idea: DAMPE's primary function is to travel into space and gather information about various high-energy particles and gamma-ray signals. We obviously cannot just zero in on the dark matter as our primary emphasis, though one possible application of the data is to search for hints about dark matter. In this era of big data, when its importance becomes so obvious, data's role as the most valuable asset is self-evident.

DAMPE is merely a start. Further research satellites and detectors will be launched in the near future to strengthen the data collection (Fig. 25.12).

Over the course of the last decade, China has made significant advancements in space technology and exploration. Chinese scientists, together with their peers from around the world, are making great strides toward dispelling those two obscuring dark clouds that have plagued the field of physics since the turn of the 21st century. For the sake of humanity's scientific progress, I fervently pray that China's DAMPE would keep up its tireless efforts to return even more crucial data.

Fig. 25.12 Gamma-Ray burst (*Credit* Depositphotos/sakkmesterke)

References

AIP: Chinese dark-matter explorer launches successfully (2015). https://doi.org/10.1063/PT.5.029455

AMS: AMS Operations. AMS - 02 (n.d.). https://ams02.space/operations/ams-operations. Accessed 16 April 2023

Bernabei, R., Incicchitti, A.: The Dark Side of the Universe (1996). https://doi.org/10.1142/3304

Chang, J.: Dark matter particle explorer: the first Chinese cosmic ray and hard γ-ray detector in space. Chin. J. Space Sci. **34**(5), 550 (2014). https://doi.org/10.11728/cjss2014.05.550

Chang, J., Adams, J.H., Ahn, H.S., Bashindzhagyan, G.L., Christl, M., Ganel, O., Zatsepin, V.I. et al.: An excess of cosmic ray electrons at energies of 300–800 GeV. Nature **456**(7220), 362–365 (2008). https://doi.org/10.1038/nature07477

Chen, N.: China's permanent magnet lift off into space—Chinese academy of sciences. Chinese Academy of Sciences (2011). https://english.cas.cn/newsroom/archive/news_archive/nu2011/201502/t20150215_140061.shtml

Chen, N.: First finding of China's DAMPE may shed light on dark matter research. Chinese Academy of Sciences (2017). https://english.cas.cn/head/201711/t20171120_186326.shtml

Close, F.: Particle Physics: A Very Short Introduction. Oxford University Press (2004)

Ghose, T.: Dark matter possibly found by $2 billion space station experiment. Space (2013). https://www.space.com/20490-dark-matter-discovery-space-experiment.html

Griffiths, D.: Introduction to Elementary Particles. Wiley (2008)

Howell, E.: Fermi telescope: studying the high-energy cosmos. Space (2018). https://www.space.com/41191-fermi-gamma-ray-telescope.html

Mazziotta, M., Cuoco, A., Luque, P.D., Loparco, F., Serini, D.: Search for dark matter signatures in the cosmic-ray electron and positron spectrum measured by the Fermi large area telescope. Proc. Sci. **358** (2019). https://doi.org/10.22323/1.358.0531

NASA: Does the Sun move around the Milky Way? NASA - StarChild Question (2000). https://starchild.gsfc.nasa.gov/docs/StarChild/questions/question18.html

NASA: Hidden Mass. NASA (2015). https://imagine.gsfc.nasa.gov/educators/galaxies/imagine/hidden_mass.html

NASA: NASA'S HEASARC: observatories - DAMPE. NASA (2020). https://heasarc.gsfc.nasa.gov/docs/heasarc/missions/dampe.html

NSF: Vera Rubin (1928–2016). National Science Foundation (n.d.). https://www.nsf.gov/news/special_reports/medalofscience50/rubin.jsp. Accessed 16 April 2023

Okada, N., Seto, O.: DAMPE excess from decaying right-handed neutrino dark matter. Mod. Phys. Lett. A **33**(27), 1850157 (2018). https://doi.org/10.1142/s0217732318501572

Passon, O.: Kelvin's clouds. Am. J. Phys. **89**(11), 1037–1041 (2021). https://doi.org/10.1119/10.0005620

Randall, L.: Dark Matter and the Dinosaurs: The Astounding Interconnectedness of the Universe. Ecco (2016)

Randall, L.: What is dark matter? Scientific American (2018). https://www.scientificamerican.com/article/what-is-dark-matter1/

Schechter, P.: Gravitational lensing: recent progress and future goals. ASP Conf. Proc. **237**, 427 (2001). https://doi.org/10.1086/316486

Swart, J.D.: Deciphering dark matter: the remarkable life of Fritz Zwicky. Nature (2019). https://www.nature.com/articles/d41586-019-02603-7

Vitillo, S.: Tracker charge identification and measurement of the proton flux in cosmic rays with the DAMPE experiment. Archive Ouverte UNIGE (2018). https://archive-ouverte.unige.ch/unige:113361

Voir en: The Higgs boson, ten years after its discovery. CERN (2022). https://home.cern/news/press-release/physics/higgs-boson-ten-years-after-its-discovery

Xinhua: Feature: DAMPE chief scientist talks hunger, hubble, high-energy particles - Xinhua. Xinhuanet (2018). http://www.xinhuanet.com/english/2018-07/23/c_137343125.htm

Yeager, A.: Cosmos incognito: Vera Rubin shines light on dark matter. MIT Libraries (2008). https://dspace.mit.edu/handle/1721.1/45424. Accessed 16 April 2023

Yeager, A.J.: Bright Galaxies, Dark Matter, and Beyond: The Life of Astronomer Vera Rubin. MIT Press (2021)

26

JWST (James Webb Space Telescope): The Most Expensive Space Telescope in History

JWST

At 5 am Beijing time on July 12, 2022, numerous Chinese astronomy enthusiasts stayed up in front of their computers, eagerly waiting for a significant moment: Joe Biden, the President of the United States, was to personally release the first official photo captured by the James Webb Space Telescope. However, when the appointed time came, Biden did not appear as promised. After waiting for nearly an hour, Vice President Harris finally made her appearance in the live broadcast, followed shortly by President Biden himself. Both leaders delivered speeches, underlining the importance of this photo. Later on, to the entire world, the first official photo taken by the most expensive space telescope in human history was unveiled (Fig. 26.1).

Bill Nelson, the head of NASA, personally explained this photo, which has been named "Webb's First Deep Field." Its actual size in the universe is equivalent to holding a grain of sand at arm's length, representing the patch of sky that this grain of sand can cover. Within this minute portion of the universe, we have captured thousands of galaxies, which were formed over 13 billion years ago, depicting a scene when our universe was just under 800 million years old.

Upon careful inspection, we notice that the light of many galaxies has been stretched into arc shapes. This is not due to deformities in the telescope but is the effect of gravitational lensing. The Webb telescope utilized the gravitational lensing effect generated by a massive galaxy cluster named SMACS 0723 to take this photo. Imagine this cluster as a magnifying glass, gathering

J. Wang, *Eye Beyond the Sky*, Astronomers' Universe,
https://doi.org/10.1007/978-981-99-9818-0_26

Fig. 26.1 Webb's First Deep Field (*Credit*: NASA, ESA, CSA, STScI)

light rays from the more distant universe behind it, allowing us to capture a clearer image (Fig. 26.2).

We can compare this with the first deep-field image (the Hubble Deep Field) taken by Webb's predecessor, the Hubble Space Telescope, in 1995. These two images vividly demonstrate the progress humanity has made in astronomical observation capabilities over 26 years (Fig. 26.3).

The James Webb Space Telescope is hailed as the largest, most complex, and most powerful space telescope ever constructed by humankind. Jointly developed by the National Aeronautics and Space Administration (NASA), the European Space Agency (ESA), and the Canadian Space Agency (CSA), and designed and crafted carefully over more than 30 years by thousands of scientists and engineers, it bears many of the hopes and dreams of the scientific community.

Initially, the budget for designing Webb was 500 million USD. This budget was not only reasonable at that time but also quite affordable, especially considering that Webb was to be the successor of the Hubble Space Telescope, which had a total cost exceeding 3 billion USD.

However, plans always fall behind changes. Webb experienced several launch delays from its design to completion, with repeated additions to the budget. By the time it was launched, it had incurred a massive expenditure of 10.8 billion USD, more than three times that of the Hubble telescope.

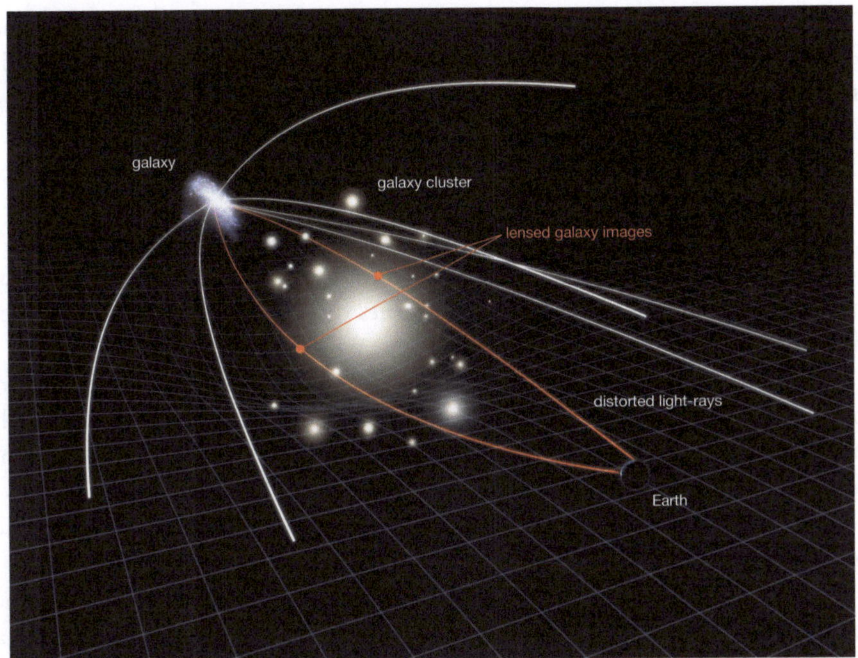

Fig. 26.2 Diagram illustrating the gravitational lensing effect (*Credit*: NASA, ESA & L. Calçada)

Fig. 26.3 Left: The Hubble Deep Field (*Credit*: Robert Williams (NASA, ESA, STScI)). Right: Webb's First Deep Field (*Credit*: NASA, ESA, CSA, STScI)

With over twenty launch postponements, Webb earned the nickname "the Perpetual Postponer." Astronomers have a love-hate relationship with this project that carries so many dreams yet has been continuously delayed, giving it another title "The telescope that ate astronomy (Billings, 2010)."

Please allow me to guide you through the complete story of the James Webb Space Telescope.

In September 1989, the Space Telescope Science Institute and NASA jointly hosted a seminar for the next generation of space telescopes, bringing together over 130 astronomers and engineers (NASA & STScl, Webb Project History, n.d.).

At the seminar, two proposals received extensive attention. One proposed that NASA should install a 10 m diameter near-infrared telescope in a high Earth orbit; the other suggested situating the infrared telescope on the moon, allowing the telescope's diameter to be expanded to 16 m.

Both proposals focused on infrared telescopes based on a specific scientific rationale.

Due to the expansion of the universe, visible light from the depths of the cosmos tends to shift more or less toward the red end of the spectrum, a phenomenon in astronomy known as redshift. The more distant a celestial object is, the more its light is affected by redshift. The first batch of stars and galaxies (Clery, 2016) formed after the Big Bang, being extremely distant, emit visible light that has shifted to the near-infrared and infrared regions of the electromagnetic spectrum (Pultarova, 2021).

The soon-to-be-launched Hubble Space Telescope mainly utilized visible and ultraviolet light for observation, rendering it incapable of capturing mid-infrared images from the early stages of the universe. Therefore, infrared observational capability became a requisite feature for the new generation of space telescopes (Fig. 26.4).

However, given that the Earth's average surface temperature is 15°C, it essentially acts as a source of infrared pollution. The James Webb Space Telescope opted for an open design without lens barrels to achieve a larger mirror surface area, allowing for the gathering of more photons. Hence, measures had to be devised to block radiation from various celestial bodies. The ideal solution was to align the sun, the Earth, and the telescope in a straight line, with both the sun and the Earth on one side of the telescope, and to position the telescope at a distant and relatively stable point away from the Earth.

Does such a point exist in the universe? Certainly.

In celestial mechanics, there exists a point where a smaller celestial body can remain stationary relative to two larger bodies, as the gravitational forces from the larger bodies balance the centripetal force needed for the smaller

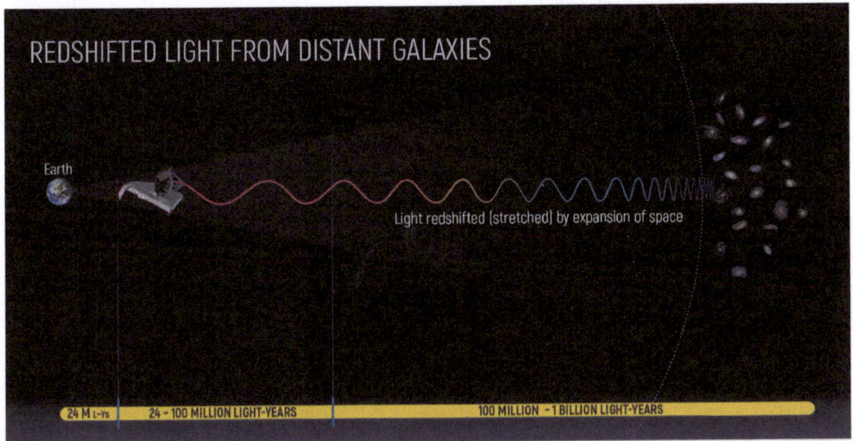

Fig. 26.4 Redshifted Light from Distant Galaxies (*Credit*: NASA, ESA, CSA, Ann Feild (STScI))

body to move along with them. This point is known as a libration point, or a Lagrange point, where a satellite can remain stably in position (Cornish, n.d.).

The Sun-Earth system has five Lagrange points, with L4 and L5 positioned on opposite sides outside the line that crosses the Earth and the sun, while L1, L2, and L3 are situated along this line. Located 1.5 million kilometers away on a direct line running through the Sun and the Earth, L2 is an ideal location for detectors and astronomical telescopes to observe the solar system, offering ease of protection and calibration with minimal fuel consumption for long-term stationing (Fig. 26.5).

In 1996, an 18-member committee led by astronomer Alan Dressler formally proposed to NASA the construction of an 8 m diameter space telescope capable of infrared observation. This design would not only enable astronomers to see through interstellar clouds to observe the internal structures of galaxies but also extend humanity's vision to more remote and ancient deep space.

Surprisingly, what ultimately persuaded NASA wasn't these scientific advantages, but the fact that the proposal was very cost-effective. Yes, you heard it right, it was due to its low cost. At that time, NASA's initiative was "Faster, Better, Cheaper," a strategy strongly advocated by the agency's chief, Daniel Goldin. Under this radical thinking, the committee quickly approved this simple, foldable, lightweight, and inexpensive proposal. In retrospect, it seems almost like a cruel joke, as the winning cost-effective proposal eventually earned the title "The telescope that ate astronomy (Billings, 2010)."

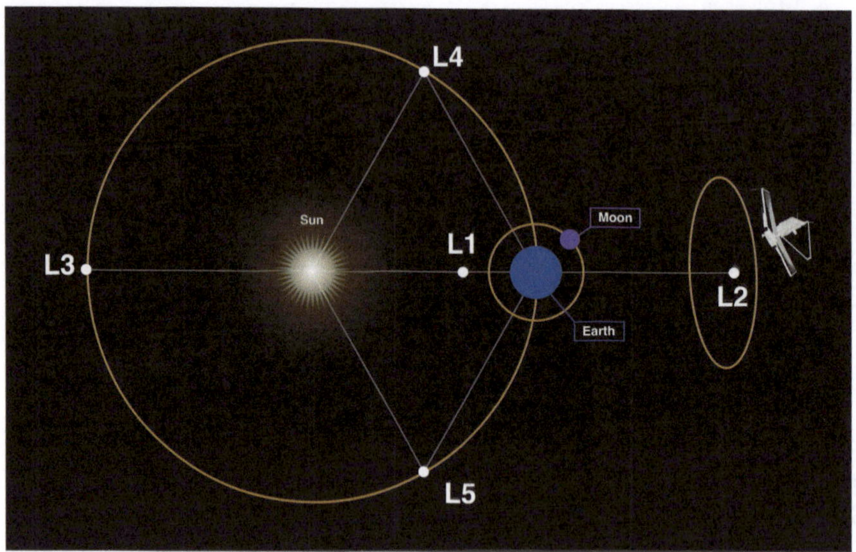

Fig. 26.5 Webb's orbit at Sun-Earth Lagrange Point 2 (L2) (*Credit*: NASA, STScI)

In 1997, NASA agreed to fund this research and collaborate with the renowned Goddard Space Flight Center in order to prepare for the development and construction of the James Webb Space Telescope. The settled total budget was 500 million USD, with a planned launch date set a decade later, in 2007.

Despite their experience from developing the Hubble Space Telescope, NASA engineers still faced numerous unprecedented challenges in their mission to capture the universe's first "dawn light" with Webb.

The first challenge was the interference caused by thermal radiation.

Every object with a temperature, including the telescope itself, emits infrared light. To prevent radiation from the telescope from overwhelming the exceptionally faint astronomical signals, both the telescope and its accompanying instruments had to be operational at a temperature of -233 °C (NASA, GENERAL QUESTIONS ABOUT WEBB, n.d.).

Contrary to the popular belief that the universe is frigid, the reality is slightly different. Spacecraft traversing the cosmos can experience rapid heating on their sun-exposed sides. To ensure that Webb could function optimally in a low-temperature environment for extended periods, engineers designed a huge sunshield for it (Fig. 26.6).

Equivalent in size to a tennis court, this grand sunshield is constructed with five layers of an ultra-thin, glistening material known as Kapton. The outermost layer of the shield faces the sun constantly, absorbing the majority

Fig. 26.6 The sunshield for the James Webb Space Telescope (*Credit*: Kevin Gill)

of the heat and operating at temperatures rising to 100 °C. Each subsequent layer progressively becomes cooler, maintaining a stable low-temperature condition for Webb's instruments (Pultarova, 2021).

However, this solution presented another challenge—how to reduce the size and weight of Webb.

The sunshield, with a thickness of only 0.1 mm—roughly equivalent to the diameter of a strand of hair, had to be contained within a compact space (Fig. 26.7).

Yet, this sunshield wasn't the hardest part of the problems. The engineers faced a pinnacle challenge in fitting the 8 m diameter mirror of the telescope into a rocket fairing with a diameter of only 4.57 m, a task demanding innovative solutions to manage the dimensions and weight critically.

To achieve this goal, engineers adopted the same approach as the sunshield by devising a mirror with a foldable structure that could be unfolded in space after launch (Fig. 26.8).

This massive mirror is composed of multiple independent hexagonal segments. To synchronize their reflective lights perfectly, precise control over these segments is indispensable, merging them into a single optical surface, a process known as phasing. Each hexagonal segment on Webb's mirror is positioned on six actuators to control its direction, coupled with a central actuator for fine-tuning the overall curvature.

Every material undergoes deformation when cooling, and Webb will endure a temperature fluctuation exceeding 200 °C from Earth to space,

Fig. 26.7 Engineers successfully stacked and unfurled a full-sized test unit of the sunshield for Webb for the first time. (*Credit*: NASA/Chris Gunn)

presenting its third major challenge—withstanding the extreme cold (Fig. 26.9).

To ensure that the deformation of each section remains within a controllable scope, engineers had to confine the margin of error to nanometers. After careful consideration, they set their sights on beryllium, a material that stood out for its strength and lightness. Moreover, beryllium demonstrated remarkable stability during repeated cycles of cooling and heating, meeting the anticipated deformation criteria and rising as the most promising candidate (Clery, 2016).

Addressing the technical issues mentioned earlier did not come without a price. In 1998, the overall budget for Webb doubled, reaching 1 billion USD. The launch schedule was also expected to be delayed by one year, moving to 2008.

By 2001, Webb's total budget had doubled twice again, skyrocketing to 1.8 billion USD, with the launch postponed to 2010. NASA scientists recognized that without containment, Webb's costs would definitely escalate further. Consequently, they proactively reduced Webb's aperture from 8 to 6.5 m and decreased the number of mirror segments from 36 to 18.

Although the aperture was reduced by only 1.5 m, which might seem insignificant, the telescope's light-collecting area decreased from 50 square meters to 25 square meters, shrinking by half (Fig. 26.10).

Fig. 26.8 Mirror assembly for the James Webb Space Telescope (*Credit*: NASA)

Despite implementing such measures, the total budget for Webb continued to rapidly soar. To further trim costs, NASA chose to utilize less precise mirrors (Wikipedian, 2023), compromising image clarity at near-infrared wavelengths of 1–2 μm. However, this outcome was deemed unacceptable since ground-based telescopes could also observe within this wavelength range (Clery, 2016).

On September 10, 2002, the next-generation space telescope was named after James E. Webb, NASA's second administrator, in commemoration of his role in realizing the monumental Apollo Moon Landings (Fig. 26.11).

By 2006, Webb's costs surged once more, compelling a re-evaluation and capping the total expenses at 4.5 billion USD, which includes approximately 3.5 billion for design, development, launch, and commissioning, as well as around 1 billion for a decade of operation. The launch was postponed to 2013.

Fig. 26.9 Engineers are preparing for the final round-the-clock cryogenic testing for Webb. (*Credit*: NASA/MSFC/David Higginbotham)

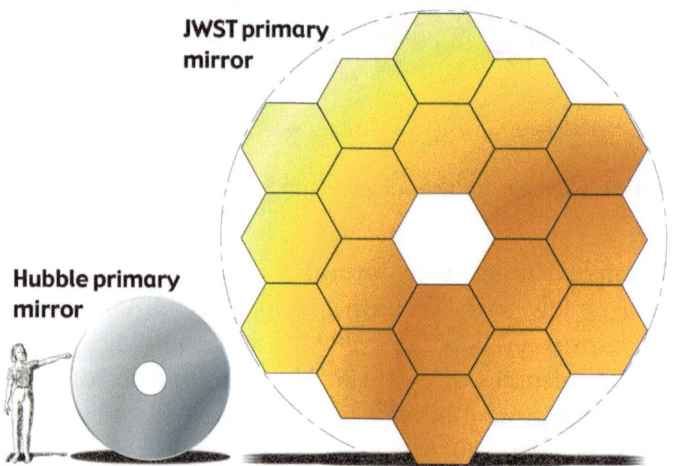

Fig. 26.10 Comparison between the sizes of the primary mirrors of Hubble and Webb (*Source*: NASA)

In the summer of 2010, the mission team underwent a rigorous review and achieved excellent scores. All key technologies were ready, except for one—cost control. The delay in progress further inflated Webb's budget to 5 billion

Fig. 26.11 James Edwin Webb (October 7, 1906 - March 27, 1992) (*Source*: NASA)

USD (Wikipedian, 2023), initiating a vicious cycle of delays fueled by cost overruns.

Anxiety began to spread within the astronomical community, as some astronomers feared that the rapidly increasing budget for the Webb project was severely draining funds allocated for other space science initiatives, thus impeding their timely completion.

A 2010 article in *Nature* titled "The Telescope That Ate Astronomy (Billings, 2010)" coined a term that widely circulated thereafter.

Despite complaints, many scientists firmly advocated for Webb, asserting that launching Webb remained a paramount objective for astronomy and astrophysics, without which most scientific goals would be unattainable.

In 2011, Webb's mirror was finally completed—a grand golden mirror resembling a giant beehive comprised of 18 hexagonal segments with a diameter of 6.5 m and a total area of 25 square meters (Fig. 26.12).

You might be surprised to learn that Webb's golden hue is not just for show; its mirror is coated with real gold. Compared to other metals, gold offers the highest reflectivity across a wide bandwidth, reflecting nearly 99% of infrared light while maintaining outstanding chemical stability (Fig. 26.13).

You might be wondering just how much gold was used? The golden layer is merely 100 nm thick (NASA, GENERAL QUESTIONS ABOUT WEBB,

Fig. 26.12 The primary mirror of the James Webb Space Telescope (*Credit*: NASA/Chris Gunn)

Fig. 26.13 Engineers are inspecting the surface of Webb's primary mirror. (*Credit*: NASA/Chris Gunn)

n.d.), equivalent to the thickness of six gold atoms or one-thousandth the diameter of a hair strand. Thinning the gold not only saved money but also reduced Webb's overall weight.

Despite such innovations, Webb's expenses soared to 8 billion USD, leading the US Congress to impose a budget ceiling of that amount.

In March 2015, NASA officials assured Congress that the Webb project was within its planned budget and schedule, pledging a launch in 2018 without further delays. By then, its budget had escalated to 8.8 billion dollars (Moskowitz, 2015).

However, fate had other plans. Three years later, on March 27, 2018 (Lewin, 2018), due to the need for further tests on Webb's intricate systems and detected flaws in the tennis court-sized sunshield, NASA once again postponed its launch, projecting it to be no earlier than May 2020—nearly a year behind schedule.

As expected, this postponement meant that 8.8 billion USD was insufficient. By 2020, the COVID-19 pandemic had spread worldwide, leading to work stoppages aimed at ensuring the safety of the workforce and consequently causing a delay in the launch until 2021 (Foust, 2020).

From 1996 to 2021, Webb's cost increased from 500 million USD to 9.7 billion USD, with 8.8 billion allocated for spacecraft development between 2003 and 2021, and an additional 861 million slated for five years of operation. According to 2020's inflation adjustment, its total cost was approximately 10.8 billion USD (Dreier, 2021).

Webb's progress transformed from initial mockery, complaints, and anxiety to become a source of jokes and tales among astronomy enthusiasts. Yet, the unchanging fact remained that everyone was eagerly anticipating the moment of its launch, carrying countless space dreams aloft with it. After all, it harbored the space dreams of countless individuals.

On September 26, 2021, a cargo carrying the James Webb Space Telescope slowly departed from the port in California, USA and set course for the Kourou launch site located in French Guiana (Wall, 2021) (Fig. 26.14).

Upon arrival, it would commence two months of preparations, with a planned launch date of December 18.

However, the launch date was postponed again a few weeks before its scheduled launch (Sun, 2021).

Initially, during the preparation phase, the straps securing Webb to the adapter "unexpectedly loosened," causing its entire apparatus to vibrate. In November, NASA announced that the launch date would be delayed until December 22 for a "check-up" of the telescope. Fortunately, after inspection, it was found that the vibration did not cause any harm to Webb.

Fig. 26.14 The James Webb Space Telescope successfully arrived in French Guiana on Tuesday, October 12, 2021, after a 16-day journey by sea. (*Credit*: NASA/Chris Gunn)

Then, there were issues with the communication cables linking the telescope and the ground system. While not a problem with the telescope itself, Webb was indirectly affected. On December 14, NASA announced another delay of two days, setting the earliest launch date to December 24. Thankfully, the issue with the communication system was quickly resolved, and the launch date was rescheduled to December 22.

However, on December 22, an extreme wind warning was issued at the launch site, pushing the launch date to Christmas Day.

On December 25, 2021, an Ariane 5 rocket successfully lifted off, carrying the James Webb Space Telescope folded and secured within its fairing (Fig. 26.15).

On January 4, 2022, Webb's sunshield, which measures 70 feet, fully unfolded. Four days later, the main mirror of the telescope unfolded, marking the successful full deployment of the James Webb Space Telescope.

By January 24, 2022, Webb had successfully reached its new home, the Sun-Earth L2 Lagrange point, approximately 1.5 million kilometers away from Earth.

Over the next decade or perhaps even longer, regardless of Earth's orbit around the sun, Webb will always follow closely, 150 km away, like a

Fig. 26.15 The Ariane 5 rocket of Arianespace successfully launched on Saturday, Dec. 25, 2021, carrying NASA's James Webb Space Telescope. (*Credit*: NASA/Bill Ingalls)

perpetual companion, until the day its propellant runs out, signaling the end of its lifespan (Fig. 26.16).

A major distinction between Webb and Hubble is the former's remote location, which makes it impossible to make any adjustments or repairs. Even a minor mishap could turn Webb into the most expensive piece of space debris. However, NASA has taken this issue into consideration during the design phase. To extend Webb's lifespan in space, there may be potential future missions to send small unmanned spacecraft for equipment replacement or propellant replenishment.

In a statement released on June 8, 2022, NASA reported that one of Webb's primary mirror segments had been struck by a micro-meteoroid between May 23 and 25, 2022 (SparkingScience, 2022). Although the micro-meteoroid was larger than anticipated, fortunately, it did not cause significant damage.

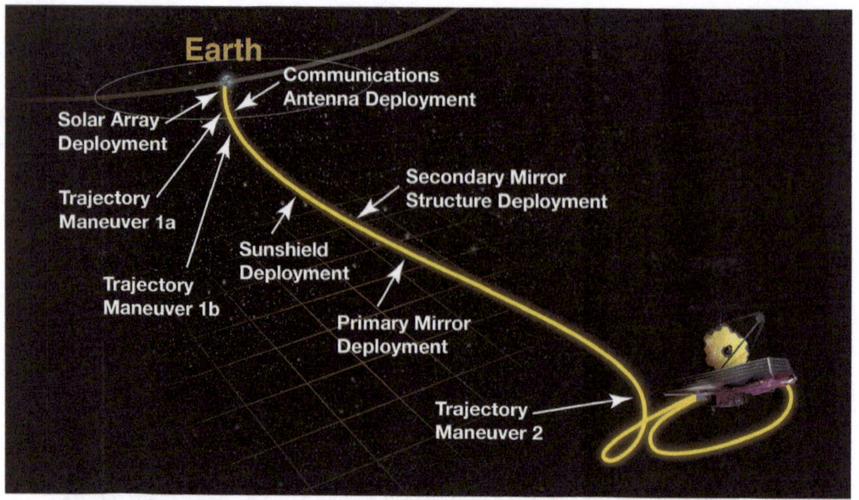

Fig. 26.16 Timeline for the post-launch deployment of the James Webb Space Telescope (*Credit*: NASA, M Clampin (NASA-GSFC))

Exploring the formation processes of ancient stars in the universe was once a primary mission of the Hubble Space Telescope and now stands as one of the main objectives for the James Webb Space Telescope (NASA, GENERAL QUESTIONS ABOUT WEBB, n.d.).

In December 1995, the Hubble Space Telescope aimed at a mere 144 arcseconds of the night sky in Ursa Major and captured the iconic Hubble Deep Field image. In this renowned photo, we caught a glimpse of celestial bodies from 13.4 billion years ago, during the early stages of the universe's formation.

In 2003 and 2012, Hubble once again targeted even deeper spaces, taking higher-resolution images known as the Hubble Ultra Deep Field and the Hubble eXtreme Deep Field, providing us with a deeper understanding of the early universe (Fig. 26.17).

However, Hubble has reached its limits and can no longer see beyond. As celestial bodies age, they emit visible light that, while traversing the cosmos and due to the universe's continuous expansion, shifts in frequency from visible light to infrared.

Astronomers predict that the James Webb Space Telescope, through its infrared observations, can extend our cosmic viewing depth by an additional 400 million years. This would potentially enable us to witness the very first rays of light that were emitted 13.8 billion years ago from the Big Bang. These are the most ancient echoes from our universe, and Webb is poised to capture and transform them into visible images for us (NASA, Early Universe, n.d.).

Fig. 26.17 Left: The Hubble Ultra Deep Field (*Credit*: NASA, ESA). Right: The Hubble eXtreme Deep Field (*Credit*: NASA; ESA; G. Illingworth, D. Magee, and P. Oesch, University of California, Santa Cruz; R. Bouwens, Leiden University; and the HUDF09 Team).

As an infrared astronomical observatory stationed in space, Webb can also analyze the formation, interaction, and evolution of galaxies. It can even reveal the internal structures of galaxies obscured from our view by interstellar clouds, thanks to infrared light that penetrates through such obstructions (NASA, What Will the James Webb Space Telescope See?, n.d.).

Stars and planets form within clouds of gas and dust. While visible light cannot penetrate these layers, infrared light can, allowing Webb to provide deeper, farther, and clearer views than Hubble ever could.

Furthermore, the James Webb Space Telescope can be utilized to study the atmospheres of distant planets. By analyzing the starlight filtered through exoplanetary atmospheres, Webb can inform us of the molecules and elements present, and even provide insights into some planetary features, including their environment and weather conditions.

In short, the James Webb Space Telescope will unlock a rich trove of infrared data. It's like an eye that we've mounted in the universe, aiming wherever we desire to see (Fig. 26.18).

For centuries, humanity has looked up at the stars, filled with wonder and curiosity about the vast universe. We've yearned to pierce the cosmic veil of mystery and uncover its most fundamental secrets. Now, with Webb, we've taken another giant leap toward this goal (Fig. 26.19).

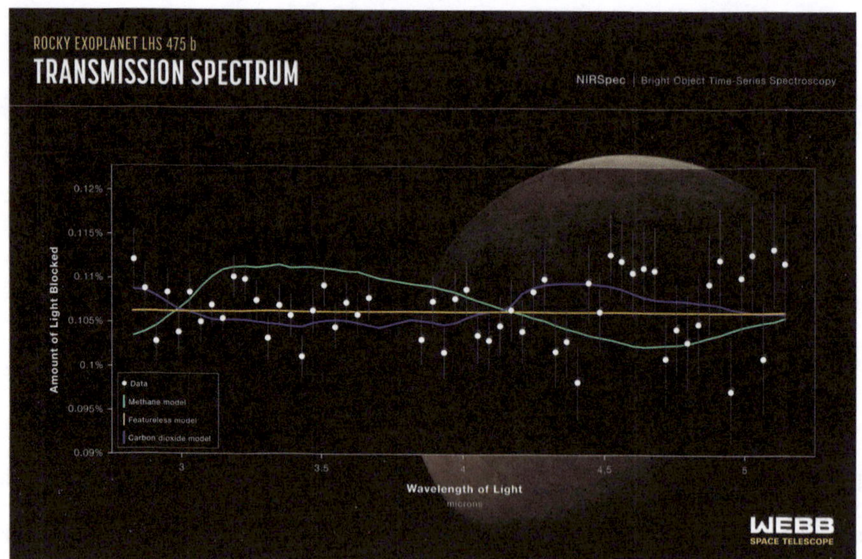

Fig. 26.18 The transmission spectrum of LHS 475 b, which was the first exoplanet observed by Webb on August 31, 2022. (*Credit*: NASA, ESA, CSA, L. Hustak (STScI), K. Stevenson, J. Lustig-Yaeger, E. May (Johns Hopkins University Applied Physics Laboratory), G. Fu (Johns Hopkins University), and S. Moran (University of Arizona))

Fig. 26.19 The James Webb Space Telescope (*Credit*: NASA/dima_zel)

References

Billings, L.: Space science: The telescope that ate astronomy. Nature **467**, 1028–1030 (2010). https://doi.org/10.1038/4671028a

Clery, D.: Building James Webb: The biggest, boldest, riskiest space telescope. Retrieved October 12, 2023 (2016), from Science.org: https://www.science.org/content/article/building-james-webb-biggest-boldest-riskiest-space-telescope

Cornish, N.J. What is a Lagrange Point? (n.d.). Retrieved October 12, 2023, from Nasa.gov: https://science.nasa.gov/resource/what-is-a-lagrange-point/

Dreier, C.: How much does the James Webb Space Telescope cost? Retrieved October 12, 2023 (2021), from Planetary.org: https://www.planetary.org/articles/cost-of-the-jwst

Foust, J.: Coronavirus pauses work on JWST (2020). Retrieved October 12, 2023, from Spacenews.com: https://spacenews.com/coronavirus-pauses-work-on-jwst/

Lewin, S. NASA Delays Launch of James Webb Space Telescope Until 2020 (2018). Retrieved October 12, 2023, from Space.com: https://www.space.com/40102-james-webb-space-telescope-launch-delay-2020.html

Moskowitz, C.: NASA Assures Skeptical Congress That the James Webb Telescope Is on Track (2015). Retrieved October 12, 2023, from Scientificamerican.com: https://www.scientificamerican.com/article/nasa-assures-skeptical-congress-that-the-james-webb-telescope-is-on-track/

NASA: Early Universe. (NASA) (n.d.). Retrieved October 12, 2023, from Nasa.gov: https://www.jwst.nasa.gov/content/science/firstLight.html

NASA: GENERAL QUESTIONS ABOUT WEBB. (NASA) (n.d.) Retrieved October 12, 2023, from Nasa.gov: https://jwst.nasa.gov/content/about/faqs/faq.html

NASA: What Will the James Webb Space Telescope See? (NASA) (n.d.) Retrieved October 12, 2023, from Google.com: https://artsandculture.google.com/story/cgVx8TmfKvBpJw

NASA, & STScl. Webb Project History. (NASA and the Space Telescope Science Institute (STScI)) (n.d.). Retrieved October 12, 2023, from Webbtelescope.org: https://webbtelescope.org/news/milestones/mission-timeline

Pultarova, T.: James Webb Space Telescope: The engineering behind a 'first light machine' that is not allowed to fail (2021). Retrieved October 12, 2023, from Space.com: https://www.space.com/james-webb-space-telescope-engineering-challenges

SparkingScience: Costing tens of billions of dollars, the James Webb Space Telescope was struck by an unidentified object, NASA: There was no way to avoid it (2022). Retrieved October 12, 2023, from Baijiahao: https://baijiahao.baidu.com/s?id=1735140899670909742&wfr=spider&for=pc

Sun, Y.: Confirmed! The James Webb Space Telescope is scheduled to launch on December 24th (2021). Retrieved October 12, 2023, from Stdaily.com: http://www.stdaily.com/cehua/Dec21th/fmxw.shtml

Wall, M.: NASA's James Webb Space Telescope arrives in French Guiana ahead of Dec (2021). 18 launch. Retrieved October 12, 2023, from Space.com: https://www.space.com/nasa-james-webb-space-telescope-arrives-french-guiana
Wikipedian: James Webb Space Telescope (2023). Retrieved October 12, 2023, from Wikipedia: https://en.wikipedia.org/wiki/James_Webb_Space_Telescope

27

CHASE (Chinese Hα Solar Explorer): China's First Solar Exploration Satellite

CHASE

On October 24, 2021, in an office at the China Center for Resources Satellite Data and Application in Beijing, the air was thick with tension. A group of scientists, including Professor Li Chuan, the chief designer of the scientific and application system for the Chinese Hα Solar Explorer (CHASE), and deputy chief designer Professor Li Zhen, gathered around a computer, eagerly eyeing the data on the screen. Data from three satellite data reception stations in Miyun, Kashgar, and Sanya was being processed in real-time, representing the first light observations from CHASE. The same data was awaited by Academician Fang Cheng, Professor Li Chuan's mentor, at the Solar Science Data Center of Nanjing University (SSDC n.d.), located one thousand kilometers away. Meanwhile, at the Xi'an Satellite Control Center, the chief designer of the satellite system for CHASE, Cheng Weiqiang, along with the head designer You Wei, and many others stood ready to send commands to CHASE.

The launch of CHASE, China's first solar exploration satellite, from the Taiyuan Satellite Launch Center just 10 days ago (Jones 2021) has captured the attention and concern of countless people (Fig. 27.1).

The data, once consolidated in Beijing, was promptly transmitted via a dedicated network to Nanjing University. Almost simultaneously, both Professor Li Chuan and Academician Fang Cheng viewed the data.

"When I saw the data," Professor Li Chuan humorously remarked during our interview, "I had good news and bad news. Which would you like to hear first?" Without waiting for my response, he continued, "The good news is,

J. Wang, *Eye Beyond the Sky*, Astronomers' Universe,
https://doi.org/10.1007/978-981-99-9818-0_27

Fig. 27.1 CHASE launch on October 14, 2021 (*Credit* Nanjing University)

the data shows that the satellite is functioning perfectly, and its condition is good."

"And the bad news?" I couldn't help but inquire further.

Professor Li replied, "The bad news is that the quality of the first set of solar images returned was not good. It appeared as if they were captured by a camera out of focus. We suspect that vibrations during the launch may have affected the optical system. Even a 1 mm displacement of the lens can result in a blurred image."

"Just like the Hubble Space Telescope when it first went into space?" I blurted out.

"Indeed, that's a fitting comparison," he agreed.

"So, do we need an astronaut to fix it as well?"

"Ha, not really," he laughed. "At the time, Academician Fang Cheng and I weren't too anxious. CHASE was designed with contingencies for such issues, allowing us to remotely adjust the optical system. So, we remained optimistic."

"That's a relief! I was worried that CHASE would be out of commission."

"However," Professor Li continued, taking a more somber tone, "in hindsight, both Academician Fang Cheng and I were overly optimistic. The issue was much more challenging than we anticipated. For the next several months, my team and I grappled with disappointment and anxiety. What we thought was a minor issue turned out to be a significant problem."

So, what became of China's first solar exploration satellite? Could the engineers of the satellite, much like their American counterparts did with Hubble, successfully repair it? Before revealing the answer, let me first delve into the history and mission of CHASE (Fig. 27.2).

Within the Milky Way galaxy, our Sun is just an ordinary star, classified by astronomers as a G-type main-sequence star (AstroBackyard n.d.). In layman's terms, it's a "yellow, medium-sized star in its prime of life." Stars of this type account for approximately 7% of the total stars in our galaxy (Wall 2020), with at least over 10 million such stars. They might seem inconspicuous; however, for our solar system, the Sun reigns supreme. It's voluminous enough to be equivalent to 1.36 million Earths (Harvey and Sharp 2022), and its mass dominates 99.86% of the entire solar system. If we imagine the Sun as a person, the droplets of water flicked from their hands after washing them would form all the celestial bodies in the solar system outside of the Sun itself—including our Earth and every living thing on it (Fig. 27.3).

Since the dawn of our civilization, humans have been fascinated by the Sun, brimming with curiosity about it, and have woven countless myths and legends around it. In ancient Chinese mythology, the deity governing the Sun

Fig. 27.2 CHASE's development team led by Academician Fang Cheng (sixth from the left) (*Credit* Shanghai Academy of Spaceflight Technology)

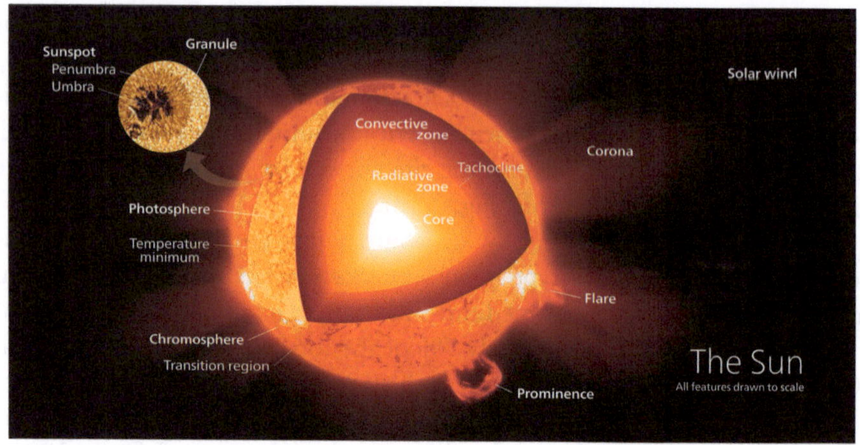

Fig. 27.3 Our Sun is a typical G-type main-sequence star (*Credit* Kelvinsong)

is a beautiful goddess named "Xihe (Wikipedian, Xihe (deity) 2023)," considered the mother of the Sun. The formal Chinese name for CHASE is given as "Xihe" in tribute to her. However, humanity's true scientific understanding of the Sun only began about 400 years ago when Galileo first observed it through a telescope. He noticed dark spots, known as sunspots, on its surface, challenging the long-held belief that the Sun was flawless.

Studying the Sun is not merely to satisfy human curiosity. The Sun has a crucial impact on our living environment. On the morning of September 1, 1859, the British astronomer Carrington fortuitously observed a solar anomaly: two bead-like white lights appeared amidst a large group of sunspots, their brightness resembling lightning—a sight he had never encountered before. Staring in disbelief, the lights grew more intense, eventually taking on the shape of two kidneys. The next evening, the most significant aurora ever recorded in human history suddenly illuminated the Earth. Incredibly, the aurora stretched as far as the equatorial regions. This astonishing celestial event introduced scientists to the phenomenon of solar flares. The Sun seemed to have "sneezed," releasing a violent solar wind rushing toward Earth at speeds more than 10,000 times faster than the most potent typhoon. As the plasma reached Earth, it triggered geomagnetic storms. These storms almost entirely paralyzed the world's telegraph communication networks of the time, as geomagnetic storms induce currents in electrical devices. In many telegraph stations across the US and Europe, massive sparks even burst from the telegraph equipment.

Since then, humanity realized the necessity of making "weather forecasts" for the Sun. Today, with over 4,000 artificial satellites (NanoAvionics 2023) densely surrounding Earth, solar activities directly influence their operational statuses. There's an increasing need for accurate predictions of solar activities.

Since the launch of the first solar observation satellite, America's Orbiting Solar Observatory-1 (OSO), in 1959, humans have launched more than 70 solar observation satellites by 2023 (Hillger and Toth 2023). The most notable among them include the Ulysses spacecraft launched in 1990, the Solar and Heliospheric Observatory (SOHO) in 1995, the Solar Dynamics Observatory (SDO) in 2010, and the Parker Solar Probe in 2018. All these solar probes were launched by the United States. It's fair to say that in the field of solar exploration, the US still leads by a significant margin. However, many countries worldwide, including China, have shown keen interest in solar exploration.

In 2015, under the leadership of the distinguished solar physicist, Academician Fang Cheng, institutions including the Nanjing University, the Shanghai Academy of Spaceflight Technology, and the Changchun Institute of Optics, Fine Mechanics and Physics of the Chinese Academy of Sciences, formally proposed China's first solar observation science and technology test satellite—the Chinese Hα Solar Explorer. Four years later, in 2019, this project was officially approved by the China National Space Administration (Jiangsu Astronomical Society 2022), and the research and development of CHASE commenced. Even though it was China's first solar exploration satellite, ambitious Chinese scientists proposed several technically challenging endeavors of international pioneering significance.

So, what are the specificities of CHASE and what pioneering capabilities does it possess on the international stage? (Fig. 27.4).

Before we delve into the technicalities of CHASE, I'd like to touch on the topic of the trajectory of solar probes. The most straightforward notion for us would be to launch a probe directly toward the Sun, believing that getting closer is always better. However, this idea is quite naive, and a foundation in astrophysics is essential to understanding the challenges.

Every celestial body in the solar system orbits the Sun. Imagine Earth as a high-speed train. If you shoot at a ground target several kilometers away from a moving train, the bullet won't hit. The bullet will inherit the train's speed and deviate from its intended path.

If a probe launched from Earth using a rocket fails to reach the escape velocity of 11.2 km/s (Wikipedian, Escape velocity 2023), it will remain in Earth's orbit and become an artificial satellite. But if the probe exceeds this speed, it will become a satellite orbiting the Sun.

However, merely getting a probe close to the Sun isn't a brute-force endeavor; it requires meticulous calculations by astrophysicists. Take, for

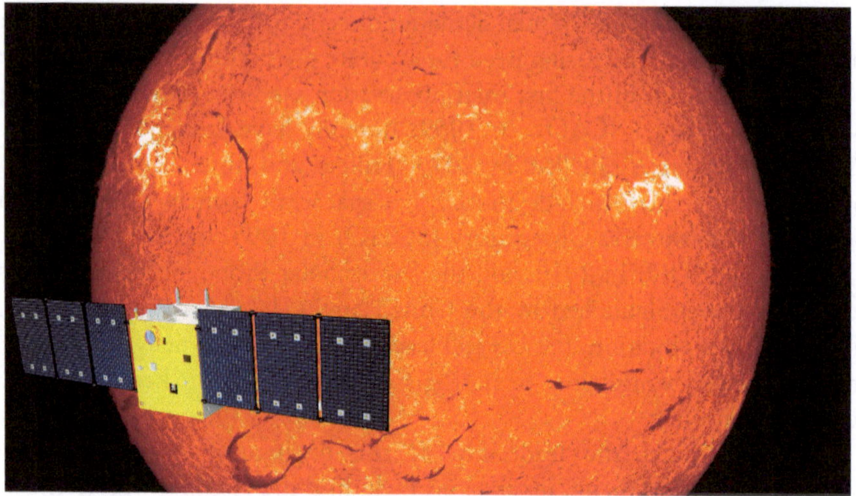

Fig. 27.4 Illustration of CHASE with the Sun (*Credit* Nanjing University)

example, the Ulysses spacecraft launched in 1990. It was the first to fly over the Sun's poles. In order to position its trajectory almost perpendicular to the ecliptic plane, scientists had to utilize Jupiter's gravity. It took Ulysses 16 months (Wenzel 1995) to reach Jupiter and then adjusted its orbit around the planet, precisely boosting its engine to harness Jupiter's gravitational slingshot, propelling itself toward the Sun at a faster speed. It took nearly three additional years to reach its closest point to the Sun, which was 193 million kilometers away (Krüger et al. 2007).

Then, there's SOHO, launched in 1995. It was the first solar probe stationed at the Sun-Earth L1 Lagrange Point (Vandenbussche 1996). Joseph-Louis Lagrange, an 18th-century French astrophysicist, identified five gravitational equilibrium points in the Sun-Earth system. Objects at these points can remain stable relative to both the Sun and Earth (Fig. 27.5).

Although illustrations can be misleading, the L1 point is about 148 million kilometers from the Sun and 1.5 million kilometers from Earth—nearly a 100-fold difference (The Editors of Encyclopaedia Britannica 1998). If drawn to scale, the L1 point would appear very close to Earth. Its advantage is that a satellite there remains relatively stationary with the Sun, providing a stable observation position.

The Parker Solar Probe, launched in 2018, holds the record for being the closest probe to the Sun ever sent by humans. To skim the Sun at a mere 8.5 solar radii away, Parker will orbit the Sun 24 times (NASA n.d.) over 7 years, each orbit closer than the last. To avoid plummeting into the Sun due to excessive speed, astrophysicists meticulously designed its path, allowing

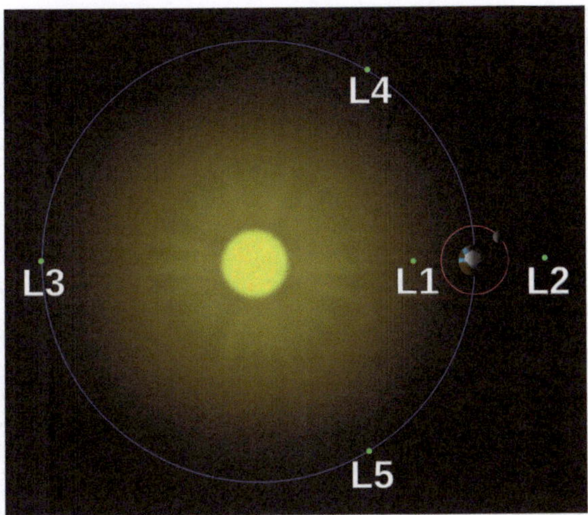

Fig. 27.5 Sun-Earth Lagrange Points (*Credit* Xander89)

Parker to utilize Mercury's gravity to decelerate 7 times (Parker Solar Probe - THE MISSION, n.d.)—a feat of elegant mathematics.

Meanwhile, CHASE operates on a Sun-synchronous orbit.

Many, hearing this term, think of the geosynchronous orbit satellites use 36,000 km above Earth. These satellites orbit Earth once every 24 h, seeming stationary above the equator. However, a Sun-synchronous satellite doesn't hover over the Sun's equator synchronized with its rotation.

This satellite still orbits Earth, with its orbital plane maintaining a consistent angle with the line connecting the centers of the Sun and Earth. To visualize, imagine wearing a hat while circling a lamp. If the hat remains fixed, different parts get illuminated as you move, which is asynchronous. But synchronizing means adjusting the hat's angle to ensure that one side always faces the light (Fig. 27.6).

CHASE is such a satellite on a Sun-synchronous orbit. Its orbital plane's angle to the equatorial plane is 97.46 degrees. Picture the Earth as your head and CHASE's orbit as a hat placed almost directly over your face while you face the Sun. From the view of observing the entire Earth-Sun system, CHASE orbits around Earth's twilight zone and completes about 15 orbits daily. This trajectory ensures that CHASE remains bathed in sunlight, never overshadowed by Earth. From CHASE's perspective, the Sun is always in view, never obscured by Earth.

On October 14, 2021, at the Taiyuan Satellite Launch Center, a Long March-2D rocket soared into the sky (Jones 2021). Aboard this rocket, in

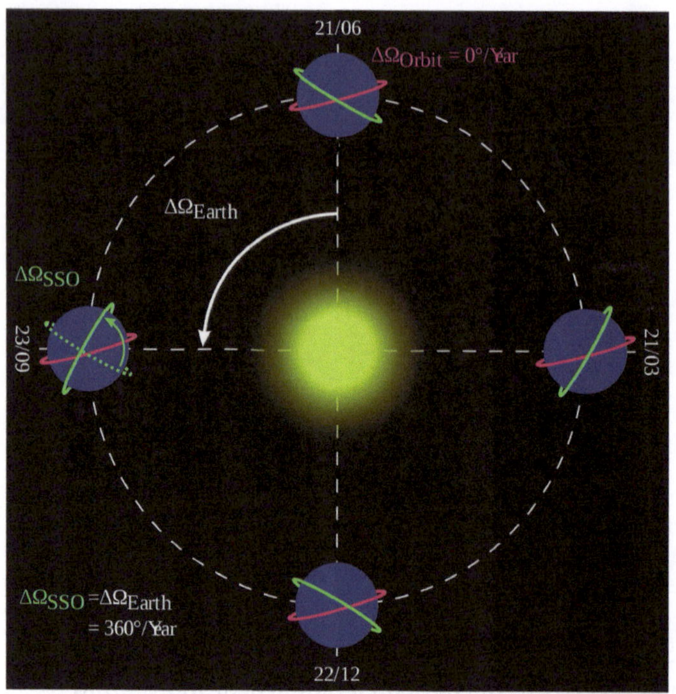

Fig. 27.6 Comparison between a Sun-synchronous orbit (Green) and a non-sun-synchronous orbit (Magenta) (*Credit* Brandir & XZise)

addition to the main satellite CHASE, there were also 11 other smaller satellites. This represented a typical multi-satellite mission with a single launch (Fig. 27.7).

By October 24, the first batch of data from CHASE was transmitted back to Earth (Jin 2022). Except for the image quality not meeting expectations, everything else seemed normal. Academician Fang Cheng and Professor Li Chuan's team weren't overly worried at first. Imperfect focusing can be adjusted, and the two CMOS (sensors that convert light signals into electrical signals) on CHASE are two adjustable components. Initially, engineers optimistically believed that just adjusting the CMOS to the focal plane would solve the problem. While CHASE orbits Earth 15 times a day, not every orbit goes over the ground control station. Hence, there are only a few control opportunities each day. Soon, Li Chuan's team realized that things weren't as rosy as they'd thought. After wrestling with the two CMOS for 2 months, the image quality showed no sign of improvement (Fig. 27.8).

A terrifying thought crept into Professor Li Chuan's mind: Had the fixed components of the entire system been displaced by vibration? Or had weightlessness led to unforeseen changes in the system? If either were the case,

Fig. 27.7 CHASE with the rocket fairing of a Long March-2D rocket (*Credit* China National Space Administration)

Fig. 27.8 Blurred imaged photos taken by CHASE (*Credit* Nanjing University)

CHASE's scientific objectives would be jeopardized irreversibly. This notion sent shivers down Li Chuan's spine. He brushed the thought aside, reminding himself that years of effort and 270 million yuan in research funds couldn't end up in a flawed satellite.

Winter 2021 silently approached. Experts from Nanjing University and the Shanghai Academy of Spaceflight Technology worked around the clock, holding meetings. Although he was 83, Academician Fang Cheng's clear thinking and computational skills often amazed the younger members of Li Chuan's team. However, as months passed, there was little progress. Before anyone realized, it was June 21, 2022. In a conference room at Nanjing University, Academician Fang Cheng, Professor Li Chuan's team, and experts from the Shanghai Academy of Spaceflight Technology gathered again. No one could recall how many meetings they'd had.

"All parameters have been tried, and every combination of CMOS positioning has been tested multiple times. It's just not working!" lamented a colleague responsible for control and measurement, unable to remember how many times he'd delivered such bad news.

The room fell silent; the weight of their predicament hung heavily in the air. Suddenly, Academician Fang, pointing to a spot on a diagram in his hand, asked, "Could it be an issue with these three mirrors of the scanning mechanism?"

A display showed the part he was referring to. "That's unlikely, isn't it? These three scanning components are constantly in motion. They can't possibly be out of focus all the time," someone remarked skeptically.

Li Chuan stared at the screen for a moment before saying, "Why not? If the scanning components have shifted as a whole, they might always be out of focus. I agree with Professor Fang's hypothesis. It's worth a try."

The previously subdued room buzzed with animated discussion. The air seemed to vibrate with the scientists' fervor. The consensus was reached: They would immediately begin adjusting the scanning mirrors.

It was as if the patient in critical condition had been given the right medication. Initial adjustments quickly showed promise. The dark cloud that had loomed over the team for over half a year finally began to lift, heralding the dawn of hope.

On August 3, 2022, at 10:08:35 Coordinated Universal Time, a perfectly imaged photo finally appeared on the computer screen (Nanjing University 2022). By then, 7 months and 20 days had passed since CHASE's launch. The project team erupted into enthusiastic applause. China's first solar exploration satellite officially began providing optimal quality solar science data. At the age of 84, Academician Fang Cheng led the team, marking a new chapter in the Chinese study of the sun (Fig. 27.9).

CHASE, which stands for "Chinese Hα Solar Explorer", weighs 550 km and is designed for a lifespan of 3 years. Based on its current operational status, it is projected to function at least until 2026 (Cheng and Chuan 2022) (Fig. 27.10).

Fig. 27.9 Perfectly imaged photos taken by CHASE (*Credit* Nanjing University)

Fig. 27.10 Fully unfolded CHASE before launch (*Credit* Nanjing University)

The scientific objectives and unique capabilities of CHASE can be summarized by two key terms: "Hα spectral detection (Li et al., The Chinese Hα Solar Explorer (CHASE) mission: An overview 2022)" and "dual-ultra platform (Zhang et al. 2014)."

The Sun's light and heat predominantly originate from the thermonuclear fusion of hydrogen atoms, a process that converts four hydrogen nuclei into a helium nucleus and releases enormous energy. Hydrogen, the most abundant element in the sun, has electrons that operate at different energy levels outside the nucleus, labeled as 1, 2, 3, and so forth. When an electron transitions from the third or a higher energy level to the second, it emits a spectral line known as the "Balmer series (Wikipedian, Balmer series 2023)" (named after physicist Johann Balmer). These lines are sequentially termed Hα (alpha), Hβ (beta), Hγ (gamma), Hδ (delta), etc. The Hα line, a prominent red spectral line, is tantamount to identifying ionized hydrogen in astronomical spectroscopy. It serves as the most straightforward method for astronomers to trace ionized hydrogen in gaseous clouds. CHASE's primary function is to scan and receive the Hα spectral lines emitted by the sun. It houses the world's most potent Hα imaging spectrometer, which can scan the entire solar disk in just 46 s, obtaining spectral information from 16 million points (Li et al., The Chinese Hα Solar Explorer (CHASE) mission: An overview 2022).

The dual-ultra platform represents both ultra-high pointing accuracy and ultra-high stability, making CHASE the most accurate and stable solar exploration satellite in space to date. To quantify, CHASE's pointing accuracy is superior to one part in fifty thousand degrees, and its stability surpasses one part in five hundred thousand degrees per second. To put it another way, if you were to target a point on the sun with a laser beam from CHASE, after traveling 149 million kilometers, the beam would deviate less than 52 m from its target and would only fluctuate about 5 m per second (Gu 2023). This precision is far greater than trying to toss a coin from Earth into a water bottle on the moon.

CHASE's extraordinary stability and accuracy stem from its unique design, which incorporates motion and still isolation with magnetic suspension mechanisms. The components used for solar observation are magnetically levitated inside the satellite, remaining untouched and virtually undisturbed (Fig. 27.11).

Since its official commencement of solar observations in August 2022, CHASE's goals have been threefold (Li et al., Chinese Hα Solar Explorer (CHASE) – a complementary space mission to the ASO-S 2019):

1. To study the thermodynamics and dynamics of solar activity in the lower atmosphere.
2. To investigate the formation and evolution of solar filaments and their relationship with solar flares.
3. To conduct comparative studies of solar and stellar flares.

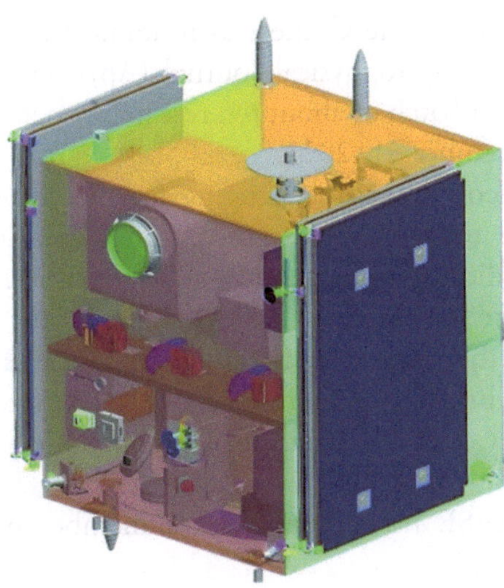

Fig. 27.11 Schematic diagram of CHASE: Upper payload bay, lower platform bay, 8 maglev actuators in the center in red. The outer shell (satellite envelope) is connected to the platform module and is not in contact with the payload module. The motion components and the payload components are completely separated and do not interfere with each other. (*Credit* Nanjing University)

To date, scientists have made numerous discoveries using data from CHASE (Chen and Yuan 2022).

For instance, for the first time, they've ascertained the distribution of physical parameters of silicon (Si) spectral lines across the solar surface and confirmed that they are formed approximately 70 km from the bottom of the photosphere. Stars are like elemental forges, where hydrogen fuses into helium, which then forms lithium, followed by beryllium and so on until iron, the 26th element. Silicon is the 14th element. This insight is pivotal for our understanding of stellar fusion processes.

Also, for the first time, it was confirmed that the sun's atmospheric rotation speed increases from bottom to top, meaning the lower you go, the slower the rotation. Further analysis by CHASE suggests this could be due to the frozen-in flux theorem between magnetic fields and plasma.

Moreover, scientists captured a white-light flare for the first time in both the iron (Fe) spectral lines and the continuous spectrum, determining its energy and how it was generated.

They also initiated the comparative study of stellar flare activities using full-disk spectral scans.

Notably, in August 2023, the internationally influential Astrophysical Journal Letters (ApJL) published a special Focus Issue on CHASE (Li et al.

2023), marking the first time Chinese astronomical equipment featured in such a capacity on an Astrophysical Journal (ApJ) series publication. This issue included the first batch of about ten research papers related to CHASE, submitted until December 31, 2023.

Perhaps these discoveries are challenging for the layperson to grasp, but that's okay. Science today has blossomed into a multitude of specialized branches, each a dedicated field of study. A scientist might spend a lifetime delving into just one narrow area, achieving significant breakthroughs. All we need to know is that each of CHASE's observations enhances our understanding of the sun, bringing us one step closer to accurately forecasting solar weather. The journey of CHASE has just begun, and it continues to surprise solar physicists.

"Over time, CHASE has become like a member of my family," says Professor Li Chuan.

"There are billions of stars in the Milky Way, and the sun is just one of them. One day, when humanity sets off for the stars, the experiences gained from the sun will be invaluable, as it's the only star we can study up close," states Academician Fang Cheng.

Fig. 27.12 CHASE and its development team (*Credit* Shanghai Academy of Spaceflight Technology)

Academician Fang Cheng, Professor Li Chuan, and their team are the modern-day Kuafu, the giant in Chinese mythology who never stops chasing the Sun (Fig. 27.12).

Currently, the CHASE 2 project is underway, set to be launched at the Sun-Earth L5 Lagrange point (Xinhuanet 2023). It will undertake three-dimensional explorations of the Sun and interplanetary space, heralding China's era of solar stereo detection. It aims to address significant scientific questions in solar physics, breakthrough key technologies in space weather forecasting, and make substantial contributions to human knowledge.

References

AstroBackyard: Types of Stars (n.d.). Retrieved October 16, 2023, from AstroBackyard: https://astrobackyard.com/types-of-stars/

Chen, H., & Yuan, Y.: CHASE's Solar Exploration Results Officially Released, Setting 5 International Firsts in Discovery (2022). Retrieved from people.cn: http://kpzg.people.com.cn/n1/2022/0831/c404214-32515817.html

Cheng, F., Chuan, L.: Introduction to the Chinese hα solar explorer (CHASE) mission. Chin. J. Space Sci. **42**(4), 546–549 (2022). https://doi.org/10.11728/cjss2022.04.yg05

Gu , H.: Biography of China's First Solar Exploration Satellite "on Duty" (2023). Retrieved from Spacechina: http://m.spacechina.com/n2018089/n2018146/c3743192/content.html

Harvey, A., Sharp, T.: How big is the sun? (2022). Retrieved from Space: https://www.space.com/17001-how-big-is-the-sun-size-of-the-sun.html

Hillger, D., Toth, G.: Solar-observing satellites (2023). Retrieved from Colorado State University: https://rammb.cira.colostate.edu/dev/hillger/solar-observing.htm

Jiangsu Astronomical Society. CHASE Satellite Science and Application System Team Wins 2021 Nanjing University Outstanding Contribution Award (2022). Retrieved from Jiangsu Astronomical Society: https://jsas.nju.edu.cn/98/d6/c21055a563414/page.htm

Jin, F.: Exploring the sun for more than half a year, what secrets CHASE brought back? (2022). Retrieved from Xinhua News: http://www.news.cn/politics/2022-07/20/c_1128846619.htm

Jones, A.: China launches first solar observatory, tests grid fins (2021). Retrieved from Space News: https://spacenews.com/china-launches-first-solar-observatory-tests-grid-fins/

Krüger, H., Landgraf, M., Altobelli, N., Grün, E.: Interstellar dust in the solar system. Space Sci. Rev. **130**(1), 401–408 (2007). https://doi.org/10.1007/s11214-007-9181-7

Li, C., Ding, M., Fang, C., Chen, P.F., Cheng, X.: Focus on Early Results from the Chinese Hα solar explorer (CHASE) Mission. Retrieved from The Astrophysical Journal Letters (2023). https://iopscience.iop.org/collections/Focus-on-Early-Results-from-CHASE

Li, C., Fang, C., Li, Z., Ding, M., Chen, P., Chen, Z., . . . Zhang, W.: Chinese Hα solar explorer (CHASE)—a complementary space mission to the ASO-S. Res. Astron. Astrophys. **19**(11) (2019).https://doi.org/10.1088/1674-4527/19/11/165

Li, C., Fang, C., Li, Z., Ding, M., Chen, P., Qiu, Y., . . . Meng, L.: The Chinese Hα solar explorer (CHASE) mission: an overview. Solar Stellar Astrophys. (2022).https://doi.org/10.1007/s11433-022-1893-3

Nanjing University: CHASE's Special Issue Articles and Updates (2022). Retrieved from Nanjing University: https://njunju.nju.edu.cn/e9/66/c3674a584038/page.htm

NanoAvionics: How Many Satellites are in Space? (2023). Retrieved from NanoAvionics: https://nanoavionics.com/blog/how-many-satellites-are-in-space/

NASA: Parker Solar Probe (n.d.). Retrieved October 16, 2023, from NASA: https://science.nasa.gov/mission/parker-solar-probe/

Parker Solar Probe—THE MISSION. (n.d.). Retrieved October 16, 2023, from The Johns Hopkins University Applied Physics Laboratory: https://parkersolarprobe.jhuapl.edu/The-Mission/index.php

SSDC: Solar Science Data Center (n.d.). Retrieved October 16, 2023, from Nanjing University: https://ssdc.nju.edu.cn/home

The Editors of Encyclopaedia Britannica: Lagrange point (1998). Retrieved from Encyclopaedia Britannica: https://www.britannica.com/science/force-physics

Vandenbussche, F.C.: SOHO—The Trip to the L1 Halo Orbit (1996). Retrieved from European Space Agency: https://www.esa.int/esapub/bulletin/bullet88/vande88.htm

Wall, M.: More than half of all sunlike stars in the Milky Way may have a habitable planet (2020). Retrieved from Space: https://www.space.com/habitable-planets-common-sunlike-stars-milky-way

Wenzel, K.: The ulysses mission — a voyage to the poles of the sun. Nuclear Phys. B Proc. Suppl. **39**(1), 59–68 (1995). https://doi.org/10.1016/0920-5632(95)00007-V

Wikipedian: Balmer series (2023). Retrieved from Wikipedia: https://en.wikipedia.org/wiki/Balmer_series

Wikipedian: Escape velocity (2023). Retrieved from Wikipedia: https://en.wikipedia.org/wiki/Escape_velocity

Wikipedian: Xihe (deity) (2023). Retrieved from Wikipedia: https://en.wikipedia.org/wiki/Xihe_(deity)

Xinhuanet: China is carrying out the demonstration of the CHASE 2 solar probe project at the L5 solar-earth point (2023). Retrieved from Xinhua News: http://www.news.cn/tech/2023-09/14/c_1129863943.htm

Zhang, W., Liao, H., Yuan, J., Zhao, H., Xu, Y., Zhao, Y., . . . Deng, C.: Motion and still isolation master-slave cooperative control double superior satellite platform (2014). Retrieved from Google Patents: https://patents.google.com/patent/CN104477410A/en